Principles and Applications of

NANOTECHNOLOGY

For Soil Health and Crop Management

Principles and Applications of NANOTECHNOLOGY For Soil Health and Crop Management

– Authors –

Vishnu D. Rajput

Abhishek Singh

Ragini Sharma

Karen Ghazaryan

Tatiana Minkina

2026

Daya Publishing House®

A Division of

Astral International Pvt. Ltd.

New Delhi – 110 002

ISBN: 978-93-5919-692-3

Published by : **Daya Publishing House**®
A Division of
Astral International Pvt. Ltd.
– ISO 9001:2015 Certified Company –
436/23, Ansari Road, Darya Ganj
New Delhi-110 002
Ph. 011-43549197
e-mail: info@astralint.com
Website: www.astralint.com

Printed at : **Replika Press Pvt. Ltd.**

ACKNOWLEDGEMENTS

We would like to express our heartfelt gratitude to the Science Committee of the Republic of Armenia and the Ministry of Science and Higher Education of the Russian Federation for their unwavering support and generous funding. Their invaluable contributions, provided under grant numbers 24FP-4C013 and FENW-2023-0008, as well as the 23PostDoc-4D007 grant provided by the Science Committee of the Republic of Armenia, and authors also acknowledge the support by the Strategic Academic Leadership Program of the Southern Federal University ("Priority 2030") which played a pivotal role in bringing to fruition the book, entitled "**Principles and Applications of Nanotechnology for Soil Health and Crop Management**". This book will be published by Astral India. Their commitment to advancing education with combination of current research and knowledge in the fields of nanotechnology and sustainable agriculture has been instrumental in making this endeavor a reality. Their support has enabled us to gather and disseminate significant insights, fostering progress and innovation in these critical domains. We express our deep appreciation for their vision and dedication, which have not only enriched the content of this book but have also contributed to the global discourse on sustainable agriculture. We look forward to continued collaboration and the positive impact it will undoubtedly have on the future of agriculture and with high quality of education and scientific research.

Vishnu D. Rajput | Abhishek Singh

Ragini Sharma | Karen Ghazaryan

Tatiana Minkina

PREFACE

In the ever-evolving landscape of modern agriculture, the integration of nanotechnology has emerged as a ground-breaking frontier, promising to revolutionize the cultivation, manage, and sustain our agricultural systems. This book, "**Principles and Applications of Nanotechnology for Soil Health and Crop Management**" delves into the multifaceted applications and implications of nanotechnology in agriculture, offering readers a comprehensive exploration of this exciting field. The journey begins with an introduction to the recent updates in nanotechnology applications within the agricultural sector, setting the stage for the detailed discussions that follow. We then navigate through the diverse world of nanoparticles, examining their types and the various approaches for their synthesis. Understanding these foundational elements is crucial as we explore the methodologies for the application, uptake, and translocation of nanoparticles in agricultural contexts. Soil and nutrient management are critical to sustainable agriculture, and nanofertilizers present innovative solutions for enhancing soil health and nutrient availability. This book addresses the impact of nanofertilizers and extends the discussion to the role of nanotechnology in managing salinity and heavy metal stress in crops. These chapters highlight the potential of nanotechnology to mitigate some of the most pressing environmental challenges faced by modern agriculture. One of the most promising areas explored in this book is the combination of genome editing tools with nanotechnology to improve crop traits. The concept of Nano-omics is introduced as a sustainable approach to crop production, emphasizing the integration of nanotechnology with other scientific disciplines to achieve more resilient and productive agricultural systems. The potential benefits of nanotechnology are immense, but so are the concerns. We dedicate a chapter to discussing the environmental, plant, and human health concerns associated with nanotechnology in agriculture, ensuring a balanced perspective on its use and implications.

Sustainability is a recurring theme, and the exploration of green or biogenic nanotechnology underscores our commitment to sustainable agricultural practices and human health. The book also examines the role of nanotechnology in zinc fortification as a strategy to combat malnutrition, showcasing another dimension of its impact on public health. Precision agriculture is another realm where nanotechnology shows great promise. We delve into how nanotechnology can enhance precision agriculture management, making farming more efficient and data driven. The final chapter expands on the synergy between nanotechnology, artificial intelligence, sensor technology, and big data, painting a picture of next-generation farming. This book is intended for researchers, practitioners, students, and anyone interested in the transformative potential of nanotechnology in agriculture. By presenting the latest research, innovative applications, and thoughtful discussions on both the opportunities and challenges, we aim to provide a valuable resource that inspires further exploration and adoption of nanotechnological solutions in agriculture.

We hope this book serves as a catalyst for new ideas, collaborations, and advancements, contributing to a more sustainable and prosperous future for global agriculture.

Vishnu D. Rajput | Abhishek Singh
Ragini Sharma | Karen Ghazaryan
Tatiana Minkina

CONTENTS

Acknowledgements v

Preface vii

1. Nanotechnology in Agriculture: An Introduction on Recent Updates 1

1. Introduction 1
2. Fundamentals of NPs 3
3. Recent Updates on Application of Agri-Nanotechnology 4
4. Concerns associated with NPs implementation 27
5. Agri-Nano-enabled products and marketing approaches for sustainable agriculture 30
6. Future suggestions and Conclusion 32

2. Nanoparticles: Types and Approaches for Synthesis 35

1. Introduction 35
2. Types of Nanomaterials 36
3. Synthesis and characterization of nanomaterials 44

3. Application, Uptake and Translocation Methodology of Nanoparticles 53

1. Introduction 53
2. NPs delivery methods, uptake, and translocation process 55

4. Impact of Nanofertilizer for Soil and Nutrient Management 59

1. Introduction 60
2. NFs Based Macronutrient Management 60
3. NFs Based Micronutrient Management 63
4. NFs base Crop Gram Nutrient Management 68
5. Function of Nanofertilizers for Mitigation of Multiple Environmental Stresses Management Under Climate change 70
6. Conclusion 79

5. **Role of Nanotechnology in Salinity Stress Management** **81**

1. Introduction *81*
2. Nanoparticles: A New Strategy for Combat against Salinity Stress *85*
3. Uptake Pathways of NPs *87*
4. Effect of NPs on Antioxidant Enzymes activities for Physiochemical Protection under Salinity Stress *89*
5. NPs Mediated Antioxidant Enzymatic Response for Mitigation of ROS under Salinity *93*
6. NPs Based Molecular Regulation for Antioxidant Defense to plant under Salinity *103*
7. NPs mediated antioxidant signaling pathways *110*
8. Conclusion *113*

6. **Role of Nanotechnology in Heavy Metals Stress Management** **115**

1. Introduction *115*
2. Global Status of HMMs Problem *117*
3. Sources of HMMs *118*
4. Impact on plant and human health *120*
5. Uptake and Translocation of HMMs in plants from the Soil *121*
6. Conventional and NPs Based Techniques for HMMs Remediation *123*
7. Nanotechnological Approaches for Remediation of HMMs *125*
8. Conclusion *127*

7. **Genome Editing Tool and Techniques for Crop Improvement with Combination Approaches of Nanotechnology** **129**

1. Introduction *130*
2. GE in Crop Improvements *132*
3. Enhancing Crop Biotic Stress Tolerance *140*
4. Enhancing crop abiotic stress tolerance *141*
5. Nutritional quality improvement *142*
6. Improvement of Economical Plants Traits *142*
7. CRISPR 2.0 *143*
8. Safety Assessment of Genome-edited Crops *146*
9. Regulation of Genome-edited crops *147*
10. Impact of Plant Genome Editing in the Post-COVID Pandemic Situation *148*
11. Future of Genome Editing *149*
12. Challenges in Genome Editing *150*
13. Conclusion *151*

8. **Nano-Omics: A Short Story for Sustainable Approaches for Crop Production** **153**

1. Introduction *154*
2. Navigating Plant Stress Responses: Insights from Nano-Omics-Based Genomics and Transcriptions *155*
3. Unlocking Plant Responses to Salinity: Insights from Nano-Omics-Based Metabolomics *158*

9. Concerns of Nanotechnology for Agriculture, Environment, Plant and Human Health **161**

1. Introduction *162*
2. Cons: Toxicity Challenges of NPs in Plants *163*
3. Challenges and Future Directions *176*

10. Green or Biogenic Nanotechnology: A Sustainable Solution for Sustainable Agriculture and Human Health **177**

1. Introduction *178*
2. Concerns about Chemical Grade NPs: Bioaccumulation and Health Implications for Plants and Humans *180*
3. Biogenic NPs: An Eco-Friendly Approach for Sustainable Agriculture *185*
4. Biogenic NPs Synthesis and Characterization Techniques *186*
5. Conclusion *194*

11. Role of Nanotechnoogy in Zinc Fortification for Combacting Against Malnutrition **197**

1. Introduction *198*
2. Agricultural zinc deficiency *199*
3. Role of Zn in Human and Plant Health's *202*
4. Need of nano-enabled approaches for biofortification *204*
5. ZnO-NPs absorption and translocation in plants *205*
6. Nano-Biofortification approaches for Zinc fortification in Crops *209*
7. Challenges of nano-biofortification *218*
8. Conclusion *221*

12. Role of Nanotechnology in Pricision Agriculture Management **223**

1. Introduction *224*
2. Geoinformatics *226*
3. Nanoparticles *229*
4. Nanobionics *235*
5. Conclusion *239*

13. Role of Artificial Intelligence, Sensor Technology, Big Data in Agriculture: Next-Generation Farming **241**

1. Introduction *242*
2. Characteristics of Big Data *243*
3. Big Data and Smart Agriculture *244*
4. Sources of Big Data *246*
5. Techniques and Tool use in Big Data analysis *251*
6. Role of Big data in Agriculture ecosystem: For Smart Farming *257*
7. Conclusion *258*

Glossary **261**

Scientific Articles for Further Reading **265**

Index **271**

ABOUT THE AUTHORS

Dr. Vishnu D. Rajput boasts an exceptional academic journey in the fields of Agriculture and Biological Sciences. With a distinction in agriculture degrees from the CS Azad University of Agriculture & Technology, India, he laid the foundation for an illustrious career. Driven by an insatiable thirst for knowledge, he pursued a PhD at the esteemed University of the Chinese Academy of Sciences, China, and further honed his expertise through a Post-doc at Southern Federal University, Russia. Currently, Dr. Rajput holds the esteemed position of Principle Scientist at Southern Federal University, Russia. His research endeavours are firmly rooted in the realm of soil contamination, with a specific focus on potentially toxic elements and metallic nanoparticles. Within this intricate domain, his work delves into the intricate web of bioaccumulation, bio/geo-transformations, uptake, translocation, and toxic repercussions of metallic nanoparticles on various facets of plant physiology. His studies span from the macroscopic aspects of plant morphology to the microscopic intricacies of cellular and subcellular organelles, including cytomorphometric alterations and DNA damage. In his illustrious career, Dr. Rajput has contributed significantly to the realm of environmental science. His research shines a spotlight on the intricate interactions between nanoparticles, heavy metals, plants, soil, and microbial communities, offering valuable insights into the broader ecosystem. His contributions have been substantial, reflected in an impressive record of 600 scientific publications. Among these are 300 peer-reviewed, highly rated full-length articles, 38 enlightening books, and 107 chapters (indexed in Scopus), all testaments to his dedication to advancing knowledge in his field. Additionally, he has presented his findings in 55 conference articles. Recognized on an international scale for his profound expertise and contributions, Dr. Rajput has also dedicated his time to peer review, evaluating 271 manuscripts for esteemed journals. His commitment to scholarly integrity has earned him outstanding reviewing certificates from prestigious publishers like Elsevier and Springer. Adding to his impressive portfolio, Dr. Rajput serves on the editorial boards of various high-impact journals, contributing to the advancement of academic discourse in his field. He is holding national and international projects including Mega-Grant, BRICS. Dr. Rajput has received "Certificate for Appreciation 2019" 2021, "Certificate of Honor 2020", Diploma Award 2021 & 2022, Letter of Gratitude 2022 and 2023 by Southern Federal University, Russia, for outstanding contribution in academic, creative research, and publication activities. He has also received "Highly Qualified Specialist" status by Russian government. In year 2023, he is included in the list of 2% worlds potential scientists by Elsevier BV publishing house. In summary, Dr. Vishnu D. Rajput stands as a beacon of excellence in the realms of academia, creative research, and publication activities, making profound contributions to the fields of Agriculture and Biological Sciences while shaping the future of environmental science.

Dr. Abhishek Singh is a dedicated researcher (scientist) currently conducting cutting-edge research at the Faculty of Biology, Yerevan State University in the Republic of Armenia Eurasia. His academic journey is firmly rooted in the fields of agriculture and biotechnology, and he has earned his doctoral degree from the esteemed Sardar Vallabh Bhai Patel University of Agriculture and Technology in Meerut, UP, India. Dr. Singh's educational journey commenced with an Integrated MSc in Biotechnology, with a specialization in Plant Studies, from the prestigious Sam Higginbottom University of Agriculture, Technology, and Sciences in Allahabad, India. At present, Dr. Singh's research efforts are primarily focused on unraveling the intricate effects of salinity stress on various aspects of plant biology. His work spans a wide spectrum, encompassing physio-biochemical characteristics, morphological transformations, intricate anatomical variations, and detailed examinations of cellular and subcellular organelles' ultrastructure. Moreover, his research delves into the complexities of DNA, RNA, and protein damage, seeking to comprehensively understand the multifaceted impacts of salinity stress on plant life.In addition to his work on salinity stress, Dr. Singh is deeply engaged in exploring the fascinating interactions between metallic nanoparticles and plants. His research touches upon the entire lifecycle of these nanoparticles, including their bioaccumulation, uptake, translocation, and the potential toxic effects they may induce. His contributions in this area are crucial for advancing sustainable agriculture by shedding light on the interactions between nanoparticles, plants, and soil, as well as their physiological and genomic responses.Dr. Singh's dedication to scholarship is evident through his extensive publication record, which includes 35 full-length research articles, 10 insightful books, and 35 chapters, all of which are indexed in the prestigious Scopus database. Furthermore, he has presented his research findings through 6 articles at scientific conferences. Beyond academia, he has shared his expertise through over 30 articles related to agriculture in various magazines, further contributing to the dissemination of knowledge in his field. Recognized for his expertise and commitment to rigorous scientific evaluation, Dr. Singh actively participates in peer review activities for reputable publications on a global scale. His contributions have earned him accolades, including a certificate as a peer reviewer from Springer and recognition for his valuable insights. In addition to his research and review activities, Dr. Singh generously lends his expertise to the editorial boards of numerous journals and societies, contributing to the advancement of knowledge in the fields of agriculture and biotechnology. In recognition of his outstanding contributions, the state government of Uttar Pradesh, India, honoured Dr. Abhishek Singh with the prestigious "Appreciation Award 2021." This award underscores his significant achievements and underscores his position as a distinguished researcher in the fields of agriculture and biotechnology.

Dr. Ragini Sharma, a dynamic researcher from Punjab Agricultural University, India, stands as an advanced author with a remarkable academic portfolio holding CSIR-NET award and experience as an academic faculty member. Her contributions to the scholarly community are evident through a prolific publication record that includes both review and research articles. Notably, seven of her papers have been indexed in Scopus, and four articles in non-Scopus indexed publications, showcasing the depth and diversity of her research output. She has authored thirteen book chapters on various aspects of sustainable agriculture, emphasizing the application of nanotechnology, vermicomposting, and geoinformatics. Beyond conventional research, Ms. Sharma has presented the abstracts at various national and international conferences, addressing the integration of indigenous earthworm species as biofertilizers, showcasing her ability to engage with the forefront of scientific research. Her active participation in diverse workshops focussing on vermiculture, wildlife, and biodiversity conservation, underscores her commitment to continous expertise and practical skills in sustainable development. What sets her apart is her multilingual expertise, demonstrated through her proficiency in Hindi and English, allowing her to communicate her findings effectively. Moreover, her commitment to knowledge dissemination is exemplified by numerous popular articles in both national and international journals and magazines. Ms. Ragini Sharma's impact extends beyond academic circles, reflecting a promising trajectory for this advanced author who seamlessly bridges specialized research and broader public understanding in the field of agricultural research.

Dr. Karen Ghazaryan stands as a prominent figure in academia, currently holding the esteemed position of Dean at the Faculty of Biology, Yerevan State University. With an impressive 16-year journey in the realm of scientific research, he has left an indelible mark across multiple domains, making substantial contributions to variouscritical areas such as as Soil Ecology, Water Resources Management, and Environmental Sciences. Dr. Ghazaryan's profound expertise is vividly evident through his substantial portfolio of more than 96 scientific research articles, showcasing his unwavering dedication to advancing knowledge and understanding in his areas of specialization. His scholarly accomplishments extend to the authorship of influential books, further solidifying his position as a thought leader and authoritative voice in his field. Beyond his scholarly achievements, Dr. Ghazaryan actively contributes to the progress of scientific publishing as a distinguished member of various editorial boards for peer-reviewed journals. His commitment to advancing research and fostering academic excellence has had a lasting impact on the scientific community. His invaluable insights continue to shape the ever-evolving landscape of environmental sciences. Dr. Ghazaryan's exemplary career, unswerving dedication, and remarkable contributions to academia position him as an outstanding and respected figure in the realm of Soil Ecology and Environmental Sciences. His profound commitment to advancing knowledge and preserving the delicate ecological balance of our planet serves as an inspiration to aspiring researchers and scientists worldwide, making him a true luminary in his field.

Professor Dr. Tatiana Minkina is an eminent figure in the academic realm, holding the prestigious position of Head of the Soil Science and Land Evaluation Department at Southern Federal University. Her impressive career also encompasses leadership as the Head of the International Master's Degree Educational Program, "Management and Estimation of Land Resources," from 2015 to 2022. Notably, this program received accreditation by ACQUIN, highlighting its academic excellence and rigour.Dr. Minkina's extensive expertise lies in the fields of Soil Science, Biogeochemistry of Trace Elements, and Environmental Soil Chemistry. Her research endeavors encompass diverse areas, including soil monitoring, assessment, modeling, and remediation, with a particular emphasis on physicochemical treatment methods.In recognition of her outstanding contributions to education and the cultivation of highly skilled professionals, Dr. Minkina received the esteemed Diploma of the Ministry of Education and Science of the Russian Federation in 2015. Her dedication to academic excellence has left an indelible mark on the educational landscape.Currently, Professor Dr. Tatiana Minkina is actively engaged in various research projects funded by prestigious organizations, including the Russian Scientific Foundation, Ministry of Education and Science of the Russian Federation, and the Russian Foundation of Basic Research. These projects underscore her commitment to advancing scientific knowledge and addressing critical environmental challenges. Minkina is a distinguished member of several esteemed organizations, including the Expert Group of the Russian Academy of Science, the International Committee on Contaminated Land, Eurasian Soil Science Societies, the International Committee on Protection of the Environment, and the International Scientific Committee of the International Conferences on Biogeochemistry of Trace Elements. Her active participation in these organizations reflects her dedication to collaborative research and global scientific advancement.Professor Dr. Tatiana Minkina's remarkable research portfolio boasts an impressive total of 757 scientific publications, with 389 published in English, emphasizing her global reach and influence. She also serves as an invited editor for the esteemed Open Access Journal by MDPI, "Water," renowned for its Impact Factor of 2.524. Additionally, she holds editorial board memberships in esteemed journals such as "Geochemistry: Environment, Exploration, Analysis," noted for its Impact Factor of 1.109, and "Eurasian Journal of Soil Science," further solidifying her role as a leading authority in her field.Dr. Tatiana Minkina's unwavering dedication and significant contributions to Soil Science and Environmental Chemistry have earned her a well-deserved reputation as a respected and accomplished researcher. Her work continues to shape the landscape of her domain, making her an influential figure in the academic community.

Chapter 1

NANOTECHNOLOGY IN AGRICULTURE: AN INTRODUCTION ON RECENT UPDATES

Short Summary

Examining nanoparticles in relation to crops offers insight into their interactions, including absorption, movement, and buildup within plant systems. While nanotechnology has made significant strides in biomedical fields, transforming treatments and diagnostics, ongoing exploration delves into its application in plant science, offering innovative, non-technical strategies for enhancing plant safety, and fostering plant growth. This chapter delves into the current uses of nanoparticles (NPs) in agriculture with the aim of enhancing plant safety, fostering plant growth, and intimately improving crop production. It also explores the potential regulatory applications of NPs, which could introduce innovative and safe approaches for delivering biomolecules intelligently and employing new strategies in managing crop nutrients, genetically engineering crops, and combating abiotic stress.

1. Introduction

Addressing the challenge of ensuring food security for a growing global population stands as one of the foremost difficulties facing our planet. With projections indicating a rise in food consumption of up to 98% by 2050, driven by a population surpassing 9 billion, the demand for increased crop production is evident. However, meeting this demand necessitates a significant allocation of agricultural resources such as fertilizers, insecticides, pesticides, herbicides, and plant growth regulators. While these resources can enhance agricultural productivity, their extensive use has adverse effects on human health due to chemical accumulation. Farmers worldwide are actively exploring novel ideas and technologies to boost crop yields, employing both intensive and extensive farming methods. Nanotechnology-based stimulants and precision farming techniques are currently being deployed to support these ongoing endeavors. Progress in nanotechnology research holds the potential to enhance critical aspects of food security, encompassing agricultural output, soil quality, efficient water usage, food preservation during storage, and food quality (Figure 1). The term "nano" originates from Greek, denoting one billionth of a whole. In terms of measurement, one nanometer equals one billionth of a meter. Nanotechnology constitutes the field of study dedicated to manipulating and manufacturing substances at the nanometer scale. Investing in advanced technology is essential for safeguarding the long-term well-being of the economy, population, and food security. Such investment aims to boost production while minimizing food wastage. Nanotechnology offers a promising avenue by enhancing nutrient bioavailability and streamlining food production processes, thereby contributing to the production of high-quality food. The exploration and advancement of nanotechnology within the food and agricultural sectors are the focus of numerous ongoing studies.

Over the past decade, nanotechnology has made significant strides in agriculture, thanks to substantial public backing. This progress can be attributed to the unregulated nature of farm production, where there's a constant exchange of energy and matter. Unlike industrial nanoproducts, the lack of control over input of nanoparticles necessitates a consistent need for input materials. Nanotechnology has facilitated the development of new agrochemical agents and delivery mechanisms, offering the potential to reduce pesticide usage while increasing crop yields. Applications in agriculture include nano-formulations of agrochemicals for more effective pesticide and fertilizer delivery, leading to improved harvest without compromising soil or water quality, alongside precision farming techniques. Moreover, nanotechnology can help mitigate nitrogen depletion caused by bleaching, emissions, or improper agricultural practices by enhancing nitrogen content and other soil nutrients through nanoparticles-based fertilizers. Furthermore, nanotechnology enables the transfer of genes or DNA using nanoparticles to develop insect resistant plant varieties. Nanotechnology also holds promise for improving the efficiency of food processing and storage, thereby extending the shelf life of various food products. Furthermore, nanotechnology could advance biomass to energy production. In agriculture, the use of nanomaterials aims to reduce reliance on pesticide spraying while boosting crop yields. Nanotechnology tools such as nanocapsules and nanoparticle-based sensors enable the detection and management of diseases. Additionally, ongoing research explores the application of nanotechnology-derived devices in plant breeding and genetic transformation.

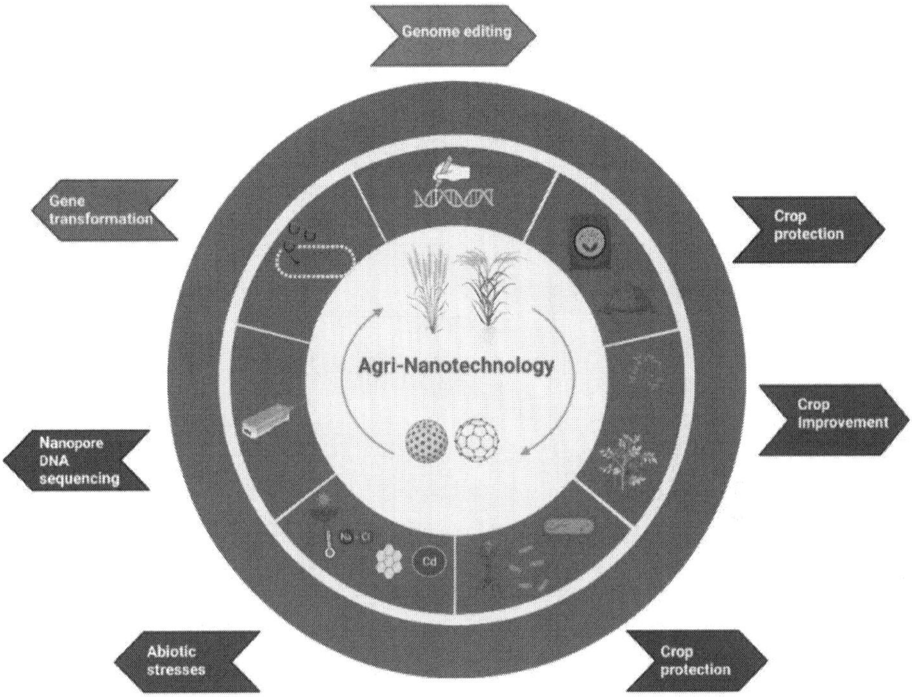

Figure 1. Diagrammatic presentation of application of Agri-nanotechnology.

While nanotechnology holds immense promise in agriculture, addressing various concerns is crucial as part of the risk assessment process. Nanoparticle enhancers made from biopolymers like proteins and carbohydrates have minimal impact on human health and the environment. Agriculture stands to benefit from nanotechnology across various facets including production, processing, storage, packaging, and transportation. Examples of areas where nanotechnology will revolutionize farming and agriculture include agricultural practices, plant nutrient uptake, disease detection, and pest management. In this chapter, we will provide an overview of non-technological aspects that contribute to enhancing crop production in agriculture with more sustainable approaches.

2. Fundamentals of NPs

NPs are commonly categorized based on their size, shape, similarity, organisation, and grouping. Circular, crystalline, and flat NPs are among the most prevalent morphologies observed. Additionally, NPs can be classified according to the dimension of electron movement, resulting in four categories:

0D, 1D, 2D, and 3D. 1D thin films, for instance, are utilized in sensor mechanisms and electronic devices. Second-generation NPs are represented by 2D-like nanotubes made of carbon, known for their excellent absorption and stability. 3D NPs encompass dendrimers and quantum dots. Furthermore, NPs are categorized based on their chemical compositions into carbon-containing, organic, and inorganic NPs. Inorganic NPs, which do not contain carbon and further divided into metals and metal oxides. Materials such as gold, silver, copper, iron, zinc, cadmium, cobalt, and platinum are commonly employed in the synthesis of nanoparticles (NPs). These NPs exhibit variations in their densities, shapes, surface area, and sizes. The synthesis of metal oxide NPs aims to enhance their performance, control their reactivity, and tailor their properties. Metal oxide NPs are produced from oxides such as Al_2O_3, CeO_2, Fe_2O_3, TiO_2, SiO_2, and ZnO. Green nanoparticles are composed of environment friendly polymers such as ferritin, lyposomes, dendrimers, and micelles. While micelles and vesicles may be susceptible to damage from electromagnetic radiation and heat, their non-toxic nature and effectiveness render them suitable for targeted drug delivery applications. Carbon based particles derived from carbon, find numerous agricultural applications including nanofibers, fullerenes, carbon black, graphene, and carbon nanotubes. Two primary methods are employed in NP synthesis: firstly, a top-down approach, where the synthesis begins with a larger form and progressively reduces it to create smaller NPs. The second method is a bottom-up approach, where the synthesis begins with the nanoparticles themselves. Various techniques are utilized for mass producing nanoparticles, employing both top-down strategies such as electron beam lithography, ion and plasma etching, anodization, photolithography, milling approaches, and bottom-up tactics such as molecular and atomic assembly or amalgamation. Bottom-up synthesis methods include chemical vapour deposition (CVD), Lazer pyrolysis, flame or plasma spring synthesis, polymer/monomer compounds, electrochemical or chemical nano-structural precipitation, and bio assisted synthesis. These synthesis techniques for nanoparticles can generally be classified into three categories: physical, chemical, and bio-assisted.

3. Recent Updates on Application of Agri-Nanotechnology

3.1 Crop production

The integration of nanotechnology into crop management offers significant potential for increasing crop productivity. Nanoparticles facilitate the controlled dispersion of chemicals, address nutrient deficiencies during fertilization, and enhance both the quality and yield of crops (Figure 3). Chemical fertilizers are commonly utilized due to their efficiency and cost-effectiveness compared to other types of fertilizers. However, improper, or excessive use of chemical fertilizers can lead to soil degradation, decreased crop yields, and environmental pollution. For instance, urea, which is highly soluble in water and prone to leaching, fulfils 80% of plants' nitrogen fertilizers needs. Any nitrogen component applied to paddy fields will usually cause leaching by runoff, volatilization, denitrification, and liquidation. The majority of nitrogen loss happens as nitrate through leaching and denitrification, with another 10-15% lost by volatilization and 2-30% through ammonia. The release of excessive nitrogen into the environment contributes to eutrophication in both water and air. Eutrophication poses various hazards, including toxic algal blooms and the depletion of marine life, endangering freshwater supplies. Nano-fertilizers play a crucial role in enhancing crop yields while addressing environmental concerns. Using nano-fertilizers can mitigate eutrophication and water pollution by improving nutrient delivery to plants and facilitating the controlled release of nutrients into the soil. Nano-fertilizers derived from banana peels have been employed in growing tomatoes, peppers, and flowers. Nano-fertilizers have been applied to cultivate and enhance various crops. For instance, chickpeas were grown using zinc oxide nanoparticles, while maize benefited from the application of silicon and iron slag powder. Tomatoes were cultivated with colloidal silica and NPK, while spinach thrived with the assistance of titanium dioxide. Grapes were grown using gold and sulfur nanoparticles. Additionally, in soil incubation experiments, calcium sulfate ($CaSO_4$) was found to decrease phosphorous loss from agricultural land. $CaSO_4$ is a commonly used soil conditioner known for its strong complexation with orthophosphate. Nano calcium sulfate has the potential to reduce soil phosphorous loss more effectively compared to conventional calcium sulfate due to its larger surface area, increased solubility, and better integration with soil and fertilizer. Utilizing nanoscale carriers in fertilizer application can help anchor plant roots to surrounding soil components and organic matter, thus minimizing chemical runoff and associated environmental issues. The use of nanoscale fertilizers can also mitigate soil toxicity resulting from excessive dosages, while slowing down nutrient release and prolonging fertilizer effects. Remarkable improvements in maize growth have been observed with the application of titanium dioxide nanoparticles has been found to enhance plant nutrient absorption and nitrate activity, thereby improving water and fertilizer management.

3.1.1 NPs based Macronutrient fertilizers

Macronutrient nano-fertilizers contain one or more essential macronutrient elements (N, P, K, Ca, and S) that provide plants with concentrated doses of these nutrients. The global demand for macronutrients is projected to reach 263 million tonnes by 2050. The widespread use of nitrogen fertilizer has contributed significantly to a 40% increase in food production per person over the past 50 years. However, excessive amounts of nitrogen, phosphorous, and potassium fertilizers ultimately find their way to surface and groundwater, causing severe damage to aquatic ecosystems. Therefore, the development of highly efficient and environmentally friendly macronutrient nano-fertilizers is crucial for achieving sustainable food production while safeguarding the environment. The primary nutrients essential for plant growth and development are nitrogen

(N), phosphorous (P), and potassium (K). These nutrients are required in various forms, including liquid and solid forms such as ammonium, nitrate, and anhydrous ammonia. Examples of NPK nano-fertilizers include apatite nanoparticles, monopotassium phosphate, and N-doped carbon dots, among others. Various fertilization methods and growth enhancers are detailed in Table 1. These nano-fertilizers typically have a small diameter of around 20nm and are water-soluble. When applied through foliar sprays or directly onto soil after being dissolved in water or fertilizer solution, they have been shown to significantly enhance plant growth. Additionally, composite nano-fertilizers that provide multiple nutrients, such as monopotassium phosphate, help migrate the adverse effects of abiotic stress, such as salt stress, on plant growth. In addition to nitrogen, phosphorous, and potassium, calcium and sulfur are also essential nutrients. Calcium oxide nanoparticles and sulfur nanoparticles have been used for foliar and soil treatments when diluted in water. Despite their wide and irregular sizes (ranging from 20 to 80 nm), these nano-fertilizers are utilized in larger quantities compared to N, P, and K nano-fertilizers (Table 1). However, it is important to note that further research is necessary to ensure their safety for widespread and long-term agricultural use, even though these nano-fertilizers primarily consist of harmless elements such as carbon, oxygen, nitrogen, potassium, phosphorous, calcium, and sulfur.

3.1.2 NPs based Micronutrient fertilizers

Micronutrients are essential plant nutrients that are required in small quantities (less than 10 mg/kg of soil) compared to macronutrients. They are crucial for the synthesis of enzymes and biomolecules involved in plant defense. Additionally, consuming meals lacking in micronutrients can have adverse effects on human health, including impaired cognition, stunted growth, and anaemia. In addition to macronutrient fertilizers, micronutrient nano-fertilizers such as zinc, molybdenum, manganese, and iron must be applied to plants. Zinc is a micronutrient essential for all living organisms, including humans. Zinc oxide nanoparticles (NPs) have been utilized as nano-fertilizers in two separate studies. These nanoparticles have particle sizes of approximately 50 and 70 nm. While it takes some time for these ZnO-NPs to dissolve in water, they and their aggregates are smaller in diameter than stomatal pores, indicating their ability to penetrate and move throughout plant tissues. Zn^{2+} ions can be continuously released from nanoparticles attached to leaf surfaces, providing plants with a steady supply of zinc that they can absorb through their stomata. ZnO NPs have been shown to enhance the growth and yield of various plant species. The absence of iron (Fe), a key nutrient crucial for chlorophyll production and the electron transport system, can negatively impact the proper physiological role and nutritional properties of plants. Maghemite (-Fe_2O_3) nanoparticles have been employed as nanofertilizers, enhancing plants' resistance to drought stress, and reducing levels of H_2O_2 while improving leaf development rate and chlorophyll content. However, Fe_2O_3 nanoparticles dissolve slowly and tend to aggregate into larger particles with a hydrodynamic size of up to 500nm, facilitating the steady uptake of iron ions (Fe_3+) by plants over prolonged periods.

Nanoscale zero-valent iron (nZVI) has been utilized to improve soil quality and increase rice yields in a separate study. Larger nZVI (100 nm) with lower coercivity (35.17 Oe) significantly increased grain yield and soil pollution removal rates. Conversely, smaller nZVI with higher coercivity exhibited increased homo-aggregation when tested in soil, negatively impacting performance.

Copper (Cu) is another crucial micronutrient essential for optimal plant and microbial development. Various Cu-based nanoparticles, including CuS, CuO, and Cu $(OH)_2$, have been applied to plants through foliar and root applications. Spindle shaped Cu $(OH)_2$ nanoparticles exhibit greater dispersibility compared to spherical CuO and CuS nanoparticles, facilitating greater Cu uptake by plants. However, the reduced dispersibility of CuO and CuS nanoparticles allows them to remain on roots longer, continuously transferring Cu to plant leaves after application.

Manganese (Mn) is also an essential micronutrient vital for plant growth and development, particularly in biochemical reactions of photosynthesis. Stable square shaped Mn nanoparticles with a hydrodynamic radius of around 100 nm can be dispersed in water without aggregation, facilitating easier absorption by plants even at higher concentrations.

Table 1: Application Agri-Nanotechnology in various fields of agriculture.

Area of Scope	NPs	Application methods/ Type	Concentration/Target Pests/ Insects/ Herbicide/plant	Mode of action
Crop production	ZnO-NPs	Foliar	5 g L^{1}	Rice (*Oryza sativa* L.) growth, yield, characters that affect yield, microbial counts, and the activity of the dehydrogenase enzyme all began to improve
		Foliar	10 mg/L	Coffee (*Coffea arabica* L.) plant was an increase in net photosynthetic rate (55%), fresh weight (37% roots, 95% leaves), and dry weight (28% roots, 85% leaves).
		Foliar	10 mg/L	Clusterbean (*Cyamopsis tetragonoloba* L.) enhancements to growth physiology and a rise in biomass accumulation and nutrient concentration
		Soil irrigation	0–1000 ppm	Peanut (*Arachis hypogaea* L.) increased rates of photosynthesis, plant growth, morphological diversity, crop yields, and overall plant efficiency
	$nCeO_2$	Soil	0–500 mg kg^{1} soil	Barley (*Hordeum vulgare* L.) increased plant functioning, increased Ce deposit in grains, and increased P, K, Ca, Mg, S, Cu, Fe, Zn, Mn, Al, amino acids, fatty acids, methionine, aspartic acid, threonine, arginine, and linolenic acid.
	Nitrogen-doped carbon-dots N-CDs	Nutrient solution	0.2 mg L^{1}	Mung bean (*Vigna radiata L.*) increased its growth rate by 200 percent (average length of shoots and roots).
	Monopotassium-phosphate (MKP)	Foliar and soil	3 g L^{1}	Tomato (*Solanum lycopersicum L.*) growth parameters were enhanced despite the presence of salt stress
Crop protection	Silver	Nano viricides	*Solanum tuberosum*	Tomato *(Solanum lycopersicum L.)* reducing TSWV infections and preventing localised damage

Area of Scope	NPs	Application methods/ Type	Concentration/Target Pests/ Insects/ Herbicide/plant	Mode of action
	Zinc oxide	Nano viricides	*Tobacco mosaic virus (TMV)*	Acceleration of TMV invasion suppression in *Nicotiana benthamiana*
	Iron oxide	Nano viricides	*Tobacco mosaic virus (TMV)*	Activated antioxidants defense system in *Nicotiana benthamiana* and induction of resistance to TMV by upregulating SA genes
	Chitosan/ tripolyphosphate	Nano herbicides	Paraquat	Eco-friendly weed control in maize and mustard using paraquat-doped chitosan/tripolyphosphate nanoparticles
	Chitosan	Nano herbicides	*E. crus-galli*	Nanoparticles of maize carboxymethyl chitosan for the controlled liberation of glutathione-sensitive herbicides
	PVC (Polyvinylchloride)	Nano insecticides	*P. vulgaris, Phaseolus vulgaris; P. xylostella, Plutellaxylostella; S. hortensis, Satureja hortensis L.; S. litura, Spodoptera litura*	Birch and southern yellow pine NPs-based biocides that are sprayed on the leaves have different effects on plant growth
	Silica	Nano insecticides	*H. armigera*(Cotton bollworm)	Compared to conventional insecticides, silica nanoparticles loaded with Brassica chinesechlorfenapyr are more effective and have a lower toxicity profile
	Polyvinylpyridine (PVPy) and polyvinyl pyridine-co-styrene	Nano fungicides	*G. trabeum*	Sapwood of a southern pine Tebuconazole and chlorothalonil, two fungicides, were slotted in into polymeric nanoparticles. Specimens of Southern pine sapwood have been bottled with biocide-containing nanoparticles and then uncovered to the wood decay fungus Gloeophyllumtrabeum in a straightforward water trial.

Area of Scope	NPs	Application methods/ Type	Concentration/Target Pests/ Insects/ Herbicide/plant	Mode of action
	Chitosan/Pectin	Nano fungicides	*F. oxysporum and A. parasiticus*	Corn, and Cucumbers Seeds of *Cucumis sativa, Zea mays*, and *Lycopersicum esculantum* benefit from a fungicide formulated for tomatoes called carbendazim nano
	Chitosan	Nano fungicides	*F. graminearum*	The maize pathogen Fusarium graminearum can be stunted by using chitosan nanoparticles encapsulated with the essential oil of the *Cymbopogon martinii* plant.
Genetic engineering	SWCNTs/ MWCNTs	Gene transformation/ Co-culture	*N. tabacum*	With *N. tabacum* leaf explants, the rate of callus formation and regrowth of shoots from the callus touched 100% by the third week of cultivation on the medium
	SWCNTs	Gene transformation/ Needleless injection	Leaves of tobacco	Silencing endogenous genes (GFP) in intact plant cells was achieved using SWCNTs, which also served to shield the siRNA from nuclease degradation
	MSNs	Gene transformation/ Spray and needleless injection	Leaves and roots of tomato	The exogenous genes were introduced into plants using the MSNs-based transient expression system, which then carried out transcription and translation in the appropriate plant tissues
	DNs	Gene transformation/ Needleless injection	Leaves of *N. benthamiana*	When compared to the control groups, the GFP fluorescence intensity dropped the most in the leaves that were injected with siRNA-DNs
	CDs	*Gene transformation/Leaf spraying*	Leaves of *N. benthamiana*and tomato	GFP transcript and protein levels were found to be reduced by over 80%.

Area of Scope	NPs	Application methods/ Type	Concentration/Target Pests/ Insects/ Herbicide/plant	Mode of action
	CaP-NPs	Gene transformation/ Co-culture	*Brassica juncea L. cv. Pusa Jaikisan*	When compared to Agrobacterium tumefaciens (54%) and naked DNA (8%), the transformation efficacy was roughly 80% when using a CaP-based delivery platform
	Gold nanoparticle	Genome editing CRISPR/Cas9 /	Wheat	Hexaploid wheat gene targeting with CRISPR/ Cas9 and DNA replicons
	Fe_2O_3-NPs	Salinity	0, 10, 20 and 30 μM	Fresh and dry weight, phosphorus, potassium, iron, zinc, and calcium content, and antioxidant enzyme activity all rise in peppermint (*Mentha piperita* L) leaves
	Cu-NPs	Salinity	50, 100, and 150 mg/L	Improved tomato (*Solanum lycopersicum* L.) yield, immunity, and mineral content
	Mn-NPs	Salinity	0.1, 0.5, and 1 mg/L	The mineral content of capsicum (*Capsicum annum L.*) plants differed between their shoots and roots.
	Cu-NP	Drought	3.33, 4.44 and 5.55 mg/L	In response to drought, maize (*Zea mays* L.) increased both total seed number and grain production.
	ZnO-NPs	Cu and Pb heavy metal stress	5, 10, 15, 20 and 25 mg/L	Lowered metal uptake in rice (*Oryza sativa* L.) and increased photosystem and thylakoid membrane strength
	Se-NP	Heat stress	10 mg/L	Improved chloroplast strength of Sorghum *(Sorghum bicolor L.)*

3.2 Crop Protection

The traditional approach to pest control involves indiscriminate application of fungicides, and insecticides, with approximately 90% of the formulations being lost to the environment rather than reaching their intended targets. Plants pests and microorganism cause an annual global crop decline averaging between 20-40%. Insecticides, fungicides, and herbicides all forms of pesticides, are crucial for modern farming's pest management strategies. It is imperative to develop effective and less environmentally harmful pesticides. Nanotechnology and other innovative approaches have the potential to render agrochemicals safer, longer-lasting, and more soluble in water, offering environmental benefits. Nanopesticides release their primary component gradually according to plant requirements, reducing the adverse effects of conventional pesticides on humans, pollinators, and the environment (Figure 3), while also enhancing the overall efficiency of conventional pesticides. Conventional pesticides often do more harm than good, including ineffective pest, pathogen, and weed control. Nanotechnology offers a method to ensure that the active ingredient is present at a threshold concentration for effective protection against insect pests, by encapsulating the key component within nanomaterials. Encapsulated pesticides, unlike conventional ones that readily dissolve and may harm unintended plants, potentially leading to resistance in target species, offer prolonged effectiveness by gradually releasing essential active ingredients at specific target sites. Inorganic nanoparticles such as ZnO, Cu, SiO_2, CaO, MgO, MnO, and Ag have shown efficacy against plant microbes. Nanopesticides have proven effective against various pests in both field and stored grain environments. Combining chitosan alginate (in nano-formulation) with paraquat has demonstrated reduced toxicity compared to using paraquat alone. Copper nanoparticles have successfully eradicated various phytopathogenic bacteria, fungi, and yeasts. The invasive aquatic weed *Eichhornia crassipes* has been effectively controlled using silver nanoparticles-chitosan encapsulated paraquat. Overall, nanoparticles can serve dual purposes in plant protection: directly shielding crops or acting as carriers for pesticides, which are then applied via spraying onto plants. Although there are a lot of possible uses for NPs, agricultural settings have just lately started to be researched as a possible use.

3.2.1 Nano-insecticides

At least 90% of pesticides that are used either spread throughout ecosystems or are ineffective in controlling insects according to research. The scenario's costs to the environment and the inflation in agriculture are intolerable. To offer sufficient protection against pest invasion and crop loss, the formulation's active components must be present at the target site in the lowest effective concentration (Table 1). Crop protection has undergone a revolution, thanks to the application of nanotechnology in the formulation and encapsulation of pesticides (Figure 2). Using a process known as nano-encapsulation, pesticide-active components may be contained in NMs of different sizes. A few of the particles used in insecticide nano-formulations can act as active substances, while others are specifically designed nanostructures with insecticidal properties. Nano-encapsulated or nano-formulated insecticides have shown increased efficacy as the active ingredients are released gradually and persist within the plant or root zone for longer durations. Traditional insecticide formulations not only lead to resistance in targeted insects but also pose a risk to non-target organisms due to decreased water solubility. Nano-encapsulation and Nano-formulation offer solutions to these limitations. Insecticides that have been nano-encapsulated or formulated exhibit numerous desirable properties which include enhanced crystallinity, stiffness, biodegradability, thermal stability, permeability, and solubility. For example, pheromones formulated in nanofibers do not significantly impact mortality in oriental fruit moths (*Grapholitamolesta* L.), indicating sustained attract-and-kill effects from the insecticide and pheromone, with controlled release mechanisms. Moreover, studies have demonstrated that insecticide nano-formulations can contribute to the development of plant mediated universal resistance. To better combat sucking or chewing insects, SiO_2 nanosphere formulations can enhance insecticide penetration into plant tissues, providing better access to cell sap. Nanomaterials in insecticides hold tremendous promise for pest control.

3.2.2 Nano-herbicides

Weeds are plants that reproduce aggressively or spread beyond their usual habitat. Herbicides are chemicals derived from biological or synthetic sources that hinder plant growth or eliminate them. Modern herbicides, often synthetic versions of natural hormones found in various plants, can hinder the growth of desirable crops. While herbicides effectively eliminate weeds, excessive use can negatively impact plant growth and human health. Synthetic or biological herbicides can be effectively transported using nanotechnology using nano-sized preparations or pesticide formulations based on nanomaterials. Herbicides are incorporated into various nanoparticles to enhance their bioavailability and effectiveness in weed control. The distinct properties of nanomaterials, such as thermal stability, biodegradability, solubility, stiffness, crystallinity, and permeability, can be exploited to develop a wide range of herbicides. Nano-herbicides, with their large specific surface area, exhibit improved affinity for the target. The efficiency of herbicide spraying is enhanced using nano-encapsulated formulations, reducing splash losses and spray drift. Soil containing nano-herbicides can efficiently eradicate weeds and weed seeds before their germination. Utilizing encapsulated herbicides such as triazine and atrazine could enhance crop protection effectiveness. Most commercially available herbicides only target the above-ground parts of plants, leaving tubers and rhizomes intact, which can give rise to new weeds. In response, researchers have developed herbicide molecules encased in nanoparticles to penetrate roots and reach areas that inhibit the glycolysis process, thus targeting specific weeds. The prolonged usage of herbicides can leave residues in the soil, obstructing the growth of subsequent crops, highlighting the significance of detoxifying herbicide residues for sustainable development. Carboxymethyl cellulose NPs exhibit the capability to remove up to 88% of atrazine herbicides from water. Ultimately, nanotechnology has the potential to expand the effective range of herbicides and prolong their efficacy.

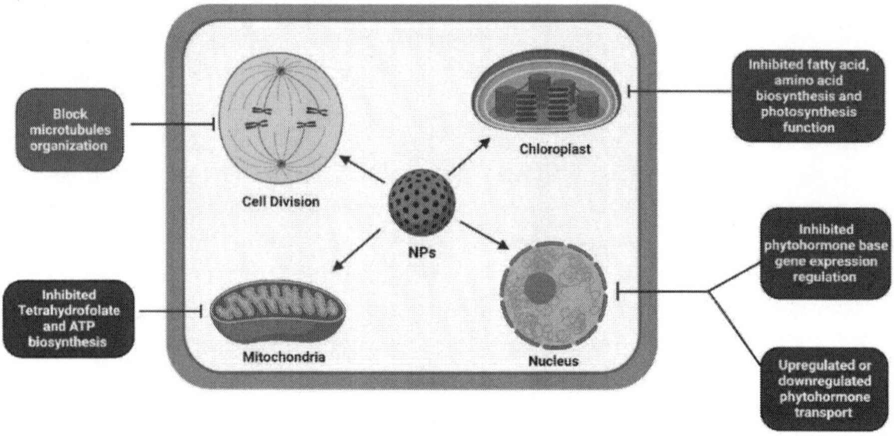

Figure 2. Mode of action of NPs mediated nano-herbicides agents use for crop protection.

3.2.3 Nano-bactericide

Any chemical that may either kill or stop bacteria from multiplying is a bactericide, regardless of whether it was created in a lab or was discovered in nature. One of the hardest challenges to overcome in farming is the emergence of drug-resistant bacteria due to the misuse of bactericides, which presents a major worldwide danger (Figure 3). The solution to this problem that nanotechnology has made feasible is promising. The way that NPs interact with the bacteria is essential to their effectiveness. For nanoparticles (NPs) to effectively evolve as nanomaterials (NMs), a comprehensive understanding of their physiochemical characteristics as well as their interaction with microbial life is necessary. Some of the microbes that cause soil-borne illness include *Ralstonia solanacearum, Fusarium oxysporum, Verticillium dahlide, Fusarium solani, Monilinia fructticola, Colletotrichum gloeosporiodes, Botrytis cinerea* and *Alternaria alternate.* It has been demonstrated that metallic oxide NPs play an effective role in controlling the growth of these soil-borne disease-causing microbes. Furthermore, soil microbial connections with decaying crops, organic matter decomposition, and nutrient cycling. Therefore, safeguarding the extensive richness and biomass of soil microorganisms is a primary responsibility of agricultural systems. Nanoparticles composed of metallic oxides, like CuO and Fe_3O_4 NPs, exert a notable influence on the variety and quantity of soil microorganisms. The distinctive physiochemical properties of nanoparticles offer potential for the advancement of potent antimicrobial agents in the future.

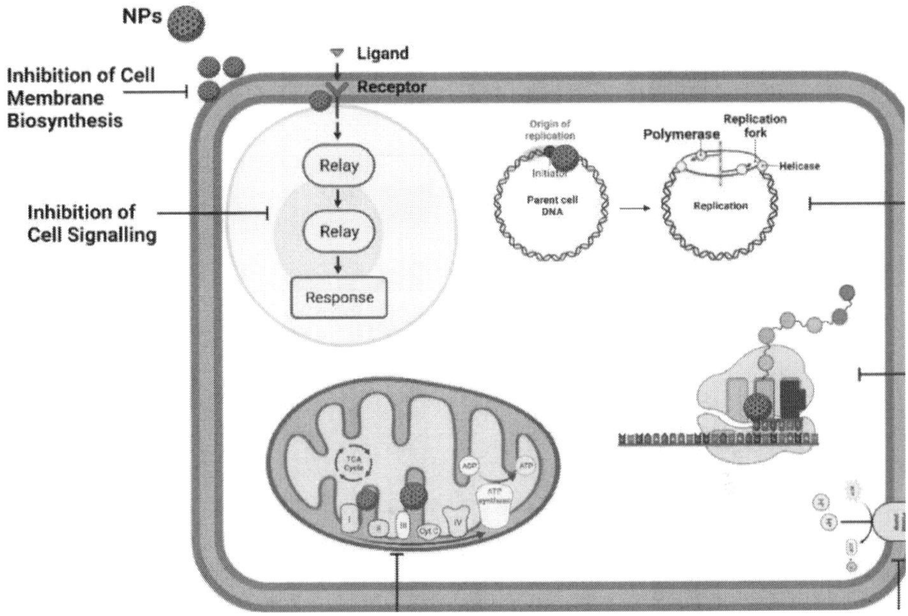

Figure 3. Mode of action of NPs mediated nano-bactericides agents use for crop protection.

3.3 Role of NPs in plant genetic engineering

3.3.1 NPs based plant transformation

Enhanced crop yields stem from factors such as climate change, a growing human population, and the utilization of biotechnology to confer desired genetic traits on plants. The plant cell walls act as barriers, preventing chemicals from entering the cells. Various methodologies, including Agrobacterium transformation and biolistic approaches, are employed to address the issue of DNA transmission in plant cells and to alter plant genetics. However, these methods have limitations due to their restricted host range and the significant damage they can cause to plants, often hindering plant growth. Despite progress in plant genetic engineering, advancements have been relatively slow until recently. There is a suggestion that nanotechnology-based approaches could offer efficient and minimally toxic means to introduce genes or other compounds into plants. Nanotechnology has significantly impacted various fields of study, particularly in medicine, energy, and industry. Nanotechnology-based technologies have been used to introduce chemicals and biomolecules into plant and mammalian cell systems. However, nanoparticle-mediated transport of plant biomolecules faces challenges due to the presence of the cell wall barrier. Four types of nanoparticles are utilized for gene transfer: polymer-based NPs, metallic NPs, silicon-based NPs, and carbon-based NPs. It is proposed that nanoparticles protect the genetic cargo from cellular enzymatic destruction, such as nucleases, facilitating effective plant transformation. The types of genetic payloads transported by various nanoparticles (NP) types differ. While metallic NPs are limited to transporting DNA as genetic cargo, carbon nanotubes (CNTs) have the capability to transport both RNA and DNA. Polymeric NPs such as PEG and Polyethyleneimine can encapsulate RNA, DNA, and proteins for delivery into cells. Silicon-based NPs are capable of transporting DNA and proteins. Certain NPs, including carbon nanotubes (CNTs) and mesoporous silica (MNPs), possess the ability to deliver genetic cargo into cells without requiring any chemical or physical pre-treatment, whereas others such as gold NPs and magnetic NPs (MNPs), do necessitate pre-treatment. NP-mediated passive delivery has been reported in tobacco, arugula, and cowpeas. Novel materials, including nanoparticles (NPs), are being explored as effective means of transportation by genetic engineering systems. A method involving plant bombardment, specifically targeting the wheat shoot apical meristems, has been investigated for this purpose. In this study, gold particles coated with a construct containing the green fluorescent protein gene were delivered to the L2 cell layer of wheat shoot apical meristems. This technology successfully facilitated the genetic modification of wheat without resorting to embryogenic callus culture, suggesting its potential applicability to other crops resistant to transformation using conventional methods.

3.3.2 Plant genome editing via NPs based tools

Gene editing is frequently employed in agriculture settings and holds promise for revolutionizing the field of plant breeding. The CRIS/Cas system, an RNA-based defense mechanism present in prokaryotes, consisting of Cas molecules and CRISPR repeat spacer arrays, has been effectively harnessed for modifying plant genomes. Traditional methods of transformation and regeneration have facilitated the successful application of CRISPR/Cas genome editing in plants. However, challenges persist in crop genome editing, including the low efficiency of homology-directed repair (HDR), and issues related to tissue culture, regeneration, and delivery mechanisms. Nanomaterials (NMs) offer the potential to enhance the precision and efficiency of the CRISPR/Cas9 technique, owing to their characteristics such as high throughput, diverse charges, high tensile strength, and small size. The targeted delivery of CRISPR/Cas9 single guide RNA by nanoparticles represents a significant advancement in genetic engineering. Mesoporous silica nanoparticles (MSNs) have been utilized to load recombined loxP sites into chromosomal DNA and transport Cre recombinase in immature maize embryos.

Following the delivery of the modified MSNs to crop cells, the loxP fragment was successfully integrated. Utilizing cationic arginine gold nanoparticles for the transportation of Cas9En (E-tag)-ribonucleoproteins (RNP) containing sgRNA in cultured cell lines demonstrated significant effectiveness, achieving around 30% active nuclear or cytoplasmic gene editing. This development marks a significant milestone in crop science research. Despite the widespread acceptance of CRISPR/Cas9 technology in many crops facilitated by current nanotechnology, additional challenges in crop genome editing remain to be addressed through agri-nanotechnology-based approaches.

3.3.3 NPs based plant gene sequencing

Nanotechnology has revolutionized gene sequencing, enabling more precise detection and utilization of plant traits, thereby enhancing their resilience to environmental pressures and diseases. Biological markers such as nanoparticles and quantum dots have demonstrated remarkable accuracy in this regard. Additionally, advancements in nanotechnology and nanoscopy have facilitated the optical mapping of DNA, providing detailed insights into the genetic and epigenetic composition of cells. Nanopore sequencing offers the potential to overcome the limitations of short-read data by allowing the sequencing of individual DNA molecules with lengths ranging from tens to potentially up to 100 kilobase pairs. Various genomics applications of nanotechnology include molecular diagnostics, genome sequencing, profiling of noncoding RNA expression, identification of transcription factor binding sites, and targeted sequencing.

3.4 Role of NPs in abiotic stresses management

3.4.1 Salinity stress

The escalating salinity levels in soil stem directly from the irrigation of agricultural areas worldwide with saline water, a consequence of water scarcity exacerbated by global warming. Salinity, characterized by an excessive accumulation of salt in the soil, poses a significant challenge to modern farming practices, and severely is compromised when the concentration of NaCl exceeds 200 nm, impacting crucial stages such as seed germination, seedling development, vegetative growth, and flowering. Numerous horticultural crops, including fruits, vegetables, and spices are adversely affected by salinity. Ionic strength imbalances induced by salt stress disrupt a wide array of biochemical, physiological, and metabolic processes, encompassing osmotic stress, water stress, oxidative stress, nutritional stress, and decreased cell division (Figure 6). To mitigate the adverse effects of salt stress on plants, nanoparticles such as zinc, silver, silicon dioxide, copper, iron, manganese, boron, cadmium, tin, titanium dioxide, cerium oxide, and potassium dioxide have been employed. In salt-sensitive Medicago sativa, tropical application of potassium nanoparticles (K-NPs) enhanced salt tolerance by elevating proline and antioxidant enzyme levels, including catalase. Treatment of pearl millet plants with silver NPs (Ag-NPs) led to reduced levels of malondialdehyde (MDA) and ROS, along with improved antioxidant activity, possibly attributed to decreased Na^{+} absorption in the leaves. Enhanced photosynthetic activity in Brassica napus was facilitated by cerium oxide nanoparticles, which modified root cells and subsequently enhanced mineral uptake. The application of nanoparticles to plants has demonstrated significant efficacy in alleviating the detrimental effects of salt stress and regulating plant responses, as evidenced by an increasing body of research.

3.4.2 Drought Stress

Drought, a significant environmental challenge, has captured the attention of both environmental and agricultural scientists alike, emerging as a prominent global agricultural issue that hampers plant development and productivity. The impact of drought stress on plant growth parameters reverberates across the economy, with crop productivity being notably affected more than by any other environmental factor. Projections by the IPCC suggest a rise in global average temperatures by year 2100, leading to widespread drought occurrences in numerous regions. In the agricultural sector, droughts disrupt regular plant growth and life cycles due to water scarcity, exacerbated by diminishing precipitation and increasing evapotranspiration demands.

The scarcity of water directly contributes to reduced crop yields, characterized by cell shrinkage, membrane disruption, oxidative stress, and leaf senescence. Drought-induced stress affects plant development by impleading cell turgor, crucial for cell expansion, and consequently impacting various aspects such as cell division rates, leaf size, stem and root length, stomatal function, and water and nutrient uptake.

Studies have demonstrated that NPs enhance plant drought resistance by improving water absorption, root hydraulic conductance, and the differential distribution of proteins involved in detoxifying ROS, stress, and hormone signalling. And oxidation-reduction pathways, among other morphological, physiological, and biochemical changes. The foliar application of metal-oxide nanoparticles like iron oxide, titanium oxide, and zinc oxide has shown promising results in enhancing the metabolic and physiological functions of crops under drought conditions.

After administrating Si NPs to water-stressed pomegranate plants, a rise in chemical or physical parameters, nutrient status, and photosynthetic pigments, particularly those associated with fruit cracking, phenolic content, and amounts of osmolytes, antioxidant enzymes, and abscisic acid, was observed. Additionally, green ZnO- NPs applied tropically at low levels have been utilized to enhance tomato drought tolerance. Fenugreek plants exposed to drought stress were reported to benefit from both nano fertilizers and green-produced Fe_3O_4 NPs. However, the effectiveness of NPs in enhancing plant drought tolerance may vary depending on certain climatic conditions.

Previous studies have demonstrated that Si NPs increase plant resistance to drought stress. For instance, hawthorn plants treated with Si NPs exhibit improved drought tolerance, accompanied by alterations in defense-related physiological indicators across various drought intensities and Si NPs concentrations. Similarly, Si NPs displayed promising capabilities for post drought plant recovery in barley through the modulation of morpho-physiological characteristics. In water-limited and salty environments, the growth and production of cucumber were enhanced. Wheat plants subjected to drought stress showed improved biomass, CAT, and SOD functions, photosynthesis rate, production, and relative water content upon treatment with chitosan NPs. Soil treatment with CeO NPs significantly improved plant growth and amplified the photosynthetic rate by regulating water use efficiency in soybean (*Glycine max* L.) plants, while foliar application of Fe NPs was found to alleviate drought stress effects on safflower cultivars. Furthermore, the use of silver NPs mitigated the negative impacts of drought on lentil (*Lens culinaris Medic.*) plants. In Arabidopsis thaliana, the delivery of abscisic acid with the aid of Si NPs was found to enhance drought tolerance.

3.4.3 Heat Stress

Heat stress can manifest in exceedingly hot environments, a trend exacerbated by global warming in recent decades. It is widely acknowledged that heat stress occurs when temperatures persist above a critical threshold for an extended period, resulting in irreversible impairment of plant growth. These extreme temperature shifts can disrupt the intermolecular interactions necessary for optimal growth during hot summers, thereby limiting plant development and fruit set. Furthermore, heat stress diminishes photosynthetic efficiency and adversely affects plant growth, survival, and productivity. In regions with tropical and subtropical climates, heat stress has the potential to emerge as a significant factor constraining the yield of field crops. Among the various environmental stressors undergoing continual change, rising temperatures are particularly concerning. Enhanced production of ROS and oxidative stress contribute to cell death in crop plants by damaging membrane lipids, disrupting cellular homeostasis, and impleading various metabolic processes. Moreover, heat stress disrupts photosystem II, induces chlorophyll breakdown, and reduces carbon fixation, collectively diminishing photosynthesis and impeding plant development. Recent advancements in modern nanotechnology have revolutionized the agricultural sector, offering the potential for enhanced plant growth and development under stressful environmental conditions. Several recent studies have explored the use of NPs to enhance crop plant's resilience to heat stress. For instance, the application of selenium NPs increased chlorophyll content, water potential, and growth rate in tomato plants, significantly alleviating heat stress. Similarly, the incorporation of TiO_2 NPs markedly reduced heat stress, as evidenced by decreased stomatal opening in tomato plants. Wheat growth was also improved upon the application of biogenic Se-NPs, and silver NPs notably enhanced the morphological characteristics of wheat plants under heat stress conditions. In conclusion, sustainable agriculture stands to benefit from increased heat stress tolerance in plants through the use of metallic NPs as nanofertilizers. Researchers have observed that Se-NPs can mitigate the adverse effects of high temperatures on sorghum plants by activating the antioxidant defense system, thereby reducing membrane damage, enhancing pollen germination, and increasing crop yields. Additionally, wheat plants treated with Ag-NPs exhibited improved morphological growth and were shielded from heat stress. Furthermore, exposure to Zn nanoparticles in wheat increased the production of antioxidant enzymes and reduced lipid peroxidation rates, contributing to better survival under heat stress conditions.

3.4.4 Flood Stress

Flooding poses a significant threat to plant life, primarily due to the saturation of soil, which impedes root respiration and nutrient uptake. Various factors such as excessive precipitation, poor drainage systems, or indiscriminate watering practices can contribute to inundation, rendering plants susceptible to detrimental effects. The ramifications of floods extend beyond mere plant survival, impacting both food security and economic stability at a national level, thereby highlighting its significance as a prominent abiotic stressor. The adverse consequences of flooding on vegetation are observed across diverse biomes, including wetland, tidal, and salt marsh ecosystems. Notably, wetland plant species exhibit a degree of resistance to submersion and waterlogging, whereas terrestrial plants are more vulnerable to flood-induced stressors. The phenomenon of hypoxia, characterized by oxygen deprivation in the rhizosphere due to waterlogging, inhibits crucial physiological processes such as root metabolism and nutrient assimilation, thereby impeding overall plant growth.

Moreover, flooding alters soil characteristics, including pH and redox potential, while also mobilizing phytotoxins, exacerbating the stress experienced by inundated plants. However, recent research suggests the potential utility of nanoparticles in mitigating the detrimental effects of flooding stress on plant physiology. For instance, studies involving silver nanoparticles (NPs) have demonstrated their efficacy in reducing stress levels and promoting growth in flooded soybean plants by modulating various biochemical pathways, including amino acid synthesis, protein synthesis, glycolysis, and wax formation. Similarly, investigations into the effects of aluminium oxide (Al_2O_3) NPs were found to enhance root length, suppress proteins associated with glycolysis, and facilitate the scavenging of reactive oxygen species by upregulating the ascorbate/ glutathione pathway (AsA/GSH). Furthermore, the upregulation of ribosomal proteins further underscores the potential of nanoparticles in Alleviating flooding-induced stress and fostering plant resilience.

3.4.5 Mode of action of NPs to mitigating abiotic stress

Nanoparticles (NPs) exhibit the capability to modify plant metabolism by facilitating micronutrient delivery, modulating gene-expression, and interfering with diverse oxidative processes. While advancements have been made in elucidating the oxidative mechanisms, the specific manner in which NPs stimulate ROS production remains not entirely understood. Notably, NP has been observed to instigate an oxidative burst and escalate ROS concentrations by disrupting the electron transport chain within both mitochondria and chloroplasts, as depicted in Figure 4.

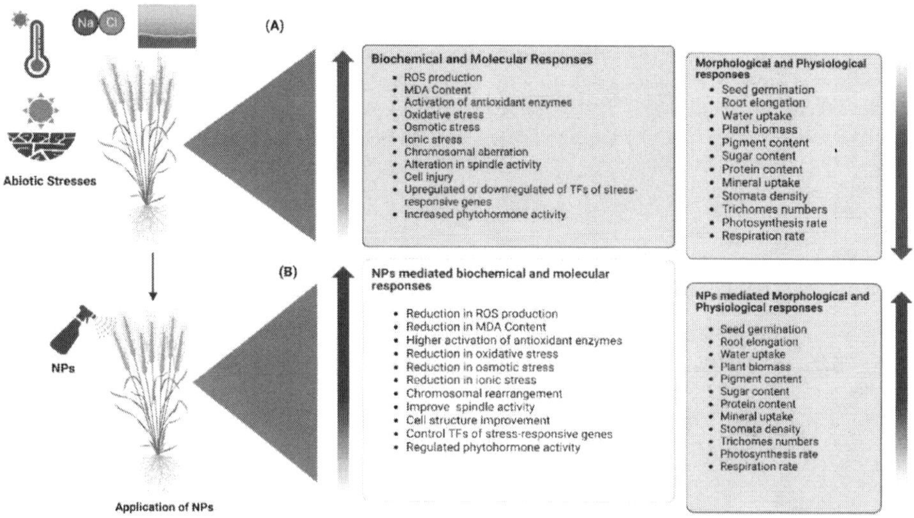

Figure 4. Effect of abiotic stresses on plants and application of NPs.

Under stress conditions, carbon fixation decelerates, leading to heightened photoinhibition, potentially resulting in increased generation of H_2O_2 and superoxide anion radicals within the photosystem. ROS produced by NPs induce protein modifications, lipid peroxidation, and DNA damage. Interactions between plants and NPs further exacerbate lipid peroxidation and DNA damage. The onset of plant cell death may occur via apoptosis or necrosis triggered by elevated levels of reactive oxygen species. However, despite their deleterious nature, reactive oxygen species play integral roles in various biological processes, including stress tolerance. The functional role of reactive oxygen species, whether as damaging agents or signalling molecules, is contingent upon the balance between ROS production and scavenging. Cells have evolved sophisticated antioxidant mechanisms to tightly regulate ROS levels. Enzymatic antioxidants such as superoxide dismutase and catalase, along with non-enzymatic antioxidants like ascorbate, glutathione, carotenoids, tocopherols, and phenolics, are integral components of the plants defense system, as discussed by some researchers. Multiple studies have demonstrated that exposure of plants to NPs results in enhanced production of antioxidant molecules. Phytohormones are recognized as key players in plant stress response signalling. These naturally occurring chemicals regulate various aspects of plant growth, development, and stress resilience. Abiotic challenges can modulate several hormone pathways, either stimulating or inhibiting their activity. Stress induced by silver nanoparticles (Ag NPs) has been observed to elevate cytokinin levels in red pepper (*Capsicum annuum* L.), whereas copper oxide nanoparticles (CuO NPs) have been found to decrease auxin and abscisic acid (ABA) levels in cotton. This evidence suggests that NPs exert an influence on hormonal balance and metabolism in plants. Additionally, numerous investigations have reported that NPs can impact the composition and function of photosynthetic pigments in plants. Plants exposed to elevated levels of NPs may exhibit inhibited growth or even mortality due to the interference of these particles with photosynthesis.

3.5 Nanobiosensors

3.5.1 Nanopore DNA Sequencing

After more than four decades since its inception, DNA sequencing technology has undergone significant advancements, enabling the collection of vast amounts of data at an accelerating pace. The commencement of first-generation sequencing was marked by the publication of the bacteriophage X174 genome sequence by Sanger and Coulson in 1977. Initially, first-generation sequence techniques dominated the field until the advent of high-throughput sequencing technology in the mid-2000s. Despite their shorter maximum read lengths compared to Sanger sequencing, second-generation sequencing methods challenged large-scale sequencing programs. However, it has been observed that the short reads obtained from these second-generation sequencing technologies often result in suboptimal assemblies, particularly when attempting de novo assembly of large and highly repetitive genomes. The exploration of basic genomes, resequencing endeavors, and RNA sequencing analyses persist in heavily leveraging these methodologies. Recent strides in sequencing technology have facilitated the acquisition of considerably elongated reads, all while sustaining data generation rates commensurate with or surpassing those of first-generation approaches. Third-generation technologies have matured to the extent that individual DNA molecules can undergo real-time sequencing, yielding reads spanning multiple kilobases (Figure 5). Such lengthened reads serve to bloster sequencing assemblies by effectively enveloping extensive repetitive segments within intricate genomes. Notably, Pacific Biosciences (PacBio) pioneered the application of single-molecule real-time sequencing for plant genome exploration, a trend later augmented by the incorporation of Oxford Nanopore Technologies (ONT) in 2014. Noteworthy advancements in nanopore sequencing technologies have significantly influenced plant pathogen identification endeavors. Researchers have formulated a standardized protocol for discerning various plant infections, inclusive of bacteria, viruses, fungi, and phytoplasma, employing a handheld sequencing apparatus developed by Oxford Nanopore Technologies, denoted as "MinION" (Figure 7). Findings from such analyses are interpreted analogously to those derived from established diagnostic methodologies like polymerase chain reaction (PCR) and enzyme-linked immunosorbent assay (ELISA). For instance, the amalgamation of nanopore sequencing with whole transcriptome amplification facilitated the prompt identification of *Camdidatus Liberibacter* asiaticus and plum pox virus in peaches within a 24-hour timeframe. Moreover, the utilization of MinION enabled the detection of multiple plant virus species within a water yam plant, including *Dioscorea bacilliform* virus, Yam mild mosaic virus, and Yam chlorotic necrosis virus. Despite constraints such as the inability to differentiate between sequences sharing identical attributes and a relatively elevated per-read error rate, advancements in nanopore technology are poised to foster the expansion of more efficient sequencing initiatives.

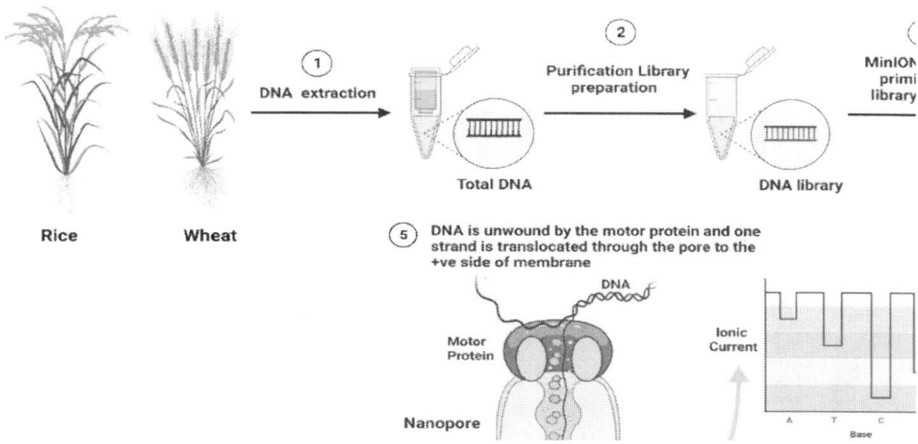

Figure 5. Diagrammatic representation nanopore mediated DNA sequencing of different crops.

3.5.2 Nano biosensors

Nano biosensors have garnered significant utility within the realm of agriculture, encompassing functions ranging from nutrient monitoring, detection of pesticides and herbicides, and assessment of insecticide efficacy, to soil moisture and pH analysis. The deployment of these biosensors, albeit modest, holds the potential to augment agricultural yields, thereby contributing to the principles of sustainable agriculture. Precession farming, leveraging modern sensor technologies, presents opportunities for yield enhancement through more effective fertilizer management, reduced input costs, and mitigation of environmental impacts. In the domain of precision agriculture, nanosensor-based smart delivery systems hold promise in enhancing the efficiency of water, fertilizers, and agrochemicals. Nanosensors facilitate the detection of various agricultural parameters including plant viruses, soil nutrient concentrations, and crop diseases. Notably, the utilization of nanosensors has enabled the detection of the acetolactate synthase inhibitor metsulfuron-methyl, accomplished through the functionalization of an atomic force microscope tip with the enzyme. Furthermore, the integration of bio-nano sensors has contributed to the enhancement of food safety by enabling rapid detection and accurate quantification of bacteria and viruses. A highly sensitive biosensor for detecting organophosphorus pesticide in vegetable samples has been developed by scientists, wherein carbon nanotube surfaces were functionalized with amino groups to facilitate the efficient immobilization of acetylcholinesterase into a glassy carbon electrode. Furthermore, researchers have introduced a multiwall carbon nanotube biosensor, employing liposome bioreactors, for the detection of acetylcholinesterase, enabling the identification of organophosphate pesticides. This biosensor exhibits a sensitivity of 1.903×10^{-12}mol/dm^3, achieved by monitoring subtle alterations in local pH around an electrode surface resulting from the inhibition of the acetylcholinesterase-acetylcholine enzyme interaction. Additionally, single-walled carbon nanotubes (SWCNTs) with electrochemical functionality serve as promising nanosensors for gases such as ammonia, nitrogen oxides, hydrogen sulfide, sulfur dioxide, and volatile organics. These sensors hold potential applications in pollution monitoring in agriculture, assessing their impact on living organisms or human health, and promoting higher crop yields. Notably, it was reported the high-sensitivity (1.0 mol/dm^3) screening of organophosphorus compounds using fluorescence spectroscopy and gold nanoparticles.

3.5.3 Nano barcode

A semi-automated, highly scalable method is used to create nano barcode particles. It entails electroplating inert metals, such as gold, silver, etc., onto templates that specify the particle diameter. This causes the templates to discharge nanorods. The ID used in multiplexed gene expression analysis are these nano-barcodes. One benefit of integrating biotechnology and nanotechnology is enhanced plant resilience to disease, salt and drought. Gene sequencing based on nanotechnology enables quick and affordable identification and utilization of plant gene trait resources.

3.5.4 Soil Nanosensors

Soil plays a vital role in sustaining plant life by supplying essential nutrients. Among its ecologically significant functions, the transformation of contaminants stands out prominently. Assessing soil quality involves evaluating various indicators such as water content, pH levels, organic matter content, and pollution levels. Identifying pollutants is crucial as they not only hinder plant growth and crop productivity but also contribute to contaminating the food chain through crop absorption. Traditionally, soil pollutant levels are determined through crop absorption. Traditionally, soil pollutant levels are determined through methods like chromatography, electrochemical detection, and mass spectrometry, following standard procedures of field sampling, plant tissue digestion, extraction, and sample purification. However, this process is labor-intensive, time consuming, and requires sophisticated equipment, and expert handling. Moreover, it lacks the capability for real-time and on-site monitoring.

Consequently, there is a pressing need for reliable technology to enable swift, on-site, and real-time assessment of soil contaminants. In addressing this need, a novel plant nanobionic sensor has been developed for selectively monitoring soil arsenic levels. The presence of arsenic and its compounds poses significant threats to both and ecosystems. In this approach, self-powered detectors crafted from specially engineered nanomaterials are integrated with plants. Researchers have devised a non-invasive technique to monitor the internal processes of plants as they uptake arsenic from the soil, employing single-walled carbon nanotubes (SWCNTs)-based near-infrared (NIR) fluorescence nanosensors embedded with plant tissues. This innovative plant nanobionic approach, juxtaposed with conventional methods, represents a significant leap forward in soil monitoring capabilities. The utilization of optical nanosensors within living plants offers a novel avenue for assessing soil quality.

The advent of this new nanobionic sensor category holds immense promise for future applications in soil quality monitoring and is anticipated to catalyze the development of additional nanosensors tailored for soil condition detection. By harnessing portable and cost-effective electronic equipment in conjunction with living plants, precise and real-time evaluation of soil quality can be achieved directly on site.

3.5.5 Metal ion detectors Nanosensors

Plants depend heavily on metal ions like calcium, potassium, sodium, and magnetism for signaling various physiological changes, which can include transient increases, waves, and oscillations of these ions in response to different stimuli. Monitoring shifts in intracellular metal ion concentrations is vital for assessing plant nutrition, emphasizing the importance of investigating the dynamic variations of metal ions within plants.

The development of ion-selective microelectrodes has been a key advancement in plant-based metal ion detection technology, allowing for the non-destructive measurement of metal ion fluxes in a plant's root elongation zone or suspension cell. However, this method requires a high level of precision and sophistication. Consequently, there is a pressing need for the development of a simpler and faster technique. Addressing this challenge, bioristor was introduced- an in vivo organic electrochemical transistor sensor based on poly (3,4-ethylenedioxythiophene): poly (styrenesulfonate)- to detect variations in the quantities of metal ions (Na^+, K^+, Ca^+, and Mg^+) in the sap of drought-stressed tomato plants. After subjecting these plants to a 30-hour water deprivation period, the device implanted in the stem demonstrated a steady increase in metal ion concentrations, enabling continuous monitoring of the plant's physiological status. This innovative gadget holds significant potential for improving drought-resistant tomato varieties. However, further research is needed to ensure its long-term stability, and biocompatibility, and to understand the interaction mechanism between the employed nanomaterials and the plant system. Moreover, reliance on wired data transmission remains cumbersome for extended usage, underscoring the importance of future integration with wireless devices.

3.5.6 ROS Detector Nanosensors

Superoxide anion, hydrogen peroxide, carboxyl radical, nitric oxide, and other ROS play vital roles in transmitting signals across various organisms. Originally formed as byproducts of aerobic metabolism, ROS have evolved to fulfil crucial functions within the complex cellular communication network. It is widely acknowledged that the release of ROS serves as a common and rapid defense mechanism in plants, activated in response to interactions with pathogens and environmental stresses. ROS are recognized as essential regulators of growth in plants influencing hormone signaling pathways and the synthesis of cell wall polymers. Hence, it is imperative to explore the generation of ROS and their intercellular activities throughout the developmental stage of plants. The most commonly utilized method for tracking H_2O_2 levels in plant leaves is the fluorescent probe, although alternative methods have also been developed. However, conventional florescence probe technology requires less time and effort for isolating plant samples. Therefore, electrochemical approaches, offering simplicity, low cost, and high sensitivity, were widely employed in studying plant biochemical process before the introduction of advanced H_2O_2 fluorescence detection technologies. According to a study conducted, haemoglobin (Hb) and single-walled carbon nanotubes (SWCNTs) modified carbon fiber ultramicroelectrode (Hb/SWCNTs/CFUME) were utilized to electrochemically monitor changes in H_2O_2 levels in aloe leaves under salt stress. By utilizing the catalytic effect of Hb in reducing H_2O_2, amperometric in vivo monitoring of H_2O_2 was achieved by inserting Hb/SWCNTs/CFUME into the leaf's central vein. Additionally, utilization of a platinum (Pt) disc microelectrode to perform in situ monitoring of H_2O_2 generated in *A. tequilana* leaves following bacterial inoculation. The electrochemical data indicated that H_2O_2 production was proportional to the number of inoculated leaves. A highly sensitive electrochemical nanosensor for ROS has been developed, utilizing electrodes with exceptional catalytic activity that spontaneously adhere to ultrathin metallic molybdenum disulfide (MoS_2) nanosheets. These nanosheets are then integrated into a freestanding paper electrode. This flexible nanosensor offers several advantages, including probability, high selectivity, and stability, and a wide linear range. Its efficacy has been demonstrated in monitoring H_2O_2 levels in aloe plant extracts. These electrochemical nanosensors promise high selectivity and stability for real-time field monitoring, eliminating the need for extensive data processing. However, most studies are conducted in laboratory settings, and the biocompatibility of these nanosensors remains unexplored. Moreover, the development of highly integrated, miniaturized, and portable monitoring systems required the advancement of electrochemical nanosensors capable of conducting in situ monitoring on living plants and seamlessly integrating with wireless devices.

3.5.7 Phytohormone Detectors Nanosensors

Plants produce various hormones such as auxins, gibberellins, abscisic acid, cytokinins, salicylic acid, ethylene, and peptide hormones. These are small organic molecules or active chemicals that regulate physiological responses in plants, induced by specific environmental signals. They play a crucial role in plant adaptation to different stressors. Techniques including high-performance liquid chromatography, gas chromatography-mass spectrometry, and capillary electrophoresis (CE) have been utilized to detect plant hormones. However, these methods are limited to in vivo applications, requiring extensive preparation of plant samples. Optical nanosensors enable the detection of hormones in living plants. Quantum dots (QDs) are particularly effective due to their strong fluorescence, broad excitation spectrum, adjustable emissions, and excellent photostability, making them widely used as bio-labels and bioimaging probes. Moreover, QDs can be conjugated with aptamers and other target-specific ligands to enhance targeted molecular imaging. A fluorescent aptasensor was developed using aptamer-functionalized zinc ion (Zn_2+)-doped cadmium telliride quantum dots (QDs) to enable in vivo detection of tomato systemin (TomSys), a type of peptide hormone found in plants. The study revealed that fluorescence was suppressed due to non-covalent interactions between aptamer-functionalized QDs and graphene oxide (GO) nanosheets in the absence of TomSys. However, in the presence of TomSys, the aptamer-functionalized QDs were released from the GO surface, resulting in the restoration of fluorescence. Metal-based fluorescent nanosensors have also been extensively developed for tracking signaling molecules. For instance, a curcumin-copper ion (Cu^{2+}) based fluorescent nanosensor was created for selective and sensitive in vivo detection of salicylic acid. The study demonstrated that Cu^{2+} initially binds to the diketone portion of curcumin, leading to the paramagnetic "turn-off" of curcumin's luminescent properties. Upon the addition of salicylic acid, a highly sensitive "turn on" fluorescence pattern towards Cu^{2+} was observed. These fluorescent nanosensors surpass electrochemical nanosensors in terms of selectivity and sensitivity, and the detection process can be visualized more intuitively due to the complex variation in fluorescence signals. However, challenges associated with optical sensors underscore the ongoing need for research on their long-term durability and biocompatibility in plants.

4. Concerns associated with NPs implementation

The term "nanotoxicity" encompasses the potential risk posed by nanoparticles to living organisms. Inflammation, oxidative stress, and damage to DNA or cellular structures are among the ways in which it may impact human health and agricultural productivity (Figure 6). In agriculture, nanoparticles can originate from both natural and anthropogenic (man-made) sources. Natural sources of nanoparticles in agricultural settings include volcanic eruptions, forest fires, and the weathering of rocks and minerals. Living organisms such as plants and animals can absorb these natural nanoparticles. Conversely, human-made sources of nanoparticles in food production include nanomaterials used in pesticides, fertilizers, and other agricultural products. Industrial processes such as manufacturing and mining can also release nanoparticles into the environment. While nanotechnology holds promise for improving agricultural practices, there are legitimate concerns regarding the long-term consequences of using nanomaterials in food production on human and environmental health. Therefore, thorough evaluation of the risks and benefits of employing nanomaterials in agriculture is imperative before widespread adoption. Regulations concerning the use of nanoparticles in farming vary among different countries and regions. Oversight of the application of nanotechnology in food and agriculture is carried out by the US Department of Agriculture (USDA), while regulation of pesticides containing nanoscale materials falls under the jurisdiction of the Environmental Protection Agency (EPA). Risk assessment and nanotoxicological evaluations are prerequisites for the successful commercial application of nanoparticles (NPs) and will significantly influence the development of safer NPs in the future. While toxicological evidence is necessary to substantiate this claim, nanostructures are generally believed to be more toxic than their non-nano counterparts. To effectively manage the interaction of NPs with the environment and biological systems and mitigate their potential toxicological effects, logical, science-based management strategies must be developed. It is evident that biogenic NPs form a protein corona (PC) around their surface, which mediates and regulates NP interactions with living systems or cells, potentially resulting in cytotoxic, genotoxic, and pathophysiological effects if the NP-cell interaction is incompatible. The PC can have either a beneficial or detrimental impact on biocompatibility, depending on various factors influencing the interaction, such as the type of protein comprising the PC, hydrodynamic size, and associated charges that can either increase or decrease NP toxicity. The morphological and physiological effects of NP phytotoxicity can manifest in chlorophyll damage, impaired root tips, reduced root length, diminished biomass, and other oxidative damage-related alterations. However, it is important to note that not all NPs exhibit these effects. For example, in *Pisum sativum*, ZnO nanoparticles (NPs), and in *Solanum lycopersicum*, Ag NPs, led to a reduction in chlorophyll levels, while in *Cucumis sativus* and *Cyamopsis tetragonoloba*, TiO_2 NPs and ZnO NPs respectively, resulted in an increase in chlorophyll content. Upon penetration by NPs, biological systems often trigger the production of reactive oxygen species (ROS), which can disrupt normal biophysical, biochemical, and stress-related activities, as well as the regulation of stress resistance genes, resulting in NP-specific genotoxic effects. ROS-induced lipid degradation leads to abnormalities and oxidative stress in cell membranes, causing additional toxic effects such as cell death and ion leakage. For instance, CeO_2 NPs in *Zea mays* induced lipid peroxidation due to ion leakage, whereas Oryza sativa did not exhibit the same effect at comparable NP concentrations (0-500 mg/L). Interactions between plants and NPs have the potential to adversely affect secondary plant metabolism, hormonal balance, and plant growth and development. Transcriptome analysis of Arabidopsis thaliana revealed that NP exposure downregulated the expression of specific genes associated with phosphate deficiency, pathogens, and stress responses, potentially compromising plant defense mechanisms and root development. Nutrient distribution may be disrupted by NPs, impacting healthy growth and development. CeO_2 NPs inhibited rhizobacteria from fixing N_2 for soybean plants, diminishing the plant's nitrogen availability and impeding normal growth and function. Conversely, TiO_2 NPs in *Cucumis sativus* increased the accessibility of phosphorus and potassium, resulting in 35% more potassium and 34% more phosphorus with a treatment of 500 mg/kg on plants. Moreover, some nanomaterials, such as TiO_2, exhibit resistance, potentially leading to the accumulation of metal components in the environment over time. Additionally, continual nutrient application, such as copper, can lead to an excess of the nutrient in the soil, which may be detrimental to plant health. One potential strategy for mitigating NPs stress involves the upregulation of antioxidant compounds and the downregulation of genes encoding metal transport. Recent research employing omics data in a systems biology approach on plant varieties such as rice, tobacco, and wheat indicates that metal NPs induce a generalized stress response, particularly the oxidative stress response. This suggests that investigating the genetic and metabolic processes triggered by NPs exposure through high-throughput analyses is essential, even if no toxic effects are evident at the phenotypic level.

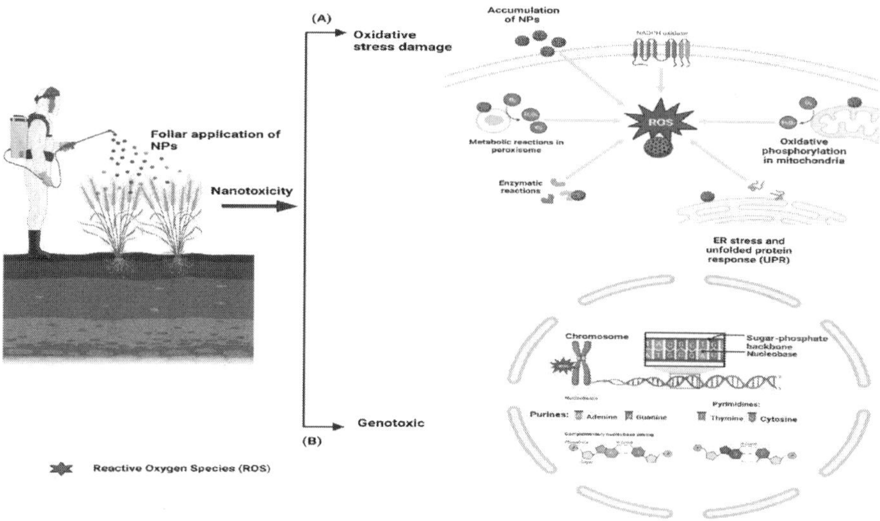

Figure 6. Mode of action NPs mediated phytotoxicity that induced (A) oxidative stress that affected various cellular organelles and biological processes also (B) genotoxic mediated damage of genetic materials that affected molecular processes.

It remains unclear whether the type of NPs and its interactions have a specific impact on nanotoxicity or if the detoxification systems activated in response to NPs stress can effectively alleviate stress at the biomolecular level. Therefore, understanding the properties of synthesized NPs is crucial before examining their effects on plant systems to minimize potential risks, both to human health and the environment, associated with their applications and other functions.

The proteome-level toxicities caused by NPs will be clarified by proteomic research that finds protein markers. Thorough in vivo and in vitro phytological testing is required before any nano agriproducts are released onto the market to ensure effective nutrient consumption with little to no associated toxicity.

5. Agri-Nano-enabled products and marketing approaches for sustainable agriculture

"Industrial Agriculture" denotes the intricate network of systems accountable for fuelling the global food supply. With the rapid expansion of consumer markets for nanotechnology, fresh prospects have emerged for the agricultural industry to exploit premium products. Despite the profusion of food in today's world, many individuals still grapple with daily hunger, often attributable to environmental factors. The aim of cultivating drought- and pest-resistant crops is to amplify yields. As per the NPD database, a staggering 243 agricultural nanoproducts are currently accessible. These diverse products are being globally introduced by 87 companies headquartered across 28 different nations (Figure 7). They encompass an array of subsets, spanning animal husbandry, fertilizers, plant hybridization, plant fortification, and soil enrichment. Manufacturers persistently innovate fertilizers, chelates, and nutrients, with commercially available product categories encompassing precursors, chemicals, and enhancers (such as controlled-release nanocomposites) utilized in fertilizer compositions, alongside plant growth regulators, seed enhancements, hormones, and fruit fortifiers. In recent times, there has been a surge of interest in nanotechnology-based soil enhancers. Plant protection primarily relies on the application of chemicals such as pesticides, fungicides, biocides, and disinfectants. Nano-sized polyethylene stretch films have emerged as a novel solution for safeguarding hay from harmful sunlight and extreme weather conditions. Animal husbandry operations involve the utilization of various dietary supplements, nutrients, feed additives, enzymes, and disinfectants. Additionally, veterinary pharmaceuticals, vaccines, and wound care products play crucial roles in maintaining animal health. In aquaculture, tools like oxygen and carbon dioxide level controllers, as well as fish counters, are indispensable. Nanotechnology holds promise for revolutionizing agriculture by facilitating disease diagnosis, enhancing plant nutrient uptake, minimizing fertilizer wastage, and ultimately, increasing crop yields.

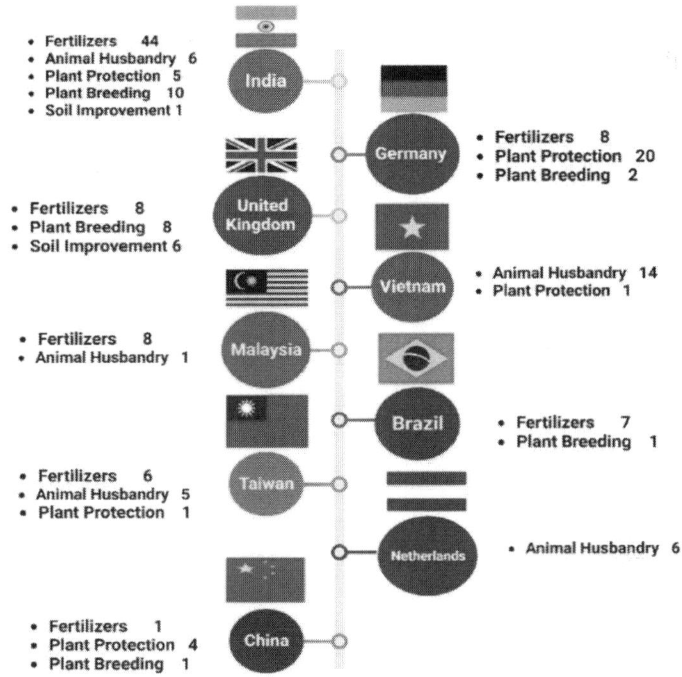

Figure 7. Various Agri-nanotechnological products produced by various countries world-wide.

Silver nanoparticles, titanium dioxide nanoparticles, silicon dioxide nanoparticles, and zinc oxide nanoparticles are widely utilized in agriculture to enhance plant health, improve stress resistance, prolong product shelf life, and optimize soil nutrient efficacy. Each of these nanoparticles possesses distinct chemical, physical, and mechanical characteristics. Nanoencapsulated forms containing essential minerals like magnesium, manganese, iron, and potassium also play significant roles in agricultural applications. Additionally, porous materials such as zeolites contribute to enhancing agricultural productivity and animal welfare. While numerous agricultural producers leverage nanotechnology, only a select few stand out as market leaders. Notable examples include Neufarm GmbH, Plant Vitality Ltd, Kanak Biotech, FRAmelco, Aqua-Yield Hub, Organic Fertilizing, Reed Mariculture Inc, Prodotti Arca S.r.l., Kimitec Group, Richfield Fertilizers Pvt., Blue Planet Environmental, Danaflex Nano, Bioteksa, AgriLife, Nano Land Baltic, Vive Crop Protection, DVS BioLife Ltd, and Samarita. Based on available data, countries like India, Germany, the United Kingdom, the United States, Vietnam, Taiwan, Brazil, China, Malaysia, and the Netherlands are among the most supportive in embracing nanotechnology's advantages for agricultural applications.

6. Future suggestions and Conclusion

The agriculture sector has seen notable advancements though the utilization of nanotechnology, which involves manipulating matter on a nanoscale level to enhance plant growth and productivity. This includes dynamic monitoring of plant growth processes, managing pests, facilitating plant transformations, safeguarding the environment, and harnessing environment micro-energy for plant use. Delving deeper into the realm of "plant nanoscience" reveals numerous unforeseen benefits offered by these nanotechnologies. However, several challenges lie ahead that must be addressed. These challenges encompass the establishment of unified international standards, fragmented policy framework, concerns regarding nano-safety, ensuring grower's health, gaining public acceptance, and effectively implementing these technologies in real-world agriculture practices. Despite the increasing scholarly work and patents related to nanotechnology in agriculture, various factors such as diverse national legislative environments, limited regulatory frameworks, and a lack of initiatives for public licensing have hindered the availability of commercial products and the progress of large-scale scientific research. Predicting the direst, indirect, and cumulative impacts of nanomaterials is challenging due to the diverse physicochemical and biological properties that distinguish them from individual atoms, molecules, or bulk materials. Moreover, understanding the movement of nanomaterials throughout the environment, organisms, and human system remains unclear. This complexity poses significant challenges in establishing consistent global policies and regulations. To address this, collaboration between international and national organizations, as well as extensive research and fieldwork, is urgently needed. Only through concerted efforts can we overcome the hurdles posed by inconsistent policies and regulations and fully harness the potential of nanotechnologies in agriculture.

There exists a gap between current studies on the hazards of nanomaterials to human and environmental health and the intricate agricultural ecosystems where their use may pose significant exposure risks. Nanomaterial applications, such as regulating agrochemical dispersion, facilitating biological molecule transport, and environmental remediation, among others, involving carbon nanotubes (CANTs), graphene, metal NPs, zero-valent iron (ZVI), and similar materials, warrant further investigation specifically, understanding the intricate interactions between these material and the complex plant system is crucial, as it will contribute to a deeper understanding of the already complex risk and exposure landscape. To address nano safety concerns, careful consideration should be given to natural material that pose minimal risk to growers' and consumers' health during experimental designs phases. Additionally, through examination of how unique properties of nanomaterials, such as shape, size, and charge, influence their interactions with various plant species is essential. Lastly, understanding the behavior and fate of nanomaterials in the environmental is imperative for accurately assessing risks associated with their use. These emerging nanotechnologies are primarily in the research phase, offering distinct advantages compared to conventional agriculture methods. While laboratory studies have repeatedly confirmed their potential, there is a notable lack of successful field implementation or commercialization examples. Furthermore, research on plant system often builds upon findings from studies on animals, suggesting the need for further exploration of their efficacy specifically in plants context. Metal nanoparticles, nano-chitosan, nano-zeolite, and other nanomaterials have shown promise in enhancing nutrient absorption efficiency and delivering active compounds to specific plant cell compartments and organelles. However, their application requires careful consideration of various soil and plant types, as well as the establishment of standardized concentration guidelines. Increasing the deployment of nanomaterials in diverse field experiments is essential to accelerate their transition from laboratory settings to practical agricultural applications. However, establishing comprehensive guidelines demonstrating the beneficial effects of these nanomaterials on plant growth and their safety for animal and human health requires extensive experimental observations over time. The transition from initial research stages to commercial applications, considering factors such as safety, affordability, and market acceptance, is a gradual process. Public perception and support play a vital role in the long-term success of agricultural nanotechnologies. Many consumers remain hesitant to purchase foods produced using genetic modification or nanotechnology. Hence, involving consumers in the production process and providing transparent and unbiased information from field trials can help enhance their understanding of nanomaterial applications. Further research is needed to assess both the risks and benefits of incorporating nanomaterials into food production, along with understanding public attitudes toward their use. By addressing nano-safety concerns and providing comprehensive explanations, public acceptance of food products utilizing nanotechnology can be improved, ultimately unlocking the potential of nanotechnology to enhance agricultural productivity. Moving forward, it will be essential to build upon existing research on interactions between plants and nanomaterials. The terminology "plant nanobionics", introduced by Strano's research group in 2014, describes the potential for enhancing plant organelle functions through the integration of non-biological nanostructures. This involves the injection of single-walled carbon nanotubes (SWCNTs) into plant leaves to achieve desired functions. The ultimate goal is to establish a symbiotic relationship between nanomaterials and plants, wherein nanomaterials are initially present during the plant's lifecycle but degrade upon the plant's demise. This concept could redefine "plant nanobionics" and offer new avenues for research in "plant nanoscience". Nanomaterials have the potential to provide insights into how plants perceive and respond to environmental cues, influencing their survival, physiology, development, and morphology. By integrating internal and external factors, plant cells can modulate gene expression, and nanomaterials play a role in this process by imparting non-native functions to plants. These dynamic changes can be observed through optical or electrical techniques, reflecting the signalling pathways within plant cells. This enables timely interventions to enhance plant growth and yield. Moreover, these nanomaterials are deemed safe for animals and humans and have no adverse effects on plant life.

CHAPTER 2

NANOPARTICLES: TYPES AND APPROACHES FOR SYNTHESIS

Short Summary

The diverse world of nanoparticles, which are central to the advancements in agricultural nanotechnology. It begins by categorizing nanoparticles based on their composition and properties, including metal, metal oxide, carbon-based, and polymeric nanoparticles. Each type of nanoparticle offers unique advantages for agricultural applications, such as improved nutrient delivery, pest control, and environmental remediation. The chapter then explores various synthesis approaches for nanoparticles, highlighting both top-down and bottom-up methods. Top-down methods, like mechanical milling and lithography, involve breaking down bulk materials into nanoscale particles. In contrast, bottom-up methods, such as chemical vapor deposition, sol-gel processes, and green synthesis, build nanoparticles from atomic or molecular precursors. Special emphasis is placed on green synthesis techniques, which use biological organisms like plants, bacteria, and fungi to produce nanoparticles. These eco-friendly methods are gaining attention due to their sustainability and reduced environmental impact. Overall, this chapter provides a comprehensive overview of the different types of nanoparticles and the innovative approaches used to synthesize them. It sets the foundation for understanding how these nanoparticles can be effectively utilized in agricultural applications to enhance productivity and sustainability.

1. Introduction

The infusion of nanomaterials into agriculture heralds a bold new era, where the infinitesimal scale of these materials holds immense promise for revolutionizing farming practices. With their unique attributes and functionalities, nanomaterials play pivotal roles across diverse agricultural domains. Take, for instance, the development of nanofertilizers, where nanoencapsulation technology enables precise nutrient delivery, optimizing plant uptake while minimizing environmental impact. Similarly, nanopesticides capitalize on this innovation for targeted pest management, enhancing efficacy while curbing overall agrochemical usage. The advent of nanosensors marks a transformative shift in precision farming, facilitating real-time monitoring of soil conditions, moisture levels, and plant vitality with unprecedented precision. Moreover, nanomaterials bolster soil health through the creation of nanocomposites, augmenting water retention and nutrient availability. In post-harvest practices, the adoption of nanomaterials-based packaging extends the shelf life of agricultural goods, mitigating losses and maximizing freshness. Additionally, animal husbandry can benefit from the uses of nanotechnology, which might lead to improvements in medication delivery, illness diagnosis, and animal feeding. The application of nanomaterials in agriculture has the potential to contribute to a future where global farming techniques are more resilient, efficient, and sustainable as nanotechnology advances.

2. Types of Nanomaterials

Nanoparticles exist in diverse form, each characterized by unique properties dictated by their constituent materials. Below are several prevalent types of NPs:

2.1 Metal Nanoparticles

Gold NPs

Gold nanoparticles, with their minute size and remarkable optical attributes, are extensively utilized across diagnostics, imaging, and drug delivery domains. Their distinctive optical properties render them indispensable in diagnostic methods like swift lateral flow assays and biosensors, facilitating precise and effective disease identification. In biomedical imaging, they act as contrast agents, enhancing techniques such as photoacoustic imaging and surface-enhanced Raman scattering (SERS). Gold nanoparticles exhibit versatility in drug delivery applications, where they can be tailored and equipped with therapeutic compounds, providing a framework for precise and regulated drug administration. Gold nanoparticles play a vital role in cancer treatment through photothermal therapy, selectively eliminating cancer cells when exposed to specific light wavelengths. They also serve as efficient nanoprobes for molecular and cellular imaging, facilitating precise visualization within biological systems. Moreover, their catalytic capabilities are utilized in organic synthesis and catalysis, driving advancements in chemical processes. Additionally, their antimicrobial properties make them promising candidates for applications in antimicrobial coatings, wound dressing, and medical devices. The versatile nature of gold nanoparticles, influenced by their size-related characteristics, positions them as valuable tools in ongoing research endeavors aimed at expanding their potential contributions to diagnostics, therapeutics, and imaging technologies.

Silver NPs

Silver nanoparticles are widely used in agriculture due to their antimicrobial properties, which help address various challenges in crop cultivation and soil management. One notable application is in the development of nanopesticides, where silver nanoparticles are incorporated into pesticide formulations to enhance their effectiveness against plant pathogens and pets. This targeted approach not only improves pest control but also reduces the environmental impact associated with traditional pesticide use.

Additionally, silver nanoparticles play a role in the field of nanofertilizers by integrating these nanoparticles into fertilizer formulations, their formulations, their antimicrobial properties can potentially help combat soil-borne diseases and promote healthier plant growth. The controlled release of nutrients enabled by nanofertilizers, combined with the antimicrobial advantages of silver nanoparticles, present a promising opportunity for optimizing nutrient utilization in agriculture.

In remediating soil, silver nanoparticles have the potential to address microbial contamination and enhance soil health by virtue of their antimicrobial properties. These nanoparticles may aid in controlling specific soil-borne pathogens and fostering a conducive environment for plant growth. However, it is imperative to thoroughly evaluate and manage the potential environment ramifications and risk associated with the utilization of silver nanoparticles in agricultures.

Although the application of silver nanoparticles in agriculture present promoting prospects for innovative solutions, continuous research is imperative to gain a comprehensive understanding of their impacts on crops, soil ecosystem, and potential implications for food safety. Striking a balance between harnessing the benefits of their antimicrobial attributes and addressing environment concerns is essential for the sustainable and addressing environmental concerns is essential for the sustainable and judicious integration of silver nanoparticles into agricultural particles.

2.2 Metal Oxide NPs

Titanium Dioxide NPs

Titanium dioxide nanoparticles, with their potential across various facets of crop cultivation and soil management, are being increasingly recognized as versatile tools in agriculture. Notably, their utilization in enhancing photosynthesis is being highlighted, as these nanoparticles can function as photocatalysts, optimizing biochemical processes within plant cells using light energy, thus potential leading to improved overall plant growth and productivity. Additionally, the promise of titanium dioxide nanoparticles as soil amendments is being explored, with their potential contributions to enhanced soil structure, water retention, and nutrient availability being acknowledged. Research into their role in soil remediation, including applications in the breakdown of pollutants, is actively ongoing. Moreover, in seed coating applications, these nanoparticles are being investigated for their ability to improve seed germination and early seedling development, providing controlled nutrient support. Pest management in agriculture can benefit from the potential of titanium dioxide nanoparticles, as they offer a controlled release mechanism for pesticides, enhancing effectiveness while reducing environmental harm.

Additionally, their capacity to offer UV protection is pertinent, particularly in greenhouse environments, where they can shield plants from excessive ultraviolet radiation. Continued research is imperative to fully grasp their impact on various crops and soil ecosystems, ensuring their responsible and sustainable integration into agricultural practices. Maintaining a balance between innovation and environmental concerns is essential as the agricultural industry explores the versatile applications of titanium dioxide nanoparticles.

Iron Oxide NPs

Iron oxide nanoparticles offer promising prospects in agriculture, demonstrating their potential to address various challenges encountered in crop cultivation and soil management. One significant area of exploration involves nutrient delivery, where these nanoparticles could improve the target could improve the targeted transportation of vital micronutrients such as iron to plants, potentially alleviating deficiencies and enhancing over all crop vitality. The magnetic attributes of iron oxide nanoparticles also present opportunity for soil remediation, as they can be directed contaminated sites, aiding in the elimination or immobilization of pollutants. The concept of "smart" fertilizers emerges through the integration of iron oxide nanoparticles, allowing for precise nutrient release in response to environmental conditions or plant needs, there by optimizing nutrient utilization efficiency. Additionally, these nanoparticles could be utilized in seed coating applications, enhancing germination rates and strengthening crop resilience by providing controlled nutrient support during initial growth phase. In the management, iron oxide nanoparticles may be utilized for targeted pesticide delivery, enhancing efficacy while minimizing environmental repercussion. While the potential advantages are evident, continuous research is imperative to thoroughly comprehend the implications, ensuring the responsible and sustainable integration of iron oxide nanoparticles into agricultural methodologies. Striking a balance between innovation and environment and environmental considerations remains paramount as the agricultural sector explores the versatile opportunities offered by these NPs.

2.3 Quantum Dots

Quantum dots, tiny semiconductor particles at the nanoscale, boast remarkable optical electronic properties attributed to their size and composition. Typically comprised of material like calcium selenide, calcium sulfide, or indium arsenide, these dots measure between 1 to 10 nanometers in diameter. Their standout feature, the quantum confinement effect, leads to size-dependent alterations in their electronic band structure. A standout attribute of quantum dots is their capacity to emit light upon exposure to external stimuli, phenomenon termed fluorescence. Their emitted light's colour is adjustable, spanning a spectrum from blue to red, based on the quantum dot's size. The adjustability renders quantum dots invaluable for a myriad of applications crossing imaging, sensing, and display technologies. Aside from their optical prowess, quantum dots boast impressive surface-to-volume ratios, making them ripe for customization with various molecules. This attribute finds purpose in the realms of biology and medicine, where quantum dots serve as fluorescent markers for illuminating cellular structures or monitoring biological functions.

While quantum dots shine with their optical and electronic prowess, apprehensions arise over their potential toxicity, particularly when crafted from heavy metals like cadmium. Researchers are fervently investigating alternative materials and coatings to allay these worries, ensuring the responsible deployment of quantum dots across myriad applications. In agriculture, where they harbor promise for meticulous monitoring and precision interventions, such efforts have paved the way for safer practices.

The unique properties of quantum dots at the nanoscale offer exciting opportunities for agricultural applications. These tiny semiconductor particles possess optical and electronic characteristics that vary with size, making them valuable tools for a range of tasks in crop management and surveillance. In agriculture, researchers are exploring the potential of quantum dots in plant imaging and diagnostics. Their minute size enables precise labelling and tracking of biological molecules within plants, facilitating the study of physiological processes. Additionally, quantum dots show promise in creating nanoscale sensors of physiological processes. Additionally, quantum dots show promise in creating nanoscale sensors for monitoring soil conditions, nutrient levels, and plant health in real time. Their high sensitivity and specificity make them suitable for detecting specific biomarkers associated with plant diseases. Moreover, investigations are underway into the use of quantum dot-based formulations for the controlled release of agrochemicals, offering targeted solutions for pest control and nutrient delivery. While the application of quantum dots in agriculture is an emerging field, ongoing research endeavors aim to fully unlock their potential, ensuring their responsible integration into farming practices and maximizing the benefits of nanotechnology for sustainable and efficient crop production.

2.4 Carbon-Based NPs

Carbon Nanotubes

Carbon nanotubes (CNTs), hailed for their extraordinary structural and material properties, are carving a niche in agriculture, offering innovative solutions to an array of challenges and ushering in a new era of farming practices. These sleek cylindrical structures, crafted from carbon atoms, boast remarkable mechanical, electrical, and thermal prowess. In the agricultural domain, CNTs are heralding a revolution in nutrient delivery mechanisms. By embedding nutrients within their nanotube framework, ingenious controlled-release systems are devised, responding adeptly to environmental cues or the specific demands of plants, ensuring optimal nutrient absorption. Beyond nourishment, CNTs are forging pathways in soil remediation, leveraging their expansive surface area and reactive nature to sequester or dismantle contaminants, thereby elevating soil quality to new heights. Carbon nanotube-infused sensors stand sentinel, providing real-time insights into soil conditions, detecting subtle shifts in moisture levels and nutrient profiles. Harnessing the power of nanoencapsulation, CNTs are spearheading the precise delivery of agrochemicals, minimizing ecological footprints while maximizing efficacy. In the realm of plant health, tailored CNTs don the mantle of nanobiosensors, enabling the early detection of disease biomarkers, ushering in a proactive approach to disease management. Moreover, CNTs champion water conservation efforts, fortifying soil structure and bolstering water retention capabilities. While these pioneering applications unveil the boundless potential of CNTs in agriculture, ongoing exploration is paramount to address safety concerns, unravel long-term ramifications, and chart a course towards responsible integration into farming practices. Embracing the possibilities of nanotechnology in agriculture promises to unlock transformative benefits while navigating any potential pitfalls with prudence and foresight.

Fullerenes

Fullerenes, characterized by their unique closed-cage carbon structures, notably represented by C60 or buckyballs, present captivating opportunities in agriculture. While their practical implementations are still in their infancy, fullerenes hold the promise of enriching various facts of crop cultivation and soil maintenance. Their distinct architecture hints at potential roles in soil enhancement, where they may enhance water and n nutrient retention capabilities. Moreover, fullerenes can be tailored as nanocarriers for precise delivery of nutrients and agricultural chemicals, offering more streamlined and controlled distribution systems for farming inputs. In the realm of environmental remediation, fullerenes' expansive surface area and reactive properties could be utilized to tackle soil pollution and promote sustainable environmental practices. Serving as nanosensors, customized fullerenes hold the potential for continuous monitoring of soil conditions and plant well-being, empowering farmers with invaluable insights for precise agricultural practices. Additionally, fullerenes may be explored for their potential in managing plants diseases, serving as carriers for targeted and regulated delivery of antimicrobial substances. Despite these alluring applications, thorough research is indispensable to grasp the environmental implications, safety considerations, and effectiveness of responsibly integrating fullerenes into agricultural methodologies.

2.5 Polymeric NPs

Polymeric Nanoparticles, known for their small size and composed of biocompatible polymers, have emerged as versatile tools with a wide range of uses in agriculture. One significant application is their utilization in the controlled release of agrochemicals, such as pesticides and fertilizers. By encapsulating these substances within polymeric NPs, a precise and targeted delivery system is achieved, enhancing their efficacy while reducing environmental harm. Additionally, polymeric nanoparticles act as carriers for delivering nutrients, ensuring a gradual and accurate release for optimal uptake by crops, thereby improving nutrients efficiency. In seed coating, polymeric Nanoparticles enhance germination rates and provide protection against environmental stressors, pathogens, and pests, ultimately supporting better crops establishments and yields. They also contribute to water management and soil health, enhancing water retention and soil structure when incorporated into the soil, which is essential for sustainable farming. Moreover, functionalized polymeric nanoparticles play a role in diagnosing and imaging plants diseases, facilitating prompt recognition and action. Nonmaterial -infused coatings contribute to post-harvest management, bolstering the longevity of agricultural yields through the augmentation provided by polymeric Nanoparticles. This leads to diminished spoilage and reduces post-harvest losses. This leads to diminished spoilage and reduces post-harvest losses. Furthermore, in environmental surveillance polymeric nanoparticles serve as miniscule sensors, delivering immediate insights into pollutants and soil conditions critical for precision agriculture and the advancement of capacity farming techniques. The multifaceted utility of polymeric nanoparticles underscores their capacity to transform agriculture methodologies, highlighting imperative for continuous research thoroughly evaluate their effectiveness, safety, and ecological soundness in crop cultivation.

2.6 Liposomes

Liposomal nanoparticles, famed for their lipid bilayer composition, emerge as dynamic agents with a spectrum of roles in modern agriculture. Once intriguing application is their prowess in delivering pesticides and fertilizers, acting as adept carriers that of offer a finely-tuned and targeted release mechanism. This not only elevates the efficacy of pest management but also optimizes nutrient utilization, fostering eco-conscious farming methods. Expanding their utility, liposomal formulations are instrumental in fortifying seeds against pathogens and environmental stressors, championing robust seed germination and vigorous crop growth. Furthermore, they spearhead the delivery of biopesticides and biofungicides, championing sustainable practices in pest and disease control. In weed management, liposomal nanoparticles bring precision and accuracy to delivery, curtailing collateral damage. Additionally, liposomes housing plant growth regulator orchestrate growth dynamics, ushering in bountiful yields and superior crop quality. Their adaptability extends to diagnostic realms, aiding in early disease detection through innovative imaging techniques. Enveloping nutrients within liposomal coating enveloping agricultural produce promise prolonged shelf life, curbing post-harvest losses and ensuring enduring food preservation. The varied utilization of liposomal nanoparticles within agriculture highlights their capacity to transform traditional methodologies, providing enhanced, focused, and eco-conscious alternatives for the sustainable cultivation of crops, Continuous research stands as imperative for the refinement of formulation and the comprehension of their enduring effects on crops and ecosystems.

2.7 Ceramic NPs

Ceramic nanoparticles, known for their distinct properties and composition, are emerging as promising element in agriculture innovation. Comprised of material like silica, alumina, or titanium dioxide, these nanoparticles offer a wide array of applications aimed at improving various aspect of crop cultivation and management. When used as soil additive, ceramic nanoparticles enhance soil structure and nutrient availability by leveraging their nano-sized particles, which provide increased surface area. This promotes optimal water retention and nutrient absorption by plants, furthermore, their capacity to encapsulate and deliver nutrients in a controlled manner enhance the effectiveness of fertilizer while minimizing environmental impact in seeds coating application, ceramic nanoparticles bolster germination rates and shields seed from environmental stressors, facilitating robust crop establishment. They also contribute to water filtration and purification, ensuring the delivery of clean water to crops. Moreover, tailored ceramic nanoparticles play a vital role is disease diagnosis through imaging techniques, allowing for timely recognition and response, the utilization of ceramic nanoparticles in post-harvest management enhance the longevity of agriculture products, significantly benefiting post-harvest practices. Moreover, in the realm of soil remediation, these nanoparticles play a pivotal role in breakdown of pollutants, effectively addressing soil contamination and promoting sustainable land management strategies. The multifaceted applications of ceramic nanoparticles in agriculture underscore their potential to tackle various challenges, paving the way for enhanced efficiency, sustainability, and eco-friendly farming methodologies. Continuous research endeavors are imperative to comprehensively grasp the enduring ramifications and refine formulations for extensive and efficacious integration into agricultural practices.

2.8 Magnetic NPs

Magnetic nanoparticles have emerged as innovative instruments in agriculture, leveraging their district attributes to enrich various aspects of crops cultivation and managements. One notable application involves their utilization in targeted nutrient delivery and soil enhancement. By incorporating essential nutrients into magnetic nanoparticles and magnetically guiding them to specific plant roots, precise and optimized nutrient delivery for plants growth is achieved. Furthermore, in the formulation of pesticides and fertilizers, magnetic nanoparticles contribute to increased efficiency by facilitating controlled release and targeted delivery. This leads to enhanced pest control and nutrients delivery efficacy while minimizing environmental impact. Moreover, magnetic nanoparticles play a crucial role in water purification, effectively eliminating contaminants through magnetic separation processes to ensure crops receive clean irrigation water. Additionally, in precision agriculture, magnetic sensors employing nanoparticles enable real-time monitoring of soil condition, moisture level, and other pertinent parameters. This empowers farmers to make informed decisions regarding precise irrigation, fertilization, and pest control strategies. Magnetic nanoparticles also find applications in seed coatings, post-harvest sorting, and soil remediation, showcasing their versatility in addressing various agricultural challenges. Continued research into novel possibilities and further integration refinement of magnetic nanoparticles offer significant potential for propelling sustainable and efficient agricultural practices forward.

2.9 Silica NPs

Silica nanoparticles have emerged as significant contributors to agricultural advancements, presenting a plethora of application that elevate crops cultivation and overall farms management practices. One notable use is in soil enhancement, while silica nanoparticles augment soil structure, moisture retention, and nutrient accessibility. Their diminutive size grants a large surface area, positively influencing water and nutrients uptake by plants, thereby fostering optimal growth conditions. Moreover, silica nanoparticles play a pivotal role in seed treatments, enhancing germination rates and safeguards against environments adversities. Within pesticide and fertilizer formulations, these nanoparticles facilitate controlled release mechanism, heightening agrochemical efficacy while mitigating ecological repercussion. Additionally, silica nanoparticles are instrumental in the development of nanosensors tailored for precision agriculture, facilitating real-time surveillance of soil dynamics and crops vitality. Employed in post-harvest preservation, especially within coatings crafted from nanomaterials, silica nanoparticles prolong the shelf life of agricultural goods by bolstering defenses against spoilage agents. Moreover, these nanoparticles play a vital role in purifying water sources, guaranteeing the provision of pristine water for irrigation purposes. As silica nanoparticles consistently exhibit their adaptability and efficacy, continuous research endeavors strive to refine their compositions, thereby nurturing sustainable and resource-conserving agricultural methodologies.

2.10 Dendrimers NPs

Dendrimers, intricate macromolecules resembling branched trees, boast a district architecture with numerous functional groups branching out from a central core. Within agricultural, dendrimers showcase unique characteristics that render them adaptable tools for a multitude of purposes.

One notable application of dendrimers in agriculture lies in their capacity for controlled and precise delivery of agrochemicals. These nanoscale structures can encapsulate pesticide, herbicides, or fertilizers, facilitating meticulous and regulate release. By employing this controlled release mechanism, the efficiency of agricultural inputs is heightened, ensuring their targeted delivery with minimal wastage. Consequently, this approach reduces the volume of agrochemicals required and lessens their environments impact.

Dendrimers also serve in soil rehabilitation. Their capability to encase and transport specific compounds can be employed for restoring contaminated soils. By encapsulating agents that degrade pollutants or enhance soil quality, dendrimers promote environmentally-friendly land management practices. Furthermore, dendrimers can be customized for particular functions, such as sensing and diagnostics in precision agriculture. Tailored dendrimers can be engineered to detects and reacts to specific environmental conditions, offering farmers valuable insight to optimize irrigation, fertilization, and pest control strategies. Dendrimers offer a promising avenue for improving agricultural methodologies by providing precise delivery systems for agrochemicals, aiding in soil remediation, and enabling precision agriculture through specialized sensing capabilities. As research in nanotechnology advances, dendrimers have the potential to play an increasingly vital role in advancing sustainability and productivity in agriculture.

3. Synthesis and characterization of nanomaterials

The creation and analysis of nanomaterials are fundamental components of nanotechnology, laying the groundwork for utilizing the distinct attributes and uses of materials on a nano-sized scale. This procedure entails producing materials with sizes typically falling between 1 and 100 nanometres, enable the emergence of innovative functionalities and properties not found in larger-scale materials.

3.1 Synthesis

Various methodologies are utilized to fabricate nanomaterials, each customized to yield particular types with specific properties. Typical approaches encompass:

- **Chemical Vapor Deposition (CVD)**: chemicals Vapor Deposition (CVD) emerges as a crucial method in nanomaterial synthesis, providing meticulous regulation over the creation of thin films and nanotubes. This technique entails the chemicals interaction of Vaporized precursor on a substance, resulting in the deposition of nanomaterials with customized characteristics. Within the CVD process, the substrate is exposed to precursor gases containing the necessary elements for nanomaterial fabrication.

Upon the substrate's surface, the precursor gases initiate chemical reactions, leading to the formation of a thin layer of nanomaterials. The selection of precursors, reaction conditions, and substrate material significantly influences the composition, arrangement, and properties of the resultant nanomaterials. A notable advantage of Chemical Vapor Deposition (CVD) lies in its adaptability and scalability, rendering it suitable for large-scale industrial production. The process can be finely adjusted to ensure precise manipulation of the thickness, uniformity, and crystal configuration of the deposited nanomaterial. Such meticulous control proves invaluable in applications requiring well-defined nanotubes or nanoscale thin films. CVD is widely applied in the semiconductor sector to create thin coatings on electronic devices. Moreover, it plays a key role in the production of carbon nanotubes, where precise control over growth is essential for electronic, mechanical, and structural uses. While effective, CVD demands meticulous adjustment of parameters to achieve the desired properties of the nanomaterial. Scientists continuously refine CVD method to expand their utility, while engineers work on improving efficiency, thereby driving progress in nanotechnology across diverse industries.

- **Sol-Gel Method**: The Sol-Gel approach epitomizes a versatile and extensively employed strategy for fabricating nanomaterials, notably nanoparticles and thin coatings. This method entails transforming a liquid precursor solution into a gel-like intermediate phase, which subsequently solidifies to produce the desired nanomaterial. The process commences with dissolving precursor compounds in a solvent to create a uniform solution. There are precursors, which may comprise mental alkoxides or salts, are chosen based on the desired nanomaterial. Subsequently, hydrolysis occurs, fragmenting the precursor molecules into smaller components that utility condenses to form a three-dimensional network known as a sol. The sol experiences gelation, wherein it undergoes a transformation into a gel with a semi-solid, three-dimensional framework. The gelation stage holds significance, as it dictates the ultimate structure and attributes of the nanomaterial. Regulation of factors like temperature, pH, and concentration is vital during gelation. Following gelation, the resultant gel undergoes a drying procedure to eliminate the solvent, resulting in a porous configuration termed an aerogel. Subsequent thermal processing, often involving elevated-temperature heating, induces the solidification of the aerogel into the ultimate nanomaterial, which can manifest as nanoparticles or thin coatings. The Sol-Gel technique is esteemed for its simplicity, affordability, and scalability. It enables precise management of the composition, morphology, and porosity of the fabricated nanomaterial. This technique finds utility across various domains, encompassing optics, catalysis, sensing, and surface treatments, owing to its capacity to generate materials with customized attributes. To sum up, the Sol-Gel approach furnishes an efficient pathway for nanomaterial synthesis, furnishing a tailored method to modulate their traits for specific applications across scientific and industrial realms.

- **Top-Down and Bottom-Up Approaches**: Nanomaterial synthesis employs two fundamental methods: top-down and bottom-up approaches, each offering distinct strategies for fabricating materials at the nanoscale. In the top-down methodology, larger structures are systematically reduced to achieve nanoscale dimensions. This process entails precisely removing or manipulating bulk materials to generate smaller entities. Lithography serves as a prominent technique in the top-down approach, where intricate patterns are etched or carved into a material's surface using specialized tools or radiation. This meticulous process enables the creation of nanoscale features with remarkable precision. The top-down approach is especially advantageous for applications necessitating intricate structures or when nanomaterials must be derived from existing macroscopic materials. On the contrary, the bottom-up strategy entails fabricating nanomaterials from atomic or molecular constituents. This approach relies on the spontaneous organization or assembly of these constituents to from the desired nanostructures. Self-assembly plays a pivotal role in the bottom-up approach, where components autonomously arrange into specific configurations. An example of this is chemical vapor deposition, where gaseous precursors react to produce nanomaterials on a substrate. Bottom-up techniques offer advantages in creating unique nanomaterials with tailored properties, often mimicking natural processes to achieve structures not easily achievable through top- methods. The selection between top-down and bottom-up methods hinges on the specific requirements of the desired nanomaterial and its intended applications. Both approaches significantly contribute to the diverse array of tools available for nanomaterial synthesis, empowering researchers and engineers to customize materials for various applications in science, technology, and industry.
- **Biological Synthesis**: Biological synthesis, also referred to as bio-fabrication, emerges as an innovative method leveraging living organisms such as bacteria and plants as natural nanofactories for nanomaterial production. This pioneering approach is distinguished by

its environmentally sustainable nature, aligning with the principles of green chemistry. In the domain of biological synthesis, microorganisms, particularly bacteria, fungi, and plants, serve as bioreactors capable of facilitating the production of nanomaterials. This process typically entails harnessing the inherit biological processes and metabolic pathways of these organisms to reduce and stabilize metal ions, thereby inducing nanoparticle formation. A notable example involves the utilization of plant extracts or bacterial cultures to synthesize metallic nanoparticles. In this methodology, the extracts derived from plants or bacterial cultures serve dual roles as both reducing and stabilizing agents for metal ions, thereby facilitating nanoparticle formation. The selection of the organism and its environmental conditions can significantly influence the size, shape, and properties of the resultant nanomaterial. The fabrication of nanomaterials is characterized by its ecological sustainability, as it circumvents the necessity for harsh chemicals and energy-intensive procedures commonly associated with traditional synthesis techniques. Furthermore, employing biological agents confers a degree of compatibility with living organisms upon the resulting nanomaterials, rendering them applicable in fields such as medicine and environmental restoration. The advantages of biological synthesis transcend its environmentally friendly nature. This approach enables the production of nanomaterials endowed with distinctive properties, often challenging to achieve through conventional synthetic methodologies. As scientists delve into unravelling the biological mechanisms underlying this synthesis, the scope of potential applications for biologically derived nanomaterials continues to broaden. In essence, biological synthesis epitomizes the forefront of sustainable nanotechnology, harnessing the intrinsic capabilities of living organisms to fabricate nanomaterials with multifaceted applications. This methodology not only fosters the progression of refined materials but also corresponds with the worldwide endeavor for eco-friendlier and more enduring technological resolutions.

3.2 Characterization parameters of NPs

3.2.1 Size

Describing nanoparticles (NPs) primarily relies on understanding their particle size and distribution, which determines whether they fall within the nano or micro scale. Electron microscopy serves as the standard technique for assessing particle size and distribution. Both Scanning Electron Microscopy (SEM) and Transmission Electron Microscopy (TEM) methods are utilized to capture images and measure particles and clusters. Laser diffraction techniques are preferred for analysing bulk samples in solid phases. In the case of liquid particles, light scattering is measured through centrifugation and photon correlation spectroscopy. Scanning Mobility Particles Sizers (SMPSs) are favoured for their efficiency and accuracy in measuring gaseous phase particles, which are notoriously challenging to image using other methods.

3.2.2 Surface area

Nanoparticles (NPs) can also be evaluated based on their surface area, which significantly influences their performance and properties due to the surface area-to-volume ratio. BET analysis is commonly employed to quantify surface area. Assessing particle surface area in liquid form can be achieved through titration, although this process is time consuming. NMR spectroscopy is utilized for this purpose. In the gas phase, nanoparticle surface area is determined using a differential mobility analyser (DMA) and an adapted Scanning Mobility Particle Sizer (SMPS).

3.2.3 Surface morphology

The diverse shapes and surface configurations of nanoparticles (NPs) play a crucial role in dictating their properties. NP surfaces can exhibit various characteristics, including crystalline or amorphous structures, as well as uniformity or irregularities. These shapes may range from spherical to flat discs, cylinders, tubes, cones, and other irregular forms. Electron microscopy techniques such as SEM and TEM are commonly employed to analyse NP surfaces, providing valuable insights. In electron microscopy imaging, liquid-phase particles are typically captured electrostatically or via filtration, while gas-phase particles are first accumulated on a surface before analysis.

3.2.4 Composition

The purity and functionality of a nanoparticle depend on its chemical or elemental composition. Elevated levels of secondary or undesirable elements within the nanoparticle can lead to reduce efficiency and increase the risk of secondary relations and contamination.

In many cases, X-ray photoelectron spectroscopy (XPS) is employed for determining the composition. Wet chemical analysis includes atomic emission spectroscopy, ion chromatography and mass spectrometry which forms an integral part of some methods. Filtration or electrostatic collection is used to collect gas phase particles along with spectrometric and wet chemical methods used for analysis purpose.

3.2.5 Crystallography

Crystallography involves the systematic investigation of various components, including atoms and molecules, that constitute crystals. Nanoparticle crystallography is achieved through techniques such as electron or neutron diffraction and powder X-ray analysis.

3.2.5 Characterization tools and techniques of NPs

X-ray diffraction (XRD), energy-dispersive spectroscopy (EDS), UV-visible spectroscopy, scanning electron microscopy (SEM), transmission electron microscopy (TEM), fourier transform infrared (FTIR) spectroscopy , atomic force microscopy (AFM), dynamic light scattering, extended X-ray absorption fine structure spectroscopy (EXAFS), and nanoscale infrared spectroscopy are among the analytical techniques commonly utilized for the characterization of de-novo synthesized nanoparticles (Figure 1).

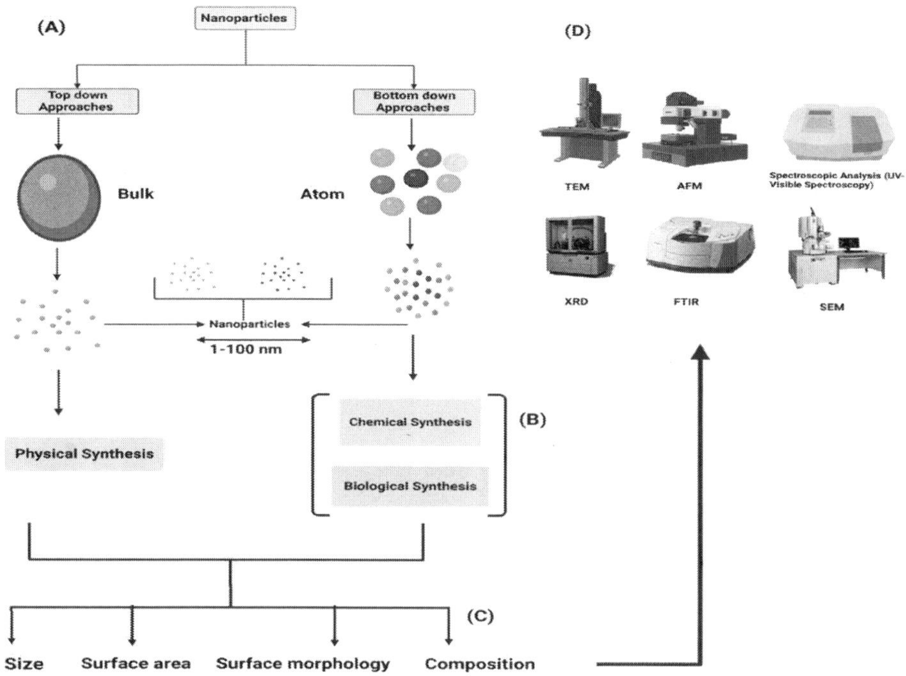

Figure 1. The figure illustrates the formation and characterization tools used for NPs. NPs are formed through (A) methods like bottom-up synthesis, top-down approaches, with (B) physical, chemical, and biological approaches with unique (C) characters like size, surface area morphology and composition. The figure also highlights the (D) characterization tools, including microscopy technique (such as scanning electron microscopy and transmission electron microscopy) and spectroscopic methods (such as UV-Vis-spectroscopy and X-ray diffraction), which are used to examine the size, shape, composition, and structural properties of NPs.

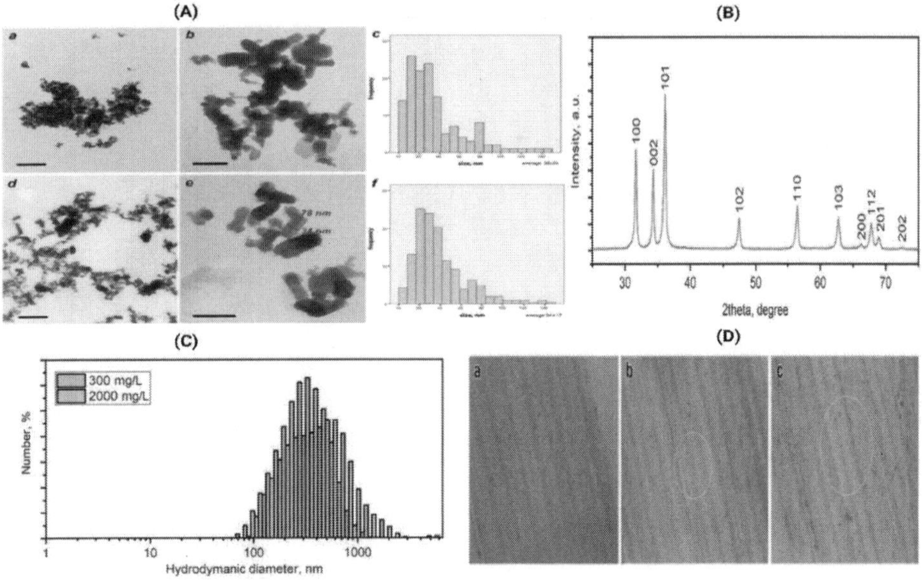

Fig 2. Characterization of ZnO-NPs with the help of (A) TEM, (B) X-ray diffraction, (C) DLS and (D) X-ray fluorescence spectroscopy. The ZnO-NPs is hexagonal lamellar particles with a major axis of 50-150 nm and a minor axis of 20-70 nm.

3.2.6 FTIR (Fourier Transform Infrared Spectroscopy)

The FTIR technique is highly efficient and user-friendly, particularly valuable in biological contexts for assessing chemical concentrations, identifying various functional groups, determining the atomic composition of biological nanoparticle samples, and analysing their surface chemistry. FTIR analysis can ascertain whether biomolecules are involved in nanoparticle synthesis and identify the types of biomolecules present in the sample. This method relies on the diverse types and frequencies of bond vibrations within molecules. Notably, FTIR offers a customizable workspace, high spectral resolution, enhanced sensitivity, and employs a high cube corner interferometer.

3.2.7 UV-Visible Spectroscopy

UV-visible spectroscopy is a widely used technique for assessing the stability and production of metallic nanoparticles. By examining nanoparticles ranging in size from 1 to 100 nm, synthesis can be confirmed through absorbance measurement across wavelengths from 230 to 800 nm. The absorption band observed in surface Plasmon resource is attributed to the presence of free electrons. Apart from confirming synthesis, UV absorption spectroscopy enables precise quantification of particles in colloidal solutions. Absorption spectra are utilized to quantify and characterize particle size, providing qualitative analysis as well.

Spectroscopy involves the measurement and interpretation of electromagnetic radiation, as transitions in energy states occur within atoms, ions, and molecules present in the sample. The absorption of ultraviolet light by molecules results in electron excitation from the ground state to higher energy states, facilitating the analysis of nanoparticle samples. UV-Vis spectroscopy is notable for its ease of maintenance and features such as double choppers, sealed optics, and adjustable spectral bandwidth, which can be as low as approximately 0.2 nm.

3.2.8 IR (Infrared spectroscopy)

Infrared spectroscopy (IR) utilizes precise instruments to analyse the molecular structure of substance by detecting their different resonance and vibrational modes. This technique has the capability to unveil the chemical composition and bonding arrangement of materials. To ensure accuracy, a specialized spectrograph is employed to compare radiation intensity and spectrum frequency against international standards for precise molecule identification. Infrared spectroscopy enables the characterization and identification of diverse chemical species, albeit with a limitation in resolving features between 5 to 10 micrometres in size. Nano-infrared spectroscopy (NanoIR) serves as a nano-scale analytical tools for chemical analysis. NanoIR merges the key attributes of Atomic Force Microscopy (AFM) and infrared spectroscopy, enabling the acquisition of infrared spectra at spatial resolutions ranging from 50 to 200 nanometres. A notable feature of nanoIR is its multifunctional capabilities. This instrument provides a comprehensive depiction of samples by integrating their mechanical, thermal, spectroscopic, and topographic attributes.

3.2.9 XRD (X-Ray diffraction)

X-Ray Diffraction (XRD) stands out as an effective technique for visualizing the crystalline structures of both organic and inorganic substances. What makes X-ray diffraction particularly appealing is its non-destructive nature, ease of use, sensitivity, depth profiling capabilities, and straightforward setup. Apart from its diverse application in forensic sciences, pharmaceuticals, geology, and polymer science, XRD proves valuable in chemical composition analysis, providing qualitative and quantitative insight such as degree of crystallinity, which offers precise details about atomic arrangements at interface. The crystals structures are elucidated based on atomic arrangements, position, and diffraction peak intensities. With X-rays possessing atomic-scale wavelengths. XRD serves as powerful tool for investigating nanomaterial architectures. Each atom in a powdered sample scatters a single X-rays beam, generating various diffraction patterns when interacting with other crystals. The intensity of peaks sampled at specific angles where X-rays scatters reflect the crystal's structure and physicochemical properties. Bragg's Law serves as fundamental principles underlying X-rays diffraction, while the joint committee and powder Diffraction standards maintains a database (JCPDS) used for cross-referencing individual diffraction patterns. Elemental analysis of metal NPs commonly employs Energy-Dispersive X-ray Spectroscopy (EDS).

3.2.10 Scanning Electron Microscopy (SEM)

Scanning Electron Microscopy (SEM) stands out as a versatile and non-invasive analytical method employing a microscope featuring a generous specimen chamber and an impressive 8.5 mm working distance, achieved through angled detectors and sharp conical objectives lenses. SEM serves as a powerful imagining tool for assessing the morphology and dimensions of nano-or microstructures present on sample surface. It offers valuable insight into the organization of components and provides swift, straightforward, and precise results. Moreover, SEM finds applications across diverse fields, including the examination of biological molecules, biochemical research, and materials science. Biomedical sciences frequently utilize SEM for the characterization of cell surfaces, organs and tissues, thereby enhancing our understanding of cellular membranes. The operational principle of SEM involves the generation of magnetically-charged electron beams, with detection of secondary and backscattered electrons resulting from their interaction with the sample's magnetic field. Additionally, SEM proves valuable for elemental analysis and determination of sample concentration, further augmenting its utility in various research endeavors.

3.2.11 Transmission Electron Microscopy (TEM)

Transmission Electron Microscopy (TEM) stands as a widely utilized and indispensable technique for the characterization of nanoparticles (NPs). Its primary advantage lies in its capability to qualitatively and quantitatively analyse the properties of synthesized NPs and their interaction within cells or tissues, focusing on their morphology, crystal structure, and dimensions. Conventional TEM methods are employed for comprehensive sample analysis. However, TEM is not without limitation, including its inability to effectively characterize NPs smaller than 10 nm, limited magnification for particles exceeding 100 nm, and the intrusive nature of sample preparation procedures. Techniques such as immune and energy-filtered electron microscopy play a pivotal role in enhancing TEM capabilities.

CHAPTER 3

APPLICATION, UPTAKE AND TRANSLOCATION METHODOLOGY OF NANOPARTICLES

Short Summary

This brief chapter delves into the essential element of nanoparticles absorption and movement within plants, with a specific focus on their relevance in agriculture. It underscores the significance of effective nanoparticle uptake by plants, exploring factors like nanoparticle size, change, and application technique that influence absorption. Furthermore, the narrative delves into the subsequent translocation within the plant system, highlighting its contributions to bolstering growth, nutrient absorption, and stress resilience. This brief overview aims to provide a quick grasp of nanoparticle application methods, uptake mechanisms, and translocation processes, thereby facilitating future advancements in agriculture.

1. Introduction

Soil-borne NPs navigate through soil pores towards plants roots, their primary entry points (Figure 1). Within the rhizosphere, lateral roots and root hairs emerge as highly efficient sites for NP absorption. Upon contact with the root surface, NPs are initially takes up by the roots from the soil. Notably, positively changed NPs exhibit a stronger affinity for root absorption, facilitated by the negative charge of the root surface, induced by the release of chemicals from roots hairs. These chemicals, which include mucus and organic acids, expedite NPs attachment to the root surface. Once inside the plant, NPs undergo unhindered movement through the plant tissues via apoplastic and symplastic pathways, symplastic pathway particularly aids in radial transport of NPs into vascular tissues, facilitating their dissemination. Symplast serves as a conduit for the transport of water and other substances between the cytoplasm of adjacent cells. NPs have the potential to infiltrate the root system through a novel adsorption interface facilitated by the development of lateral roots, as depicted in Figure 1. Although structurally and functionally akin to the leaf surface, the root epidermis lacks full development in the root hair surfaces and root tip epidermis of primary and secondary plants roots. NPs make direct contact and penetrate the root epidermis upon exposure. While root epidermal cells permit the passage of some water and nutrients through their cell wall, they possess tight cell walls capable of sieving out undesirable smaller and larger debris. Acting as a physical barrier between the root and soil, the plasma membrane of root cells selectively allows the passage of small molecules like steroids while blocking hydrophilic molecules and macromolecules. These compounds undergo transportation across the membrane via channels and carriers such as ATPase and the GLUT1-4 transporter/carrier. Additionally, these membrane channels and transporters may play a role in facilitating the transport of NPs from the soil to root cell. In instances where the root lacks an exodermis, NPs have the capability to penetrate the central column, or xylem, of the root. Specific NPs have been observed to cause damage to the plasma membrane, resulting in the formatting of new pores in the epidermal cell wall, thus enabling the entry of larger NPs, as demonstrated in other studies. Various mechanisms are implicated in the absorption of NPs by cells during their introduction to plant tissue, including the ion pathway, endocytosis, protein binding to cell membranes, and physical damage. Nanoparticles (NPs) deposited on the surface of leaves are absorbed by the plant through various structures such as trichomes, hydathodes, or stomach. The waxy cuticle of the leaf epidermis, primarily composed of wax, cutin, and pectin, acts as a crucial natural barrier against NP entry while also preventing water loss in developing leaves. Nonetheless, within the waxy cuticle, there exist two distinct channels: one hydrophilic and the other lipophilic, each with diameters ranging from 0.6 nm to 4.8 nm. Hydrophilic nanoparticles (NPs) within a diameter smaller than 4.8 nm diffuse through hydrophilic channels. Conversely, lipophilic NPs are absorbed through diffusion and infiltration via the cuticle's lipophilic channels. Using high-resolution confocal fluorescence microscopy, it is demonstrated that carbon dots smaller than 2 nm can penetrate cotton leaves via the cuticular pathway. However, the absorption of nanoparticles (NPs) by plants through the epidermis is hindered by the small size of the pore channels in the cuticle. When NPs are applied to the leaf surface, they tend to accumulate in the vascular area and epidermis. Additionally, numerous studies have indicated that NPs can be transported to various plant organs. Based on these findings, the research team hypothesized that NPs could be absorbed via the stomach pathway, ad depicted in Figure 1.

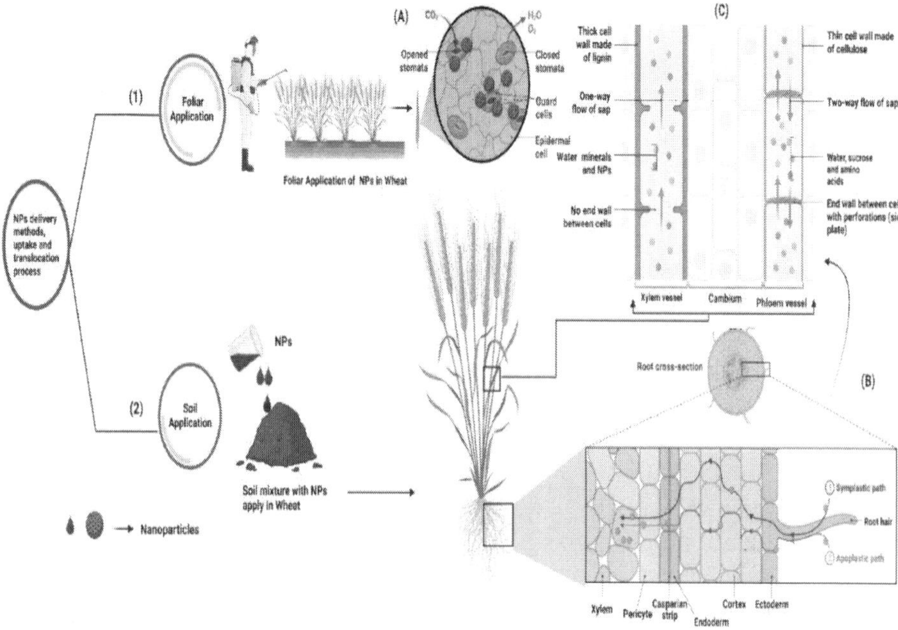

Figure 1. Application methodology of NPs though foliar (figure 1) and soil application (figure 2). The NPs which apply through (A) foliar method uptakes by plants through stomata and soil mixture NPs transport via (B) symplastic and apoplastic pathway through (C) xylem and phloem in different parts of plants.

2. NPs delivery methods, uptake, and translocation process

2.1 Soil Application and uptake of NPs

In soil environments, nanoparticles (NPs) undergo a range of bio/geo-transformations that influence their bioavailability and toxicological effects. Following interactions with root systems, NPs migrate to aboveground segments and accumulate within cellular or subcellular organelles. The uptake to NPs from the soil via plant roots represents the initial phase of bioaccumulation. Additionally, it has been observed that very small NPs, typically with diameters ranging from 3 to 5 nm, can penetrate plant roots either through capillary forces or by directly traversing root epidermal cells. Primary epidermal cells construct a semipermeable cell wall with minute pores that effectively hinder the passage of larger NPs. The formation of initial pores in the epidermal cell wall facilitates the uptake of certain NPs. As nanoparticles (NPs) penetrate cell membranes, they traverse extracellular gaps until reaching the central vascular cylinder, allowing fir vertical ascension via the xylem. Access to the central vascular cylinder necessitates the crossing of the Casparian strip boundary through the symplastic pathway, as illustrated in Figure 1. This process involves endocytosis, pore formation, and delivery subsequent to attachment to carrier proteins in the membrane of endodermal cells. NPs embedded in the cytoplasm can move across cells via plasmodesmata. The Casparian barrier serves to trap NPs that fail to penetrate the plant, while those that reach the xylem are accepted by shoots and roots. Absorbed NPs may be localized in the outer layer cell membrane, the inner environment of cortical cells, or within plant centers, whereas non-absorbed NPs on the root surface of soil aggregates can alter nutrient absorption. Upon seeding on soil, seeds can be promptly absorbing soil-bended NPs through the seed coat via parenchymatic intercellular spaces, leading to NP diffusion in the cotyledon.

2.2 Foliar Application and uptake of NPs

In contemporary agriculture, there is a growing trend toward utilizing foliar sprays containing engineered nanoparticles (ENPs) for various purposes such as nano-fertilizers, nano-pesticides, nano-sensors, and nanocarriers. Compared to traditional soil- root treatment, applying NPs directly onto the leaves enhances the effectiveness of plant protection method. Spraying foliar solutions containing NPs allows them to penetrate the plant primarily through the stomata, subsequently traveling through the plant via both apoplastic and symplastic pathways (Figure 1A). The adoption of foliar NPs has been shown to enhance crop yield and quality, while also bolstering plant defense against pests and diseases. The mechanisms underlying the potential harm caused by foliar NPs remain incompletely understood. Moreover, it is imperative to investigate the chemical and physical properties of NPs, as well abiotic factors such as temperature, humidity, and light, enhance the effectiveness of this technology in facilitating foliage uptake of NPS. NPs applied to leaves can enter through stomata or cuticles. The cuticle serves as the leaf's primary defense, preventing particles smaller than 5nm from entering the plant. NPs enter the plant through stomata, and cells transport them into the vascular system via apoplastic and symplastic pathways. NPs ranging from 10 to 50 nm tend to traverse through the cytoplasm of adjacent cells (symplastic route), while NPs sized between 50 and 200 nm migrate between cells (apoplastic pathway).

NPs taken up by the plant are transported through the phloem sieve tubes alongside the sugar solution. Various plant parts, including roots, stems, fruits, grains, and young leaves, act as significant reservoirs for the sap, allowing NPs to move bidirectionally within the plant's vascular system. The apoplastic pathway, known for its non-selective and least resistant path, plays a crucial role in this transport. It is widely recognized that the apoplastic route efficiently transports various water nutrients and non-essential metal complexes. The effective adsorption of NPs following foliar application depends on factors such as the application method, NP size and concentration, and environmental conditions. Various factors, including leaf morphology, chemical composition, presence of trichomes, leaf exudates, and waxes, influence the deposition of NPs on the leaf surface.

By retaining NP on the plant's surface, the trichomes on plant organs can alter surface activity and increase the amounts of time that external materials are present in tissues. Plants above ground and in their hypogeal sections may have wounds and impairments that serve as efficient routes for NP internalization. A variety of leaf morphological factors, such as the shape and chemical composition of the leaf, the presence of trichrome, leaf exudates, and waxes, influence the trapping of NPs on the leaf surface.

2.3 Translocation of NPs

The plant's internal nutrient transport is categorized into two crucial pathways: apoplast and symplast. Apoplastic translocation involves the movement of nutrients through interconnected cellular membranes within plant cells. Conversely, symplastic translocation occurs via protoplasts of different plant cell connected by thin cytoplasmic connections. These pathways facilitate the uptake and release of dissolved ions into and out of root cells. Specifically, ions can only enter root cells through symplastic pathways using membrane-specific channels and transporters that facilitate their passage across the plasma membrane (Figure 2). It has been demonstrated that apoplastic transport promotes the radial movement of nanoparticles (NPs), potentially enabling them to reach the core cylinder and vascular tissues of the root, facilitating their transports to the aerial parts of the plant. This mode of NP translocation proves particularly advantageous for applications requiring systemic NP delivery. However, the presence of a layer of lignin-like structures known as the Casparian strip impedes the radial migration of the root endodermis. Consequently, NPs must ultimately enter cells to undergo symplastic transport. Delivering NPs intracellularly into plant cells pose greater challenges compared to animal cell, primarily due to the robust cell wall that serves as an external barrier to cell entry. Plant cells have been observed to employ alternative strategies for NP uptake, including those involving hole formation, membrane translocation, or carrier proteins. Plasmodesmata, which are membrane-bound cytoplasmic bridges with adjustable diameter (20-50nm), facilitate the migration of NPs from one cell to another after penetrating the cytoplasm. Studies on Arabidopsis, rice, and poplar have contributed to our understanding of NPs transport through plasmodesmata, encompassing NPs of various sizes.

Tiny particles have the ability to move throughout the entire plant via both symplastic and apoplastic pathways, ultimately reaching the xylem and phloem vessels. Notably, NPs have a tendency to accumulate in organs with high fluid uptake capacity from the phloem, such as flowers, fruits, and seeds, due to their skink activity. Worries regarding NP accumulation in specific organs are as signification as concerns about toxicity of NPs to plants.

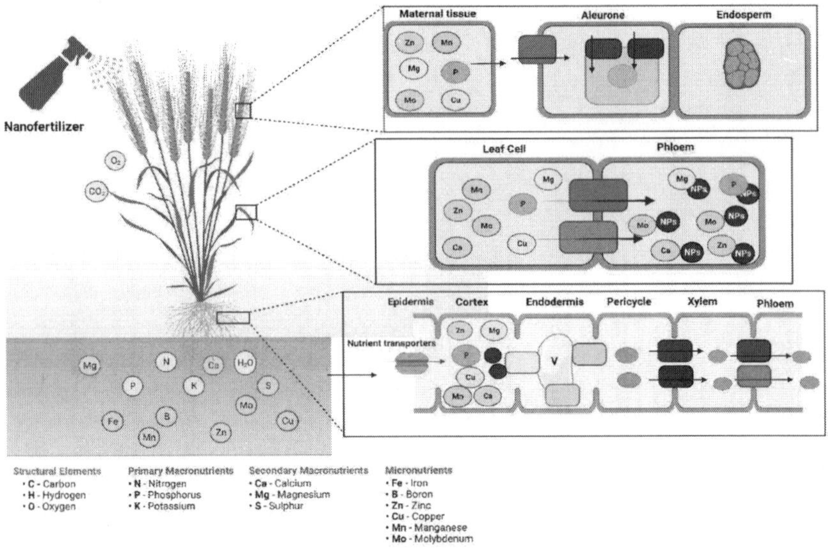

Figure 2. Diagrammatic repreparation of nanofertilizers as sources of plant nutrients that help crop production. This nanofertilizers help in plants growth and developments and improve nutrients contents in grains.

CHAPTER 4

IMPACT OF NANOFERTILIZER FOR SOIL AND NUTRIENT MANAGEMENT

Short Summary

This chapter explores the transformative potential of nanofertilizers in soil and nutrient management, highlighting their role in enhancing agricultural productivity and sustainability. Nanofertilizers, due to their nanoscale size and unique properties, offer significant advantages over conventional fertilizers, including higher efficiency, targeted delivery, and reduced environmental impact. The chapter begins by discussing the different types of nanofertilizers, such as nano-encapsulated nutrients, nano-emulsions, and nanoscale carriers. These formulations allow for the controlled release of nutrients, ensuring that plants receive a steady supply over time. This leads to improved nutrient uptake and utilization, minimizing nutrient loss through leaching and volatilization. The application of nanofertilizers also improves soil health by enhancing microbial activity and nutrient availability. This is crucial for maintaining soil fertility and structure, which are essential for sustainable crop production. Additionally, nanofertilizers can be engineered to address specific nutrient deficiencies and soil conditions, offering customized solutions for various agricultural settings. Furthermore, the chapter examines the environmental benefits of nanofertilizers, such as reducing the overuse of chemical fertilizers and minimizing pollution. By providing a more efficient and sustainable nutrient management strategy, nanofertilizers hold the promise of significantly boosting crop yields while mitigating the adverse environmental impacts associated with traditional fertilization practices.

1. Introduction

Plants necessitate essential nutrients, encompassing both macro and micronutrients, to flourish and attain their utmost potential (Figure 4). The successful germination of seeds may be impeded in the absence of any of these constituents. Nonetheless, an excess of nutrients can also inflict harm upon plants. Ensuring the provision of requisite nutrients for plants to uphold normal cellular function presents a challenge. A consistent supply of nutrients in appropriate proportions enables plants to complete their life cycle proficiently. Upon absorption of minerals, plants expeditiously employ these elements in a diverse array of metabolic processes, fostering the development and arrangement of plant tissues. These components constitute crucial constituents of numerous enzymes, pigments, and proteins, playing pivotal role in progresses such as cellular signalling and metabolism. Seventeen essential nutrients have been identified as critical for plant development, including nitrogen, phosphorus, potassium, hydrogen, boron, carbon, oxygen, magnesium, sulfur, chlorine, calcium, manganese, iron, nickel, copper, zinc, and molybdenum. Among these, phosphorus, nitrate, and magnesium are pivotal for plant health and growth. Since these elements cannot be directly sourced from the atmosphere, plants rely on root absorption for nutrient uptake. Nano-fertilizers have emerged as a technological solution to address nutrient deficiencies due to their minute size. Carbon-based compounds, metal oxides, and other nanoporous materials are utilized in the production of nano-fertilizers. There nanoparticles exhibit a diverse range of properties. Research findings indicate that nano-fertilizers can enhance plant surface area, facilitate slow-release nutrient functions, and regulate the distribution of macro and micronutrients among plant of varying sizes and concentrations.

2. NFs Based Macronutrient Management

In order to lessen the mass required to nourish plants while still satisfying their nutritional demands, nanomaterials have been mixed with macronutrients such as calcium (Ca), nitrogen (N), phosphorus (P), magnesium (Mg), sulfur (S), and potassium (K). These plant NFs are composed of a selected nanomaterial encasing one or more macronutrients. Through 2020, an extra 265 million tonnes of NPK are anticipated to be consumed by the agriculture industry.

2.1 Nitrogen NFs

One of the vital major macronutrients that is deficient in agricultural soil is nitrogen. Though the air contains 78% nitrogen, plants can only use this nitrogen in the form of nitrates and ammonium. Manufacturing companies use the Haber-Bosch technique. It is well recognized that nitrogen fertilizers affect plant growth and development, resulting in a healthy supply of nutrient-dense food. The fact that a large portion of synthetic N fertilizers evaporates and washes of during and immediately after field application is one of their main issues. In recent years, an improved N nano-fertilizer for gradual, targeted release was created using urea-coated hydroxyapatite nanoparticles. Because of their availability of N and P source and chemical compatibility, these compound NPs were selected. Thus, due of reduced consumption and N leakage into the environment, nitrogen fertilizers induce eutrophication and excessive greenhouse gas emissions. It is discovered that maize plants developed more quickly, assimilated more nutrients, and generated more grain of better quality when they were planted using zeo-urea nanofertilizer (N-NF). When N-NF was applied to starflower, the plant grew faster and produced more essential oil. When urea-modified zeolites were compared to synthetic fertilizers, it was discovered that the soybean seed production increased. The application of urea to the nanofilm enabled an N-NF that was beneficial to Brassica napus L. Corresponding advantages from applying nano-N and chelated nano-N included decreased nitrate leaching and enhanced production in a potato crop. In a recent study, it is discovered that greenhouse-grown coffee seedlings were fertilized with NPK- coated nano-fertilizers (NFs). The authors observed that the application of NPK NFs led to an increase in leaf number and photosynthetic plant area, thereby enhancing nutrient uptake and growth rate in coffee plants compared to the control group, the authors noted a 30.7% increase in overall chlorophyll content, a 71.7% increase in net photosynthesis rate, and a 17.1%, 16.3% increase in NPK content (zero NF). Due to their ability to slow down N release, reduce volatilization and leaching rates, enhance nutrient uptake, and promote crop growth and yield, nano-fertilizers come highly recommended.

2.2 Phosphorus NFs

Similar to nitrogen, phosphorus is essential for all living forms and is utilized nearly solely (nearly 90%) in agricultural activities. Rock phosphate serves as the primary raw material for the majority of commercial phosphate fertilizers. Among the main issues brought on by P fertilizers is eutrophication, which is produced by a rise in the P and N levels in water. It also results from the fertilizers' lower usability and absorption capability, which cause the useful P from to quickly transition into the less available from. In the past, scientists have looked at several biological techniques to boost plant P absorption. These comprise phosphatase, fungi and bacteria that produce phytotase, bacteria that solubilize P, and vesicular-arbuscular mycorrhiza. Because of the low levels of organic carbon in the soil and their high propensity for evaporation, the P-mobilizing bacteria proved to be limited use. It restricted the amount of energy available to the microbes that mobilize p. These variables lower the mortality rate of microorganisms. Researchers studying nanotechnology are looking at way to increase the delay release and absorption effectiveness of P fertilizers in order to preserve phosphorus in a form that is more useable for extended periods of time and allow native phosphorus to be mobilized to plants. Tricalcium phosphate is used as a precursor salt in the process of fungus-mediated P NPs. Soybeans were then impacted by hydroxyapatite

NPs stabilized with carboxymethyl cellulose for the processing of seeds, the application of organic crude phosphorite by ultrasonic dispersion as a P nano-fertilizer produced promising outcomes. Phosphorus (P) may be present in adequate amount in the soil, but because P is unstable, framers need P fertilizer often. By using zinc nanoparticles, the rate at which native P is transformed into plant-available P was increased. Numerous enzymes, such as phosphatase and phytase, require zinc as a cofactor. ZnO NPs have increased the capacity of these enzymatic processes together with P absorption of 11% in cereals and legumes, which has led to better growth, biomass, nutritional importance, and yield. It has been shown that cotton plants benefit from exposure to ZnO NP as well as P supplements. ZnO with phosphorus increased the biomass, protein content, and photosynthetic pigments of the cotton plant, improve native P mobilization, which raises P absorption. This will not only boost agricultural yields but also lessen the ecological consequences of environmental phosphorus shortage and eutrophication caused by phosphorus.

2.3 Potassium NFs

Potassium (K) is the third most essential major nutrient because it may control a wide range of metabolic activities, including the opening and closed of stomata, the transfer of water and nutrients, the activation of over 60 enzymes, and the pH of the cytoplasm. The immunological system of plants is strengthened by potassium. Since K+ comes from the disintegration and decay of potassium mineral rocks, the plants are absorbing the K found in the soil. It is a known uncommon element that, even at larger accumulations, does not produce any harm. Normally, normal tissues contain 1-5% of this element. K+'s crystal ionic radius is 0.138 nm, whereas its hydrated ionic radius is 0.2798 nm. While soil alkalinity raises the concentration of phosphate and potassium, soil acidification decreases their availability. A lack of K inhibits the growth of roots and shoots as well as the quantity, shape, size, color, flavor, and yield of seeds in fruits. It is possible to remove between 50 and 70 percent of the potassium fertilizer, which would result in a large financial loss and have a detrimental effect on the quality of the water and soil. A K fertilizer's maximal potential usually ranges from 30% to 50%. Kubavat conducted research and created a low-rate K discharge nano-potassium fertilizers formulation. The scientists also came to the conclusion that applying a fertilizer containing nano-potassium might reduce soil K loss while preserving a steady supply of K for the crops. Using K-loaded zeolites improved the hot peppers (*Capsicum annuum L.*) overall yield, K concentration, harvest index, and chlorophyll content. The foliar application of a nano-K fertilizer also significantly improved the biomass, growth, and quantity of Cucurbita pepo. Comparably, an analysis demonstrated that NFs K applied topically can increase wheat production and growth. By increasing the translocation of NPs-based K, NFs K can improve agricultural yield. The detrimental effects of water deficiency on morphological characteristics and yield components, especially during the seed filling stage, are reduced by about

25% when 3 kg/ha of K NFs are used. It is known that NFs K increases the activity of related enzymes (GS, GOGAT, etc.) in nitrogen metabolism, which in turn increases the expression of genes belonging to the GLN family.

2.4 Magnesium NFs

Since magnesium is a necessary component of chlorophyll, plants depend on it for survival. In actually, it is a ribosomal structure component (bridging element), a cell's protein maker that effectively contributes to protein synthesis by including the production of polypeptide chains, which is turn forms amino acids. It functions as a co-factor for several enzymes and promotes the absorption and translocation mechanism of macromolecules and nutrients, such as phosphate sugar, include plants. The disintegration of rocks such as biotite, olivine, dolomite, epsomie, etc. produces soil magnesium. The soil has this element in an exchangeable and water-soluble form, which is what the plants are absorbing as Mg^{2+}. Mg fixation results from a drop in soil pH for the reason Mg^{2+} ions interact with $A1^{3+}$. In acidic soils, 65-70% of this magnesium is soaked and is inaccessible to plants. Depletion of base cations such as potassium, calcium, and magnesium results in a rise in soil acidity. A foliar spray investigation demonstrating that a 0.5 g/L dosage of Mg NPs generates a considerable increase in biomass and photosynthesis in black-eyed pea plants was conducted. The Mg NPs also altered the permeability of the cell membrane. Scientists found that wheat plants treated with a foliar spray of biosynthesized Mg-NPs (5.9 nm, 20 mg/L) showed increases in root length, root tip number, and root biomass. The possibility that Mg-NP-induced enhancement of leaf solar absorption explains the occurrence was put up. MgO-NPs (50, 150, and 250 mg/L) have been shown to enhance tobacco cultivated on matrix media by increasing SOD and POD activity. Because magnesium is crucial for increasing antioxidant enzyme activity, Mg-NP can improve plant stress tolerance and alleviate symptoms of magnesium insufficiency. *Ralstonia solanacearum* bacteria were found to be substantially reduced in tobacco cultivated in soil when MgO-NPs (150 and 250 mg/L) were administrated rather than bulk MgO.

3. NFs Based Micronutrient Management

A plant requires trace quantities of these key components. The pH of the soil greatly affects the quantity and accessibility of micronutrients. Among the key elements in this group (Ni) are iron (Fe), manganese (Mn), zinc (Zn), copper (Cu), boron (B), molybdenum (Mo), chlorine (C1), and nickel (Figure 1).

3.1 Fe NFs

Iron (Fe) plays a crucial role in the growth and development of plants, being the fourth most important nutrient for them. Its significance lies in its involvement in various metabolic processes essential for plant life. These include cellular respiration, electron transfer, replication, lipid metabolism, photosynthesis, nitrogen fixation, and gene regulation. Many proteins, enzymes, and organelles within plant cells, such as chloroplasts, mitochondria, and vacuoles, rely on iron for their proper function. Studies have shown that the application of iron nanofertilizers (FNFs) can enhance germination and improve the growth of various crops compared to control or synthetic iron sources. For instance, it has been demonstrated that spinach (*Spinacia oleracea* L.) exhibited better growth when exposed to iron pyrite nanoparticles. Scientists observed that peanut (*Arachis hypogaea* L.) plants treated with iron nanoparticles showed healthier root growth in field conditions compared to untreated plants. The studies by researchers further support the potential benefits of using iron nanoparticles (Fe NPs) as a replacement source, particularly in iron-deficient soils.

The effects of Fe NPs (2-6 nm) were investigated on the germination of green gram (*Vigna radiate* L.) and observed that the application of Fe NPs led to increased radical length and fresh biomass compared to the control treatment using ferrous sulfate ($FeSO_4$). This suggests that Fe NPs could enhance early growth stages and biomass production in green gram, indicating their potential as a more effective source of iron for promoting plant growth. Similarly, experiments were conducted with rose periwinkle (*Catharanthus roseus*) exposed to Fe NPs at various concentrations. They found that plants treated with Fe nanoparticles exhibited higher levels of chlorophyll and protein compared to those treated with a control solution. This indicated that Fe nanoparticles can serve as excellent replacement sources for traditional iron fertilizers, particularly in soils deficient in iron.

3.2 Zn NFs

Zinc (Zn) is a micronutrient that plays a critical role in various physiological and biochemical process essential for proper plant functioning. It is involved in controlling the plant's defense mechanisms by activating antioxidant enzymes like superoxide dismutase, glutathione reductase, and peroxidase. These enzymes help protect the cell membrane's lipids and proteins from oxidative damage caused by reactive oxygen species. Furthermore, zinc regulates the remobilization of photo-assimilates, which is crucial for reproductive processes and grain development in plants. Zinc ions (Zn^{2+}) have a crystal ionic radius of approximately 0.07 nanometres (nm) and a hydrated radius of 0.075 nm. The form of zinc that plants uptake in Zn^{2+} ions, which also plays a role in regulating phosphorus translocation and controlling toxicity caused by excessive phosphorus absorption. The availability of zinc to plants is influenced by various factors, including soil pH. Zinc exists primarily as cationic Zn^{2+} ions, which can from complexes with other ions such as calcium, reducing its availability, especially at higher pH levels. Conversely, soils with lower pH levels are more conducive to zinc availability-Zinc nanofertilizers (Zn NFs) have been demonstrated to have positive effects on crop germination, seeding development, and overall harvest quantity and quality, suggesting the use of Zn NFs could be beneficial in improving crop productivity, particularly in soils where zinc availability may be limited. Zinc oxide nanoparticles (ZnO NPs) have shown improved protein content, sugar content, and antioxidant activities in cabbage (*Brassica botrytis* L.) and tomato (*Lycopersicon esculentum* L.). Conversely, cucumber plants were found to absorb more ZnO NPs (100 mg kg^{-1}) than their synthetic bulk counterparts but observed suppression of cucumber growth characteristics at this concentration. Other studies have shown benefits of Zn NFs on various crops including sunflowers (*Helianthus annuus* L.), pearl millet (*Pennisetum Americanum* L.), rice, maize, sugarcane (*Saccharum officinarum* L.), and potatoes. Researches have reported improvements in shoot growth, leaf area, dry weight, final yield, and protein contents of these crops. Furthermore, ZnO NPs have been found to enhance germination and growth rate in seeds of Cicer arietinum and rice. However, optimal concentrations of ZnO NPs enhanced root and shoot growth in chickpeas and mung beans, with negative effects observed at higher concentrations. Peanuts treated with ZnO NPs exhibit elevated chlorophyll content, reaching levels as high as 100 mg/kg. Furthermore, studies have shown alterations in PSII fluorescence kinetics, with increases observed in parameters such as Phi-Eo, and PIABS, alongside decreases in energy flux parameter values including ABC/RC, TRO/RC, and ETO/RC, in plants treated with varying concentrations of ZnO NPs. Additionally, tomato plants subjected to treatment with ZnO NPs at a concentration of 8mg/L for approximately 30 minutes have exhibited notable enhancements in photosynthesis rate, chlorophyll content, and carbonic anhydrase activity. In summary, Zn NFs in the form of ZnO offer a more cost-effective alternative to synthetic zinc fertilizers commonly utilized in modern agriculture, particularly in foliar, soil mixing, and seed priming applications. It is well-established that zinc fertilizers play a pivotal role in promoting plant development and augmenting both crop yield and quality.

3.3 Mn NFs

Because manganese (Mn) is a co-factor for many different enzymes, it is necessary for a variety of physiological and metabolic processes as well as increasing a plant's tolerance to external stressors. It is necessary for photosynthesis, ATP, chlorophyll, fatty acids, protein synthesis, and secondary metabolites including lignin and flavonoids. It has been demonstrated that adding Mn to wheat, maize, sugarcane, soybeans, and common beans greatly boosts their growth and output. When mung beans are treated with nano-Mn fertilizers or Mn NPs, their root and shoot development increases by 52% and 38%, respectively, in comparison to a control treatment, (MnSO4; commercially available manganese salt at a recommended dose). When compared to a control group, Mn treatments considerably lengthened the roots of lettuce (*Lactuca sativa* L.). Likewise, Mn ions resulted in a 22% increase in aubergine (*Solanum melogena* L.) yield. Mn NPs, nevertheless, had no effect on the germination of lettuce seeds, the growth of watermelons (*Citrullus lanatus* L.), or the length of the roots of white mustard (*Sinapis alba* L.). Mn NPs attaches to the chlorophyll binding protein (CP43) of photosystem II in a physiological context, increasing photosynthesis and the electron transport chain's effectiveness. Compared to plants cultivated in soil supplemented with traditional bulk fertilizers, plants grown in soil loaded with Mn NPs have shown noticeably greater rates of nitrogen uptake and metabolism.

3.5 Copper NFs

While it regulates so many diverse physiological activities, including the metabolism of cell walls and several signalling pathways associated with transcription, copper (Cu) is an essential element for the growth of plants. Copper is also essential for the synthesis of molybdenum cofactor and influence protein trafficking, iron mobilization, and nitrogen metabolism. It is a reliable cofactor for several enzymes and is essential for respiration, phenol metabolism, protein formation and tracking, defense mechanism modulation, and enhanced male flower fertility. Plants take up copper from the soil as Cu^{2+} ions. The majority of soils have enough copper (Cu), but because Cu is an immobile metal, plants frequently lack it. In a crystal, the hydrated radius is 0.073 nm while the ionic radius of Cu2+ is 0.072 nm. Raising the pH of the soil lower the availability of Cu. According to recent research, using CuO NPs nanofertilizers has been shown to improve root development and soybean and chickpea (*Cicer ariteinum* L.) germination. Cu, Co, and Fr nanocrystalline powder (40-60nm) were found to have a germination rate of 65%, 80%, and 80% of soyabean seeds, respectively. It surpasses then 55% germination rate seen in a control sample with 0% NF. Similarly, increased leaf surface area, chlorophyll content, number of grains per spike, and weight of the plant all contributed to increased growth and yield in wheat at different Cu NP concentrations. Moreover, the application of 5 mg L^{-1} of Cu NPs. Research demonstrated an improvement in the flavonoid levels, sulfur assimilation, and proline and glutathione production in Arabidopsis thaliana. Cu NP treatment, on the other hand, decreased the hardness of the fruit on cucumber plants, which affected the quantity and quality of water lettuce (*Pisita stratiotes* L.). In conclusion, depending on the application rate, it seems that Cu NFs may be employed to considerably increase the yield and its biochemical properties.

3.6 Titanium NFs

TiO_2 ability of transforming light energy into electrical or chemical energy in an environment with plenty of sunlight resulted in its industrial use as a pigment in paints and sunscreens. TiO_2 is useful in NPs research including plant growth and responses to seed germination, pest management, pesticidal deterioration, superior water purification, and senor technology to recognize pesticidal deterioration, superior water purification, and senor technology to pesticidal residues. TiO_2 NPs have been encountered to improve the photosynthetic activity, germination rates, nutritional value, yields/biomass, and chlorophyll levels in spinach, lettuce, wheat, tomato, and watermelon. The wheat with 10 ppm of TiO_2 germinated the quickest as in contrast to the control condition. On the other hand, compared to the control, 2-10 ppm nanosized TiO2 produced noticeably longer shoots and seedlings. Higher doses of TiO2- based NPs, however, inhibited or had no effect on wheat. According to a different study, old spinach (*Spinacia oleracea*) seedlings germinated more quickly when nano –TiO_2 treatments were administered at the proper does. TiO_2 NPs help spinach because they increase Rubisco activase's enzymatic activity, which enhances light absorption capacity and reduces oxidative stress on chloroplasts from UV radiation during photosynthesis. TiO2 NPs in their anatase crystalline form have more catalytic activity when used on *Cucumis sativus*, which can help to promote the formation of carotene and chlorophyll. Titanium dioxide (TiO_2 NPs) enhanced the light-absorbing capacity, oxygen evolution, and electron transfer rate of spinach leaves, in addition, TiO_2 NPs stimulate the Hill reaction and chloroplast reaction. Enhanced physiological activities were also influenced by elevated RuBisCO levels and nitrogen metabolism. However, plants react differently depending on the concentration and environment of TiO_2 NPs. Additionally, since they absorb more light, plants exposed to TiO_2 NPs grow more quickly.

3.7 Silica Nanofertilizers

To assist seedling withstand abiotic challenges, silica nanoparticles promote healthy growth and the activation of antioxidant enzymes. Drought and salt are two the absorption of potassium and reduce the intake of sodium. Under salt stress, Si NPs increase the absorption of potassium and reduce the intake of sodium. The growth and development that might be used to lessen the negative consequences of climate change are evidence of the beneficial benefits of silica silicon nanoparticles. For instance, silica may be able to reduce the amount of water that evaporates from the soil's surface.

3.8 Carbon NFs

Carbon atoms, the main building block of every living creature, are responsible for the basic structure of biomolecules, including proteins, carbohydrates, and fats. Carbon in necessary for the growth of plants as well as the synthesis of oxygen and nourishment for other living things, Manure and decomposing plant biomass are used by farmers as soil-enriching carbon sources. Because carbon comes from either the organic material in the soil or the carbon dioxide in the air, plant growth and carbon are related. A lot of interest has been focused on manufactured carbon nanostructures because of their constant physicochemical features. Carbon nanotubes, or CNTs, are absorbed by plants and have the ability to influence their growth and development. Many studies on plants have focused on carbon nanomaterials and structure, and the findings have indicated that these materials extend roots, increase plant biomass, and speed up seed germination. According to researchers, carbon nanoparticles (CNPs) have been applied to enhance photosynthesis, water absorption, cell division, and agricultural yield. For instance, the number of fruits and blossoms that grew on tomato plants rose by 100% when carbon nanotubes (CNTs) were applied. The quality of chlorophyll, phenols, vitamin C, beta-carotene, flavonoid, and H_2O_2 in seedlings of tomato rose by 111%, 78%, 11 folds, 85%, 45%, and 215%, respectively, when improves the seed's ability to absorb and retain water. After being immersed from two days in a medium holding carbon nanotubes (CNTs), tomato seeds treated with CNTs contained 19% more moisture than untreated seeds. These findings demonstrate the role of CNTs in the uptake and retention of water. The precise method by which CNTs generate microspores and water penetration channels in seed coat is yet unknown. The idea is that CNTs can regulate the amount of AQPs in the coverings A class of membrane proteins known as APQs performs role similar to those of a water channel.

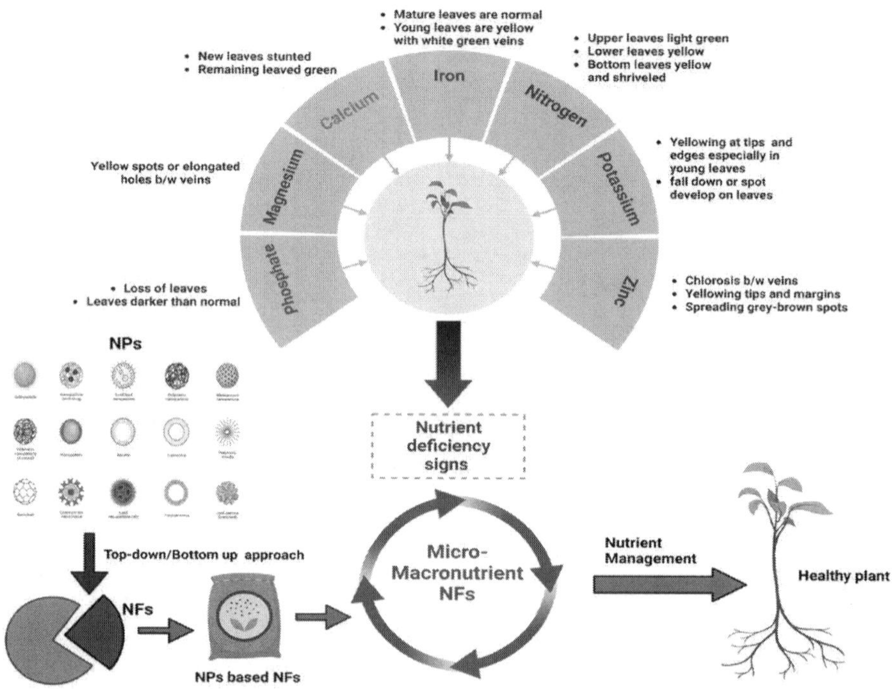

Figure 1. The figure highlights the role of NFs in nutrient management for plants. It showcases how NPs, with their unique properties, facilitate the efficient delivery and targeted release of nutrients to plants. NFs act as specialized carriers that enhance nutrient uptake reduce environment losses and optimize nutrient utilization in plants.

4. NFs base Crop Gram Nutrient Management

A developing technology that will help the agriculture sector is nanotechnology. This technique can improve crops while removing several problems related to both genetic and traditional agronomy biofortification. Nutrient translocation in nanomaterials and nano-fertilizers occurs from the roots towards the leaves, and by enhancing seedling growth, nitrogen metabolism, photosynthesis, and protein and carbohydrates synthesis, they can potentially reduce various stress conditions and enhance the productivity of crops. It could be more efficient to apply nano-fertilizers directly to the target to increase nutritional content. Because wheat is a major staple crop biofortification efforts have focused on it on several occasions. Numerous studies have been carried out to far on small and large sizes in fields or greenhouse, indicating the beneficial and negative impacts of nanomaterials on parameters such as the nutritional content of wheat. For large-scale agricultural efforts aimed effort aimed at the wheat biofortification process, nanotechnology is ineffective. To ensure the success of large-scale wheat biofortification initiatives, a deeper comprehension of the nutritional makeup of various wheat genotypes is needed. Nanomaterials have been employed in wheat biofortification albeit little study has been done to support his claim. In order to do that, this will look at how well nanomaterials are currently being used to enhance wheat seeds and crops with a variety of nutrients. The advantages and disadvantages of many nanotechnology-based biofortification techniques for wheat were examined. Wheat harvest boosted by nanobiotechnology hold potential in the battle against world hungers, notwithstanding failures. It is generally acknowledged that using agronomic biofortification to boost the mineral content of crops is a useful and cost-effective strategy. The fact is that agronomy-based biofortification does not survive for the intended amount of time because of poor drainage and poor soil quality results in maximum applied nutrients being lost. Applying nano-ZnO fertilizer topically, as opposed to conventional Zn fertilizer, can result in a notable increase in photosynthetic pigments and a twenty percent rise in gram protein content. By using nano ZnO foliar spray, wheat protein content is increased and Zn leaching into the soil after harvest is significantly reduced.

Nutrients are applied topically to wheat plants. In comparison to applying zinc to the soil for grain biofortification, it is far more economical. Applying ZnO NPs has caused Zn to accumulate in grain creases and the aleurone layer, raising the level of Zn in endosperm. Zn biofortification might be accomplished similarly to seed priming by foliar-spraying wheat plants with ZnO-NPs. ZnO NF accumulate when applied to leaves, where it is swiftly absorbed by the plant and used metabolically. It has been proposed that ZnO NPs build up on wheat leaves because they are directly absorbed by the leaf cuticles, penetrate through the lea epidermis by means of the apoplast, and adhere to the mesophyll cells regardless of the particle coating.

When used as fertilizer, silica reduces the acidity of the soil and improves nutrient absorption. The second most common element on Earth's surface dioxide. According to study, NPs of SiO2 applied to the soil as well as the foliar surface of wheat plants under drought and stress conditions has boosted wheat output by approximately 17.81% and 23.35%, respectively. Three distinct nano-chelated fertilizers-silicon, boron (B), and zinc (Zn)-have been applied topically. Their effects on grain production and protein content as well as development of wheat plants have been investigated.

5. Function of Nanofertilizers for Mitigation of Multiple Environmental Stresses Management Under Climate change

Abiotic and biotic stressors on cultivation crop can be lessened in a variety of ways by NFs because of their enormous surface area and nanoscale diameter. A robust plant defense mechanism, improves photosynthetic rate, improves nutrient absorption efficiency, and controlled phytohormone production are only a few of the biochemical, morphological, and physiological indicators that may be enhanced by applying NFs in agriculture (Fig 7). Hence, the influence of natural fertilizers (NFs) on crop productivity during stressful situations could be attributed to their capacity to: (a) lessen oxidative stress, which raises plants' stress tolerance; (b) boost several biochemical reactions within stressed plants; (c) control salt toxicity; (d) reduce the building up of hydrogen peroxide and malondialdehyde, and (d) boost the effective absorption of nutrients of the growth and metabolism of cultivated plants. In contrast, nano-fertilizers depend on the increased surface area of the nano-active constituents to produce a satisfactory reaction. "Smart" NFs are those that have the ability to release nutrients in a regulation manner, transport them as NPs or enclosed them in nanomaterials. Research has demonstrated that using NFs or NPs to raise the level of plant antioxidant molecules can greatly reduce the harmful effects caused by a range of environmental conditions.

5.1 NFs responses under Abiotic stresses

1. Salinity Stress

Due to global warming-induced water shortages, agricultural regions worldwide must be irrigated with saltwater raises the salinity of the soil. Salinity is one of the main issues with contemporary faming; excessive salt build-up in the soil can stunt plant growth and ultimately cause plant death. Most plant perish when the NaCl content rises beyond 200 mM. Salinity has an impact on a wide range of horticultural crops, including fruits, vegetables, and spices. Salt stress result in an imbalance of ionic strength, which impacts a number of physiological, metabolic, and biochemical processes. It can also induce nutritional, osmotic, and water oxidative stress, all of which reduce cell proliferation.

The reduced soil osmotic potential of salt-affected soils leads to insufficient plants feeding and be a good way to address this issue. The usage of SiO2 NPs was observed to improve leaf dry weight, chlorophyll content, proline content, and level of antioxidants in the presence of salt stress. Additionally, it was discovered that, in comparison to untreated plants, plants treated with SiO_2 NPs were able to grow significantly better under salt stress conditions. Likewise, maize plants treated with SiO_2 NPs and cultivated in salt stress produced higher biological yields than plants treated with a control treatment. Researcher used nano-calcium to *Solanum Lycopersicon* growing under salt stress, and saw similar outcomes. Additionally, they discovered that compared to plants fertilized with synthetic monophosphate, those treated with nano-calcium produced 76% more fruits per plant. In order to improve growth and production while lessening the negative impacts of salt, researchers sprayed foliar applications of nano-NPK (50-100 ppm) onto pea plants. According to reports, in conditions of salt stress, nano-chitosan dramatically increases the rate of germination and development of bean plants (*Phaseolus vulgaris* L.).

2. Drought Stress

Due to its most detrimental effect on agricultural productivity, drought is one of the most destructive environmental stresses. By 2100, global average temperatures are expected to rise by 1.8 to 4.0 C, and many regions would be affected by extensive drought, according to the IPCC ("AR6 Climate Change 2021: The Physical Science Basis – IPCC,"). When there is insufficient water for crops to grow and develop normally, this is referred to as a "drought". As precipitation declines and evapotranspiration needs increase simultaneously, drought conditions get worse). Drought stress prevents plant development because water is necessary to sustain cell turgor and because pressure on the cell walls might force the cells to expand.

Drought slows down cellular division and development, disorganizes stomatal oscillations, shrink levels, changes the interactions between nutrients and water, lower agricultural yields, inefficiently utilizes water, and lengthens stems and roots. By improving the soil's ability to store water, the use of stress-ameliorative materials, such as NFs, can significantly reduce the detrimental effects of drought stress on the plants. Moreover, oxidative stress can raise the creation of ROS, which harm cell membranes, permits solution to escape from cells, and eventually causes the cells to die. Additionally, studies show that NPs can raise antioxidant levels, which reduces the production of H_2O_2 and malondialdehyde. It has been discovered that hawthorn (*Crataegus sp.*) seedlings, drought tolerance is increased by nano-enabled production like Si NPs, without creating any changes to the plant's physiological or chemical processes during situation. Comparing ZnO- NPs (0.5 and 1.0 g/L) to control treatment (0 g ZnO-NPs/L), it was discovered that ZnO-NPs increased soybean germination rates under water stress. Foliar spray application of Fe NP has been found to boost oil percentages and safflower (*Carthamus tinctorius* L.) production (Table 1). After receiving a foliar spray treatment with 0.02% TiO_2 NPs, it was discovered that a water- stressed wheat crop's number of tillers, grain weight, ultimate grain yield, and harvest index all rose.

When cultivated under drought stress, lentil (*Less culinaris Medik*) plants treated with silver nanoparticles showed increase in both seed germination and root dry weight. Lastly, Fe NPs and salicylic acid can aid strawberry (*Fragaria ananassa*) plants become more drought- tolerant. In the face of drought stress, applying 41 kg ha1 of nano-chelated nitrogen fertilizer boosts wheat output. Similar finding was made by other researchers discovered that using nano- NPK and nano-zeolite loaded with N might boost the development of sage (*Salvia officinalis*) and reduce water stress.

3. Heat Stress

Elevated temperature led to oxidative stress, which diminishes plant development, chlorophyll levels, and net photosynthesis rate. One of the main causes of oxidative stress, which damages plants membrane lipids by allowing ions and solutes to seep out as results of heat stress, is an overabundance of reactive oxygen species. However, low concentrations of Se NPs were also discovered to enhance plant's hydrologic connection, chlorophyll levels, and antioxidant capabilities, therefore greatly mitigating the harmful effects of heat stress on plants. When plants are exposed to high temperature or heat stress, they create proteins known as heat shock proteins.

Investigators have discovered that employing multiwalled carbon nanotubes can increase the expression of the heat shock protein gene in plants. In comparison with the control treatment (zero Ag-NPs), treating heat-stressed wheat plants in the trifoliate phase utilizing a foliar application of Ag-NPs (50-75 mg L^{-1}) enhanced growth. There was an upsurge of about 5.0% and 5.4% in the lengths of the roots, 22.2% and 26.1% in the length of the shoots, 6.6% and 7.2% in the number of roots, 1.3% and 2.0% in the fresh weights, and 3.0% in the dry weights. The foliar application of 10 mg L^{-1} of Se NPs at a higher temperature and stress during the sorghum (*Sorghum bicolor L. Moench*) booting stage resulted in increased pollen germination, enhanced antioxidant defenses, and enhanced seed output.

4. Heavy metals Stress

In today's world, one of the negative factors affecting agricultural yields in heavy-metal (HMs) stress. Due to human action like industrialization and urbanization, HM pollution has risen globally. Chemical fertilizers and pesticides have been used more often, which increases the risk of HM stress in crop plants. Heavy metals such as Ag, Cr, Co, Cd, Hg, Ni, and Pb are detrimental to plant growth. Since plants are at the base of the food chain, it is possible that higher animals would consume these HMs and suffer long-term health issues including liver and kidney damage as a result. By altering their physiology and resulting in various problems, HMs also have an indirect impact on plants). These may adversely affect the productivity of crops and the potency of medicinal plants.

Numerous reports about the potential use of NPs in the remediation of soil polluted with heavy metals have been published. In a recent study it is shown that applying FeO NPs to lower Cd toxicity also boosted the biomass, chlorophyll content, and antioxidant enzymes in wheat plants. These nanoparticles can lessen the buildup of harmful ions within plants cells, hence preventing ionic stress. NPs can lessen plant toxicity caused by heavy metals (HM) due to their large surface area and tiny size. Similar claims have been made about how Si NPs protect wheat, rice, and peas from HM-induced phytotoxicity. Researchers found that adding Fe-NPs to wheat plant development also decreased the cellular oxidative stress brought on by Pb and Cd toxicity. Similarly, it was shown that the magnetite nanoparticles increased the development of wheat seedlings by lowering their exposure to harmful amounts of Cd and Cr. Recent studies have demonstrated the effectiveness of FeO and ZnO-NPs in promoting plant development and reducing HM and drought stress. However, in order to lessen the harmful effects of heavy metals (HMs) on the development and growth of plants, innovative nano remediation techniques are needed.

5.2 NPs and their impact upon plant response towards abiotic stress

Among the effective NPs that have been the focus of the most investigation include carbon nanotubes (CNTS), metal-based NPs (Ag NPs and Au NPs), crystalline nanoscale powders (Fe, Co, and cu), and metal-oxide NPs (iron oxide, TiO_2 NPs, ZnO NPs, SiO_2 NPs, CuO NPs, and $CaCO_3$ NPs). NPs are more reactive and show higher metabolic activity due to their high surface energy and high surface-to-volume ratio, which has a range of consequences on plants. NPs have the ability to interact with plants and swiftly trigger molecular processes. Furthermore, the NMs have a dual function in oxidative stress indication and ROS defense, both of which can trigger the plant's antioxidant defense system. NPs enable plants adapt to adverse environments more successfully and to generate high yield by reducing the harmful effects to abiotic stress and affecting different anatomical, morphological, physiological, biochemical and molecular properties connected with the plants.

1. Impact of NPs on morphological and anatomical features of plants under abiotic stresses

Plant morphology is significantly impacted by abiotic stressors such as salt, drought, excessive heat or cold, and heavy metals. These stressors can influence features such as fresh and dry weight, leaf area, shoot and root length, total plant development, and crop output. Researchers claim that priming seeds with low NP concentrations boosts root length, and seed germination rate. When temperature stress is present, all of that can be recovered. Because of plant-derived silver nanoparticles, aerial structures can enhance the leaf area and dry weight in hot weather. TiNPs have been shown to cause an increase in plant height in *Dracocephalum moldavica* under salt stress. Under drought stress, Fe-NPs have been shown to enhance plant growth and biomass production in *Brassica napus*.

There are several ways in which plants might change their morphology in the aftermath of abiotic stress. The abiotic stress encountered has an impact on the morphological alterations. Drought caused stomata to produce large sub-stomatal chambers and thicken the upper-epidermal cuticle, cuticular margins, and lignification of xylem tracheids. In response to heat stress, roots and shoots have bigger xylem vessels and greater stomatal and trachomatous densities. Transcriptional rate slows down, stomata remain open, and cell size shrinks. In *Vitis vinifera L. CV. Jingx*, heat stress results in cellular damage, particularly to the mesophyll cell membrane permeability. Increased temperature cause *Zygohyllamqatarenes Hadidi* to display bimodal stomatal behaviour, which reduces transpiration-related water loss. Plants that get NPs help their bodies adapt physiologically to harsh condition. TiO_2 NPs-induced heat stress on maize plants results in effective control of stomatal opening. The epicuticular wax layer (EWL) of strawberry plants treated with SiO_2 NPs was reinforced, which helps to lessen the negative impacts of salt. Studies showed that $A1_2O_3$ NPs prevented soybean cell death. ZnO-NPs have been demonstrated to lessen the detrimental effects of salt stress on the vascular tissues and epidermis of *Sorghum bicolor* plants. Recent research examined the impact of CeO_2 NPs on salt-stressed *Brassica napus* L. It was shown that the former significantly reduced root apoplastic barriers.

2. Impact of NPs on Physiological Response of plants under abiotic stresses

Plants have a range of physiological reactions to stress, which include modification to their photosynthetic apparatus, water intake, transpiration, material uptake, lipid peroxidation, and seed germination. Abiotic stress has a detrimental influence on seed germination, pollen fertility, and chlorophyll content, all of which are related to photosynthesis. Plant cells also experience plasmolysis, protoplasmic streaming, and electrolyte loss. Heat stress impairs plants' capacity to absorb nutrients. Because Cd stress prevents plants from absorbing water, Cd build-up in the soil affect the absorption of both macro- and micronutrients.

Drought stress has been found to have a deleterious effect on thylakoid membranes, which lower photosynthesis. Elevated salinity in the soil inhibits seed germination. Many vital physiological functions of plants, such as photosynthesis and germination, are inhibited by heavy metal. Si NPs can help drought-stressed plants because they increase stomatal conductance and hasten photosynthesis. To enhance stomatal conductance, leaf transpiration, and net photosynthesis rate in tomato plants under heat stress, a consistent concentration of nano-TiO_2 can be applied. While the energy loss from the photosystem II (PS II) was reduced, the uncontrolled energy loss was enhanced by nano-TiO_2. The quantifiable physiological reactions of nanoparticles exposure are listed in Table1.

3. Impact of NFs on Biochemical Response of plants under abiotic stresses

(A) ROS Production under Abiotic stresses

Both healthy and stressed plant cells generate reactive oxygen species (ROS), such as alkoxyl radicals (RO•), superoxide radicals ($O_2^{\bullet -}$), hydrogen peroxide (H_2O_2), hydroxyl radicals (•OH), singlet oxygen ($^1O_2$), and peroxy radicals (ROO•). ROS that are extremely reactive are created when O_2 is steadily depleted by high-energy radiation or electron-transfer processes. Energy and an inevitable electron leakage caused by the electron transport functions of mitochondria, apoplasts, endoplasmic reticulum (ER), chloroplasts, peroxisomes, and plasma membranes, as well as the cell wall, are needed for plants to active ROS. This energy leakage can also occur because of distinct metabolic pathways in various cell regions.

When light is present, chloroplasts and peroxisomes are the main source of ROS production; when light is not present, mitochondria are the main source of ROS production. The highly structured thylakoid system that houses the photosynthetic equipment for catching light is founded inside the chloroplast. ROS are primarily produced by the light-harvesting system in thylakoid, which are made up of photosystems PS-I and PS-II. In the vicinity of PSII reaction centers, oxygen molecule, O_2 may create $^1O_2$ when chlorophyll is overexcited in response to stress and Mehler reaction's electron transfer generates $O_2^{\bullet}$ from 1O2. Additionally, most of the H_2O_2 produced in peroxisomes is produced by flavin oxidases. Univalent reduction of O_2 in close proximity to the electron transport chain inside mitochondria is one potential source of H_2O_2 in plant cells. Apart from these organelles, ROS production occurs at various places inside the cell. The cytoplasmic side of the plasma membrane is where localized NADPH-dependent environmental changes. Furthermore, ER contributes to the production of via Cyt-P450. When ABA and stress signal are combined, the apoplast is ready to produce H_2O_2 in reaction to stressful situations. Diamine/ polyamine oxidases and oxalate oxidases, which are found in the cell wall, can create H_2O_2, which can be converted by class III peroxidases.

(B) Role ROS Homeostasis in Plants

1. NFs/NPs Based Antioxidants Response for ROS Homeostasis in Plants:

An enzymatic antioxidant such as superoxide dismutase (SOD), catalase (CAT), adenosine triphosphate (ATP), peroxidase (APX), peroxygenase (GPX), peroiredine (Prx), or thioredoxin (Trx) can assist lower the ROS content in a system (Fig7). The metalloenzyme SOD has three isoforms (EC: 1.15.1.1): Cu, Fe-SOD, which exists within the mitochondria, chloroplast, and peroxisomes; Mn-SOD, which is located within the peroxisomes, mitochondria, and vascular tissues; and Zn-SOD, which is located inside the cytosol, chloroplast, peroxisomes, nucleus, mitochondria, and apoplast, the primary point of defense against ROS is SODs. They catalyze the disproportionation of O2• free radicals as they transform one radical into H_2O_2 and second into O_2, preventing the production of the more harmful free radical •OH. Enzymes that catalyze the breakdown of the H_2O_2 antioxidant into water and oxygen gas are called CAT (EC:1.11.1.6), APX (EC:1.11.1.11), and GR (EC:1.6.4.2). The metalloenzymes APX and CAT are found in mitochondria and peroxisomes, correspondingly. Moreover, APX may be found inside the peroxisomes, chloroplast, microbodies, and cytosol. It is a crucial enzyme in the ascorbate-glutathione (AsA-GSH) pathway. AsA has to act as a reducing substrate for the enzymes to stay stable and function as intended. The AsA-GSH, or Foyer-Halliwell-Asada, pathway is a set of non-enzymatic (AsA, GSH) and enzymatic (APX, MDHAR, DHAR, GPX) antioxidant components that counteract excess H_2O_2. It is found in chloroplasts, plastids, mitochondria, and peroxisomes. Flavoprotein oxidoreductase GR is expressed at 70-80% level in the chloroplast, although it is also present in non-photosynthetic tissues and organelles. Its efficacy depends on pH as well as local glutathione disulfide (GSSG) and NADPH concentrations. The GSH/GSSG ratio is necessary for H_2O_2 detoxification and is regulated by GR, which uses NADPH as an electron donor to catalyze the conversion of oxidized GSH (GSSG) into the reduced form of GSH (GSH). Thiol peroxide enzymes, or Prxs, neutralize peroxidase substrates such hydrogen peroxide and alkyl hydroperoxide to eliminate oxidative stress.

Nanoparticles (NPs) derived from plant sources exert influence on antioxidant enzymes such as SOD, CAT, and POD by targeting cellular organelles and releasing various constituents. Researchers observed heightened SOD activity in plants treated with TiO_2 NPs, underscoring this regulatory effect. In agriculture, certain elemental and oxide NPs have been utilized to enhance resistance to drought stress. Si NPs have been particularly effective in mitigating abiotic stresses like drought, as evidenced by studies. Moreover, treatments involving silica and ZnO NPs have been noted for their capacity to improve growth, physiological, and biochemical traits across various crops. A study demonstrated reduced drought stress in wheat plant exposed to Si NPs. Additionally, ZnO NPs were found to detrimental effect of both salinity and drought stress on plants. Researchers discovered that excessive nanoparticles (NP) application induced oxidative stress in plants, resulting in phenomena such as cell cycle arrest, regulating of cell death proteins and induction of antioxidant (MDA) levels and free radicals (H_2O_2 and O_2) when subjected to drought conditions. Moreover, it was observed a decrease in MDA levels following the induction of the plants antioxidant system after TiO_2 application. Additionally, the antioxidant enzymatic activities of peroxidase (POD) and catalase (CAT) were found to be increased.

When plants face environmental challenges like salinity stress, they initiate a defense mechanism by generating ROS. To combat those detrimental effects of ROS accumulation, plants enhance the production of antioxidant enzymes, as observed in numerous studies. Due to their antioxidant properties, NPs contribute to plants ability to withstand oxidative stress. Some NPs such as Co, Fe, and Ce, bear structural similarities to catalase (CAT), while other like Ce, Mn, Cu, and Fe share similarities with peroxidase (POD) enzymes. Studies revealed that cerium-treated Arabidopsis plant's exhibited improved ROS scavenging through ROS-NSCC (reactive oxygen species-non-selective cation channel) activity. Under salinity stress conditions, pearl millet (*Pennisetum glaucuml.*) was subjected to priming with silver nanoparticles (Ag NPs) at concentrations of 10 mM, 20 Mm and 30 Mm, alongside varying levels of NaCl stress (o, 120, and 150mM NaCl). The plants development characteristics have been greatly enhanced by these NPs, which have also been linked to an increase in antioxidant enzymes such as SOD, CAT, and glutathione peroxidase (GPX) as well as a drop in the sodium-potassium ratio. It is discovered that antioxidant enzyme activity was increased by modest doses of Ag NPs. Tests using different concentrations of TiO_2 NPs under salinity stress (0, 50, 100, and 200mM NaCl were conducted on *Dracocephalum moldavica*. It was discovered that TiO_2 NPs at 100mg/L increased antioxidant content and decreased H_2O_2 concentration.

Enzymes that are sensitive to temperature changes and required for a variety of metabolic processes. Enzymes may decouple from metabolic pathways as a result of heat stress, which would raise ROS levels (Fig 2).

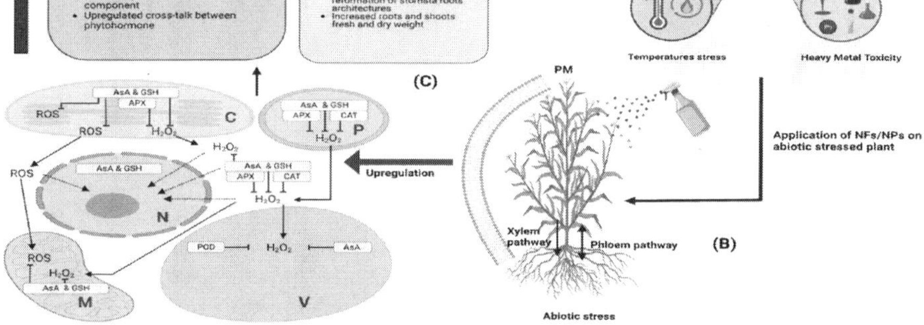

Figure 2. The effect of NFs/NPs on plant salinity stress (A) various types of abiotic stresses (B) application of NFs/NPs on plant facing abiotic stresses effect (C) NFs/NPs upregulated the antioxidant responsive pathway and (D) NFs/NPs mitigated the abiotic stresses at physiobiochemical, and molecular level on plants. Possible signalling pathways: dotted arrows. AsA, ascorbate; APX, ascorbate peroxidase; C, chloroplast; CAT, catalase; GSH, glutathione; M, mitochondria; N, nucleus; P, peroxisome; POD, guaiacol-type peroxidase; V, vacuole.

Singlet oxygen ($^1O_2$), superoxide radicals (O_2), hydrogen peroxide (H_2O_2), and hydroxyl radicals make up the majority of these ROS. Plants that employ NPs are better able to withstand heat stress. By boosting the synthesis of SOD and GPX and lowering H_2O_2 levels in plants, ZnO and TiO_2 improve defense. TiO_2 NPs enhanced photosynthetic activity while lowering H_2O_2. When exposed to high temperatures, plants treated with Zn NPs and Fe NPs exhibit an enhanced antioxidant enzyme activity. In wheat plants stressed by moderately high temperature, TiO_2 boosted PSII activity.

NPs can be used in several scenarios to reduce the levels of heavy metals in plants by upregulating, antioxidant enzymes such as superoxide dismutase (SOD), catalase (CAT), ascorbate peroxide (AspPX), AND GUAIACOL peroxides (GPx). Heavy metals are bound by NPs, which stopes them from circulating and entering the body. Because of their smaller size and increased surface-to-volume ratio, nanoparticles (NPs) may interact with other molecules more easily. NPs can even cross the membrane of a cell. Different quantities of TiO_2 NPs were used to reduce to toxicity of Cd in maize

Through the stimulation of metabolic pathways and the elevation of SOD and glutathione level, high amount of TiO2 reduce toxicity. IT was discovered that ZnO NPs reduce the amount of Cd in wheat. The content of Cd decreased as SOD and POD activity rose. Scientists looked at how little Cd was absorbed by soyabean when nano-TiO2 was present. Studies had administered Cu NPs to brassica in order to decrease Cd toxicity, and this led to increased levels of plants biomass, peroxide (POD), catalase (CAT), and superoxide dismutase (SOD). Researchers were able to drastically minimize the amount of arsenic (As)and cadmium (Cd) accumulation within the plants by applying ZnO NPs, which in turn lead to lower levels of As and Cd accumulation within the rice grains. Studies on the application of FeO NPs to rice plants revealed that the plants' antioxidant enzymes and glyoxalase system were enhanced. Moreover, the chelating agents accumulate greater when as is immobilized in vacuoles and cell walls. Following Vigna radiata's exposure to Fe_2O_3 NPs, there was an increase in total antioxidant capacity along with increase in SOD and CAT and decrease in guaiacol peroxidase. It was discovered that when CuO NPs were added to rice, plants biomass and antioxidant activity and as levels fell sharply.

2. NFs/NPs Based Non-Antioxidants Response for ROS Homeostasis in Plants

Among the most harmful ROS, such as $^1O_2$ and •OH, cannot be eliminated by enzymatic antioxidant systems. As a result, plants' only means of removing ROS are non- enzymatic methods. ROS often results lipid peroxidation and a sharp rise in MDA concentrations. In order to counteract the unchecked creation of ROS that happens during stress, nonenzymatic substances such as glutathione (GSH), suitable solutes, phenolics, alpha-tocopherol (vitamin E), carotenoids, flavonoids, and proline collaborate. GPX uses the ubiquitous thiol tripeptide glutathione (GSH) to accelerate the breakdown of H_2O_2 and contributes in the formation of a conjugate through a process catalysed by glutathione S-transferase (GST) and glutathione peroxidase (GPX), which is responsible for the breakdown of lipid peroxides and H_2O_2 and its scavenging. In the AsA-GSH pathway, ascorbic acid, often known as vitamin C, serves as a co-factor of POXs and an electron donor for APX. AsA regenerates both tocopherols and xanthophylls, which both contribute to the quenching of this excitation energy. Enzymatic and non- enzymatic antioxidants acquire vital avenues for tightly controlling ROS in plants cells. These pathways are responsible for efficiently reducing oxidative damage of NPs increases the effectiveness of non-antioxidants, which may assist to lessen oxidative stress in situations when salt is a problem.

6. Conclusion

In conclusion, nanofertilizers represent a groundbreaking advancement in soil and nutrient management, offering substantial benefits for agricultural productivity and sustainability. Their nanoscale properties allow for precise and efficient nutrient delivery, significantly enhancing nutrient uptake and utilization while reducing losses due to leaching and volatilization. By improving soil health through increased microbial activity and nutrient availability, nanofertilizers help maintain soil fertility and structure, which are vital for long-term agricultural sustainability. Moreover, their ability to provide tailored solutions for specific nutrient deficiencies and soil conditions makes them adaptable to various farming environments. The environmental advantages of nanofertilizers, such as minimizing the overuse of chemical fertilizers and reducing pollution, further underscore their potential to revolutionize traditional fertilization practices. As the agricultural sector continues to seek sustainable solutions, nanofertilizers offer a promising pathway to achieving higher crop yields with lower environmental impact.

CHAPTER 5

ROLE OF NANOTECHNOLOGY IN SALINITY STRESS MANAGEMENT

Short Summary

Globally, salinized land is gradually becoming more widespread. The capacity to feed a rising population is threatened by reduced food yields and quality brought on by salt stress. We talked about the processes that underlie nano-enabled antioxidant enzyme- mediated plant tolerance, including preserving the homeostasis of ROS, improving plants ability to hold on to K^+ and excrete Na^+, boosting nitric oxide production that include signaling pathways, and reducing lipoxygenase activities to minimize oxidative damage to membranes. Frequentlv employed strategies were emphasized, such as maintaining ionic state equilibrium and shielding cells from oxidative damage. Nanotechnology enabled salt tolerance in plants is also explored as well as the possible contribution of physiobiochemical and molecular pathways. The objective of this chapter to provide an overview of salinity stress with mode of action of NPs which help scientists, young researchers, students to studying anything from plant science to nanoscience to better understand the possibilities for creative solutions to salinity problems for sustainable agriculture.

1. Introduction

Plant maturity and development are impeded by abiotic stresses, which are closely related to changes in the climate and have a negative impact on agricultural yield and quality. Abiotic stressors therefore represent a risk to the sustainability of agricultural output on a worldwide scale. The Food and Agriculture Organization (FAO) reports that food constraints affect millions of people and that by the end of 2030, food demand would have increased by more than 40% when compared to 2011's production. Salinity stress is one of the most important abiotic variables affecting crop yield, plant development and efficiency is salt stress. Soil salinization is a significant issue affecting large portions of the world's land area. According to research, its impact is widespread, affecting approximately 53% of Asia and Australia 13% of Latin America 10% of Europe 9% of Africa and 2% of North America.

The presence of salt in the soil affects plants in various ways. When plants are grown in saline conditions, their internal levels of chloride (Cl^-) and sodium (Na^+) may rise. This increase in Na^+ and Cl^- ions can be detrimental to plant life, with Cl^- being particularly harmful. Salinization alters the physiological, biochemical, molecular, morphological, and metabolic processes of plants. It also lowers the osmotic potential (to 0.2 MPa) leading to drought stress and ion toxicity. An excess of Na^+ in the soil can cause several issues for plants, including ionic and osmotic imbalances, malfunctioning of cell membrane, increased formation of ROS and negative effects on plant cell proliferation and growth.

Plant cells produce ROS during salinity stress, which result in ionic, osmotic, and nutritional stress. The most prevalent reactive oxygen species (ROS) generated by plant cells are superoxide anion radical ($O2^{\bullet -}$), singlet oxygen ($^1O_2$), hydrogen peroxide (H_2O_2), and hydroxyl radicals ($OH^\bullet$). These ROS are produced by the mitochondria, peroxisomes, apoplast, and chloroplast, respectively. ROS are produced by chloroplasts 10 times more quickly than by mitochondria in the presence of light. ROS can be a double-edged sword in plant cells; while they can cause damage at larger concentrations, they are unable to act as stress signaling molecules. On the other hand, ROS may function as a supplementary messenger for basic cellular activities. ROS serve as secondary messengers, governing essential cellular processes such as the cell cycle and programmed cell death in plants. Furthermore, they engage in intricate interactions with various signaling molecules, including jasmonic acid, ethylene, nitric oxide, and calcium. It has been observed that rice, subjected to salt stress, demonstrates increased production of both ROS and ethylene.

When high-energy exposure processes occur, electron transfer result in the generation of very reactive ROS, which gradually reduce the molecular oxygen(O_2) (Figure 1). Plants constantly produce reactive oxygen species (ROS) due to metabolic activities occurring in different organelles or because electrons from the mitochondria, plasma membrane, and electron transport system in chloroplasts always flow onto O_2. Exposure to salt disturbs a plant's cellular equilibrium, which leads to a spike in ROS production (Figure 1). All living things are poisoned by ROS at high concentrations. We refer to a state as "oxidative stress" when a cell's defense against ROS production is overwhelmed. Oxidative stress puts cells at risk due to protein oxidation, nucleic acid degradation, lipid peroxidation, initiation of programmed cell death, and enzyme inhibition caused by the heightened production of ROS influenced by environmental conditions. These factors collectively contribute to cell death.

In reaction to ionic imbalances and hyperosmotic stressors, ROS molecules build up at large concentrations in plant tissues, which can lead to subcellular damage. A consequence of lipid oxidation brought on by an excess of ROS, malondialdehyde (MDA) is detrimental to cellular metabolism and physiological functions (Figure 1). The biological activities of membranes, such as cellular homeostasis, molecular transport, cell signaling, and process compartmentalization, are disrupted by a loss of membrane integrity. By altering photosynthetic pigment, the nutritional imbalance brought on by salinity, besides ionic disruption and hyperosmotic stress, decreases photosynthetic activity.

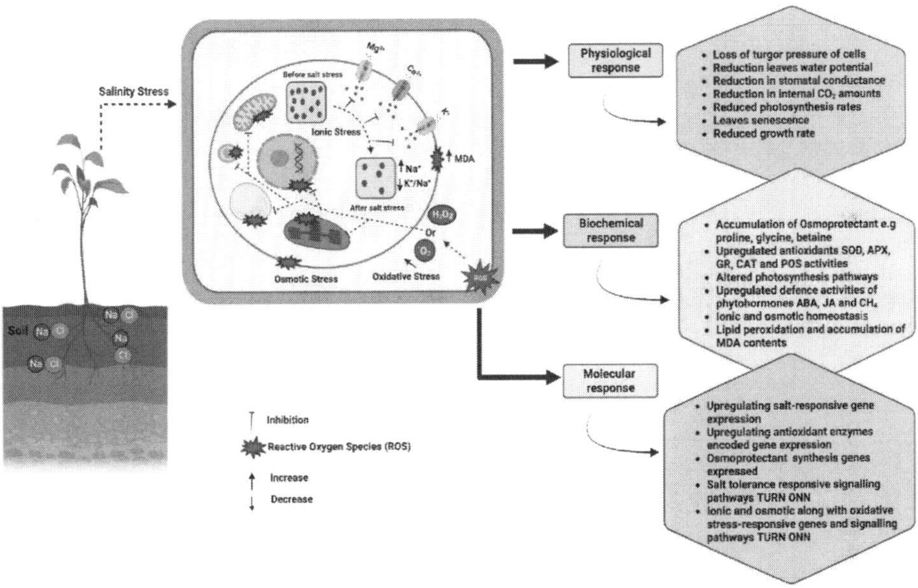

Figure 1. Diagrammatic representation of physio- biochemical and molecular response of salinity stress on the plants. Salinity stress induced ionic and osmotic stresses that developed oxidative stress and ROS production. ROS accumulation can alter the physio-biochemical and molecular cellular process of the plants.

Additionally, changes in gene expression and the overproduction of ROS in plant peroxisomes, mitochondria, and chloroplasts are closely related. One well-understood method of resistance to salt stress is the production of antioxidant enzymes genes. During salinity stress, an efficient antioxidative pathway that incorporates both enzymatic and nonenzymatic antioxidant machinery detoxifies an excess of ROS. The ascorbate-glutathione cycle (AsA-GSH) is mediated by the enzymes ascorbate peroxidase (APX0), glutathione (GR), dehydroasorbate reductase (DHAR), and monodehyroascorbate reductase (MDHAR). Ascorbate (AsA), glutathione (GSH), carotenoids, phenolics, and tocopherols are examples of nonenzymatic antioxidants found in cells. However, the quantity of ROS within the cell determines whether they are destructive or signaling. Moreover, a variety of plant morphological, physiological, biochemical, and molecular mechanism, such as seed germination and maturation, root, shoot, and flower maturation, as well as a decrease in stomatal conductance, trichome count, photosynthesis rate, leaf water potential, turgor pressure, and an enhancement of osmoprotectant built up, antioxidant enzymes phytohormones (ABA, JA, CH_4) and ions (Na^+ and Cl^-) are linked to ROS-mediated osmotic, ionic, and oxidative stress.

These days, there are several methods for improving plant salt tolerance including genetic engineering and the exogenous application of antioxidants. However, nanoparticles (NPs) have demonstrated especially encouraging outcomes. Using nanotechnology to increase plant tolerance to salt might offer a workable substitute for the approaches used now to enable efficient agricultural output. The enhanced resistance of plants to salinity stress is attributed to the ability of nanoparticles (NPs) to shield photosynthesis, facilitate reactive oxygen species (ROS) detoxification, and alleviate osmotic and ionic stress.

Because of their morphology (size, forms and structure ranging from 1 to 100 nm) and strong surface reactivity, NPs can effectively replace traditional crop protection approaches with reduced eco-toxicity and minimum input requirements. A number of recent investigations have confirmed that, under saline settings, NPs decreased ROS stress by including the enzymatic antioxidant system, or biogenic SiO_2 on *Pisum sativum* L. and *Lycopersicon esculentum Mill,* Au-NPs on *Triticum aestivum* L, CeO_2 on *Zea mays* L, ZnO-NPs on *Lycopersicon escultentum,* Ca- phosphate NPs on *Vicia faba* L., carbon NPs on *Lactuca sativa* L., zeolite NPs on *Solanum tuberosum* L, urea- functionalized hydroxyapatite NPs on *Helianthus annuus L.,* biogenic ZnO-NPs on *Abelmoschus esculentus* L., biochar NPs on *Triticum aestivum,* chitosan NPs on *Momordica charantia,* and silver NPs on *Pennisetum glaucum*. The application of Ag-NPs and C-NPs has been shown to increase catalase (CAT) peroxidase (POD) and ascorbate peroxidase (APX) activities. This leads to an improvement in soluble sugar and osmoprotectant content, helping to maintain redox balance and ion homeostasis. On the other hand, the application of Fe-NPs increases CAT and peroxidase (POX) activities, leading to enhanced soluble carbohydrate levels, reduced malondialdehyde (MDA) content, and increased levels of reactive oxygen species (ROS). These improvements contribute to an increase in shoot dry weight and seed yield.

Furthermore, NPs regulate genes such as *MAPK2, DDF2, P5CS, A10G2288, Ao3G1517, D09G1420, GhD11G2183, GhA09G1415, TAS14* and *ZFHD* that are important to produce enzymatic and non- enzymatic antioxidants such as proline-CAT, ascorbic acid (AsA), POD and SOD. The quantity of ROS and the likelihood of oxidative stress in plants can be reduced by upregulating these genes in saline environments. In summary scientists have reached a scientific consensus that NPs applied at low concentrations or in a dose-dependent manner may stimulate the ROS detoxification system by increasing the expression of signaling genes. Hence, the utilization of a diverse array of

NPs has significantly enhanced the potential for successful crop cultivation in salt- affected soils, resulting in increased yields. This chapter underscores the pivotal role of nanotechnology in augmenting plant tolerance to salinity stress. It presents a comprehensive summary of the physiological, biochemical and molecular mechanisms involved with a particular focus on NPs' mediation of antioxidant enzymatic activities for plant adaptation under salinity stress. We envisage that this overview will establish a framework for the integration of nanoscience with plant science, thereby facilitating the development of innovative solution to agricultural challenges such as salt stress.

2. Nanoparticles: A New Strategy for Combat against Salinity Stress

There appears to be encouraging proof in recent years that a plant nano-biotechnology strategy can increase plants stress tolerance. According to research, enhancing plant stress tolerance by using NPs has emerged as a feasible, long-term approach to boost agriculture productivity that is sustainable and low-cost. Plant cells acquire Cl^- and Na^+ ions because of their saline environment which causes an ionic imbalance and toxicity. Salinity enhances the expulsion of K^+ ions out of mesophyll cells of leaves and increases the deposition of Na^+ in cytosols. Salinity and development reduce the soil's ability to absorb nutrients, which also lowers production. Examples of nano- enabled techniques that are developed to promote plant growth include the application of NPs through a variety of strategies (seed coating, foliar spray, and soil irrigation) genetic engineering of plants to increase defense- related phytohormones and efficiency for photosynthesis, and the development of nano-sensors for real-time plant health tracking. Because of their unique physiochemical properties--which include small size, stability, high reactivity, specific surface area and structure relative to their respective bulk equivalents NPs are useful for wide range of antioxidant application. CeO_2-NPs, or nanoceria exhibited catalytic activity in scavenging free radicals in the chloroplast of plant cells. CeO_2-NPs thereby lessen the damage that an excess of ROS causes to a plant's photosynthetic activities as because of abiotic stress. While catalase, a form of nanoceria with a lower Ce^{3+}/Ce^{4+} ratio, lessens the harmful effects of H_2O_2 by displacing it into water and oxygen, high Ce^{3+}/Ce^{4+} nanoceria work like SOD and prevent green algae development by creating H_2O_2. Because of the loose-fitting Ce^{3+} bonds that form in the defect sites of the nano-active ceria when O_2 vacancies are generated, nano-ceria coated with polyacryl acid can catalyze the detoxification of reactive oxygen species. It has been demonstrated that maintaining the k+ ion homeostasis of the mesophyll cells via hydroxyl radical detoxification using nanoceria enhances the photosynthetic process and increases the salt tolerance of plants. Mn-NPs (MnO, MnO_2, and Mn_3O_4 are also able to participate in catalytic activities and scavenge ROS. The SOD-like activity of MnO-NPs was found to greatly excel over the intrinsic manganese-dependent SOD.

MnO_2 can imitate the activities of POD, CAT, and OXD enzymes in addition to SOD. The redox exchange between manganese's two oxidative sites (Mn^{2+}/ Mn^{3+}) illustrates the multifunctionality and enzyme-like activity of Mn_3O_4-NPs. MnO-NPs are more efficient ROS scavengers that CeO-NPs. The introduction of CeO-NPs stopped the reduction in PS II (photosystem II) quantum yield and CO_2 assimilation that was brought on by abiotic stress-induced ROS scavenging, particularly hydrogen peroxide. TiO_2-NP application enhanced the levels of antioxidant enzymes SOD, CAT, and GPOX and reduced oxidative stress in Lemna minor plants. After being exposed to NPs, several plants species, including as *Gossypium herbaceum* L., *Arabidopsis thaliana* L., *Triticum aestivum* L., and others have shown a high degree of saline tolerance. NPs such as nano-zinc, cerium oxide, carbon nanotubes, put- carbon quantum dots (put-CQD), titanium dioxide and silica have been used to increase plant resistance to salt (Figure 2). A small number of NPs, such as MWCNTs, Se-NPs and ZnO-NPs can considerably lessen the effects of salt on Brassica napus seeding growth. When seeds of Brassica napus are exposed to salt, their rate of germination is greatly increased by seed priming them with NPs of zinc oxide, selenium oxide, and cerium oxide. *Culinaris Medic* L. cultivated on Si-NPs shown improvement in growth and seed germination even when they were exposed to salt. Manganese NPs were utilized to prime the seeds of *Capsicum annuum* L. plants, shielding them against the damage induced by salt spray. *Brassica napus* L. exhibited increased salinity tolerance when amended with multi-walled carbon nanotubes, leading to a reduction in the Na^+/K^+ ratio ROS production and thiobarbituric acid content. Nevertheless, further investigation is warranted at the molecular, physical, and chemical levels to elucidate the mechanisms through which nanoparticles enhance plant salinity tolerance.

The NPs can function as enzymes and offer the plants resistance. Zinc, for instance, is essential to plant survival, and studies using ZnO-NPs have demonstrated improved SOD and GPX gene expression in tomato plants cultivated in salinized soil. It has also been discovered that in plants exposed to abiotic stress, TiO_2 and SiO_2-NPs boost the expression of antioxidant enzymes and photosynthetic activity. Researchers discovered that whereas large concentration of NPs were shown to induce oxidative injury, low quantities of stabilized nanosized zero-valent iron (Fe) were found to minimize damage caused by oxidative strain in plants. Additionally, bioengineered FeO-NPs have been demonstrated to enhance the growth of Triticum aestivum by increasing the level of essential nutrients (N, P, and K) while reducing the level of salt (Na^+) and chlorine (Cl^-) within the plant. This mitigates the harmful impacts of ROS formed due to salinity stress. Fe_3O_4 NPs were show to exhibit catalytic activity that was much higher with decreasing particle size and to have activity comparable to that of POD enzyme. However, DNA damage, decreased cell viability, and apoptosis (programmed cell death) may result from oxidative stress in mitochondria brought on by FeO-NPs. Higher quantities of Fe_3O_4-NPs have been demonstrated to have cytotoxic effects in addition to oxidative damage.

Hence, date from the scientific literature demonstrates that usage of NPs with varying concentrations have opposing effects. It has been demonstrated that a smaller dosage of NPs is more beneficial.

3. Uptake Pathways of NPs

3.1 Uptake of NPs by plants

3.1.1 Pathways for NPs uptake from Soil

The main entrance sites for NPs into soil are the roots, where they move from soil pores (Figure 2). The rhizosphere's lateral roots and root hair are the most effective location for NP absorption. The root surface is where NPs first come into touch with plant roots because roots absorb NPs from the soil. The root surface is negatively charged because of chemical discharge from the root hair, and the positively charged nanoparticles are more likely to accumulate and are absorbed on it rapidly. Organic acids and mucus are some of these substances. Once within the plants, NPs have unrestricted access to apoplastic and symplastic mobility within the tissues. The apoplastic route facilitates NPs' radical trafficking into vascular tissues. Transport of materials, including water, occurs between the symplast and the cytoplasm of the next cell. The formation of lateral roots creates a new adsorption contact through which NPs can reach the root column (Figure 2). The root epidermis and the leaf surfaces are similar in structure and function. On the other hand, the root hair surfaces, and root tip epidermis of primary and secondary plant roots are not fully formed. When NPs are exposed here, they pierce and come into direct touch with the root epidermis. A little amount of nutrients and water can flow through the cell wall of root epidermal cells. Tight cell walls of root cells allow them to filter out bigger and smaller undesirable material. Small molecule compounds like steroids can flow across the physical barrier that the root cell's plasma membrane creates between it and the soil, but hydrophilic and macromolecules are blocked. ATPase and the GLUT1-4 transporter/carrier are two examples of the channels and carrier that move these molecules through the membrane. The movement of NPs from the soil to the root cells may also include these membrane channels and transporters. NPs may enter the root's xylem, or core column, if the root lacks an exodermis. Further research has demonstrated that specific NPs can cause damage to the plasma membrane and that bigger NPs can enter the epidermal cell wall through the formation of new hole. Various techniques facilitate the absorption of nanoparticles (NPs) by plant cells upon their introduction into plant tissue. There methods include the ion pathway, endocytosis, protein binding to cell membranes, and physical damage.

3.1.2 Pathways for NPs foliar uptake

NPs are foliated on the leaf surface, where they settle and are absorbed by the plant via the stomata, hydathodes, or trichomes. The primary constituents of the cuticle of the leaf epidermis are wax, cutin, and pectin. It acts as the main natural barrier preventing NP penetration and stops water loss in developing leaves. Nonetheless, there are two unique channels in the waxy Carneum: one is lipophilic and the is hydrophilic. The width of hydrophilic and lipophilic channels can range from 0.6 nm to 4.8 nm. Diffusion of hydrophilic NPs with a diameter of than 4.8 nm is made possible by the hydrophilic channels. The process of diffusion and infiltration via the lipophilic channels in the cuticle facilitates the absorption of lipophilic NPs by leaves.

Recent research showed that carbon dots smaller than 2 nm may penetrate cotton leaves by means of the cuticular route using high-resolution confocal fluorescence microscopy. However, the tiny size of the pore channels in the cuticle limits the number of NPs that plants may absorb via the epidermis. It has been demonstrated that applying NPs to the leaf surface causes them to accumulate in the epidermis and vascular region. In this regard, a number of studies have discovered that nanoparticles are capable of reaching various plant organs. The study team postulated that the stomatal channel might facilitate the uptake of nanoparticles (Figure2).

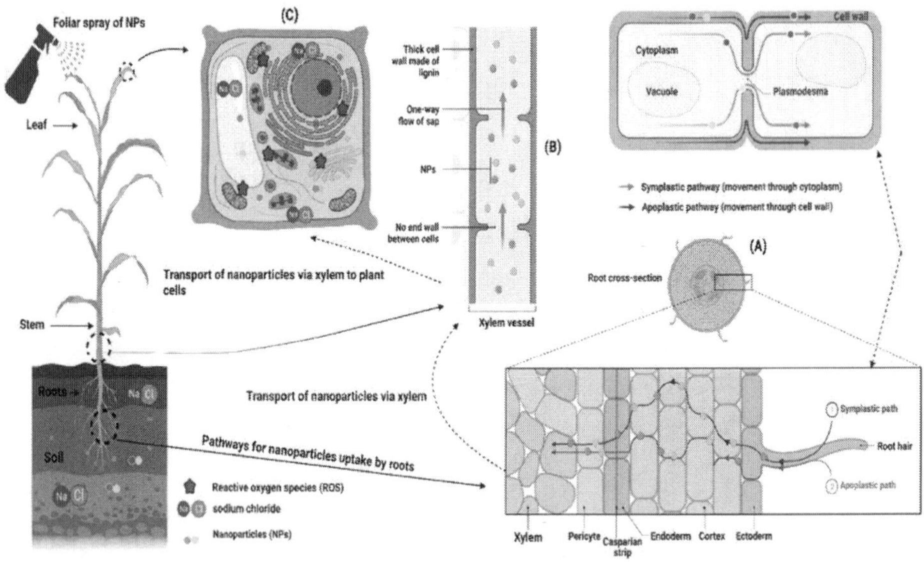

Figure 2. Schematic illustration of NPs addition techniques, plant absorption, and translocation through (A) Symaplastic and apoplastic pathway that after transport to other part of plants via (B) Xylem vessel. (C) To protect plants from salinity stresses, NPs and reduced ROS can be applied to the soil, coated on seeds, or sprayed on the leaves.

3.2 Translocation of NPs

When NPs are translocated in plants, it implies that they go from their original position to the roots, stems, and leaves of the organism. NPs (pores on the surface of leaves) might enter a plant by foliar spray, root absorption, or even the stomata. The size, shape, surface charge, and coating of the NP define its entrance point. Once inside, NPs have the freedom to travel to various areas of the facility. This applies to the fruits as well as the leaves and stalks. Plants translocation of NPs is influenced by several variables. Relevant factors include NPs' size, shape, surface charge, and concentration. Additionally, the kind of plant and the exposure pathway may have an impact on NPs translocation. Greater surface area NPs prefer to collect in the roots of the plant, whereas smaller surface area NPs are more prone to translocate inside the plant. The surface charge of the particles can also influence their reception and translocation; positively charged particles are more readily absorbed by plants. The number of NPs in the surrounding environment can also have an impact on the absorption and transport of NPs within plants. NPs translocation in plants can be influenced by a wide range of circumstances, and further research is needed to completely comprehend this process. Plants roots can transfer accumulated NPs to other aerial sections of the plants, which include developing seeds. Plants characteristics and other variables have a major role in the translocation of NPs. For example, studies have shown that gold (Au) can accumulate inside *O. sativa* shoots. But the shoots of *Raphanus raphanistrum* and *Cucurbita pepo* are resistant to its buildup. In addition, the positively changed Au^- NPs are quickly absorbed by plant roots. On the other hand, the plant's roots and shoots transfer negatively changed Au NPs more efficiently. The most persistent NPs, SiO_2 and TiO_2, are found in full speciation in plant tissues. When ZnO NPs in various forms, such as Zn-phosphate, are exposed hydroponically, zinc mostly accumulates in the roots and shoots of Z. mays plants. Increased dissolution in the rhizosphere, plant absorption, and Zn ion translocation are a few possible causes. Wheat crops cultivation is soil showed comparable zinc buildup and speciation. It was discovered that ZnO and CeO_2-NPs moved from the soil into *Glycine max* L. Zn was shown to have experienced biotransformation into Zn-citrus inside plant tissue, and CeO_2-NPs were identified to be transported as Ns. CUO-NPs were transferred from the roots to the shoots in Z mays by xylem and from the shoots to the roots by phloem. Earlier investigation has demonstrated the involvement of TiO_2-NPs in translocation in *Triticum aestivems*. They postulated that particles cannot collect in the root if their diameter is more than 140nm. Moreover, particles that are too large to fit into the stele or be transferred to the shoot can be stored in the root parenchyma of T. aestivum by a threshold diameter of 36nm.

4. Effect of NPs on Antioxidant Enzymes activities for Physiochemical Protection under Salinity Stress

NPs have been shown several studies to increase levels of antioxidant enzymes. The author of the investigations tried to assert that a small number of NPs have traits in common with enzymes, which help plants better withstand and adjust to oxidative stress. Few researches on NPs claim, the plant in fending against the oxidative environment because of their characteristics that resemble those of some antioxidant enzymes. For example, the behaviour of the POD and CAT enzymes are comparable to those of iron, cesium and cobalt NPs. Ag-NPs pretreatment of *T. aestivum* L. seeds resulted improved plant growth, proline content, soluble sugars, and POD activity in *T. aestivum* L. plantlets facing salt stress, as demonstrated by the plants augmented with zeolite and zinc. Under salinity, silicon and boron nanoparticles demonstrated notable improvements in photosynthesis, water and nutrient retention, antioxidant enzyme activity, and nutrient consumption capacity.

4.1 Protection of plant growth process

Cu is a crucial metal- based micronutrient that impact many important metabolic processes in plant cells. The most prevalent and important and Cu based is plastocyanim which is necessary to sustain photosynthesis through the transfer of electrons in the lumen of thylakoids. There has been much research on the role that copper metabolism. Furthermore, because many enzymes involved in redox reaction need copper in their structural compositions, Cu has an impact on the metabolic activity of plants. Cu, an element present in the SOD isoform, provides plants with resistance against oxidative stress. According to finding on maize plants, adding Cu-NPs to plants improves the level of amino acids and osmo- protectant as well as the antioxidant' defense mechanism. This decreases the negative effect on nutrition, photosynthesis, and water connection. Compared to controls, fruit containing Cu-NPs exhibited higher levels of GSH by 81%, phenols by 16% and vitamin C by 80%- Cu treatment boosted APX activity in leaves by 140%, GPX activity by 26%, SOD activity by 8%, and CAT activity by 93%. Numerous studies have shown that Fe-NPs play a critical mediating function in the establishment of plant resistance to salt stress. When plants of the *Dracocephalum Moldavia* species were subjected to salt stress, Fe_2O_3 in the form of NPs effect was seen on the plants. The scientist discovered that by boosting antioxidant enzymatic activity, plants treated with 60 mg L-1 with nano Fe_2O_3 under salt stress were able to grow more quickly. Fe_2O_3 (3mm) on *Trachyspermum ammoniacum* under salinity and discovered that the combination of Fe_2O_3 NPs and SA treatment improved the K^+/NA^+ ratio and the antioxidant of the antioxidant system (polyphenol oxidase, POD, CAT, and SOD) to decrease salinity stress. The change improved the plant's Fe content, root and shoot growth, leaf pigmentation, membrane stability index, and seed yield. Furthermore, the function of CAT, POD, and polyphenol oxidase was improved when Fe-NPs were applied as a foliar supplement to Helianthus annuus cultivated in a saline regime. Low concentrations of MWCNTs appeared to have a superior reaction of *Ocimumbasilicum* toward salt by increasing photosynthetic pigment and triggering antioxidant defense system enzymes (CAT, GP, and APX) and non-enzymatic components (the quantity of phenol). Their research showed that TiO_2 at 100mg L-1 improved plant development under salt stress, and they connected this to an increase in the activity of many important enzymes.

4.2 Protection of Photosynthetic pigments and process

The primary technique by which plant on earth transform light energy into chemical energy is called photosynthesis. The ultimate energy source for all living things is photosynthesis, which the organism uses either directly or indirectly. The basic ingredients needed to transform light energy into chemical form that can be further stored in sugar molecules in chloroplasts include chlorophyll, water, and carbon dioxide. Numerous studies have shown that these chlorophyll complexes and processes were impacted by salinity stress, and that foliar applications of NPs greatly increase plant chlorophyll content. This enables plants to generate more light-harvesting complexes, which improve their capacity to absorb and use light energy during photosynthesis.

In order to investigate the effects of TiO_2-NPs on the physiological and biochemical response of *T. aestivum* to TiO_2-NPs in phosphorus-enriched soil, a through life cycle research on development under salinity stress was recently observed. The applications of 50 mg L^{-1} TiO_2-NPs promoted *T. aestivum* growth in salty environment by increasing the activity of antioxidant enzymes SOD (2.06-33.97%) and POD (up to 13.19%). Additionally, it has been shown that when subjected to salt stress, *T. aestivum* primed seeds treated with ZnO-NPs at a concentration of 50 mg L-1 exhibit improved leaf water potential and reduced Na concentrations at the seeding stage. By upregulating SOD, plants preconditioned with ZnO-NPs exhibit a more effective oxygen scavenging system and photosynthetic electron transfer chain. Plants preconditioned with ZnO-NPs exhibit a more efficient oxygen scavenging system and photosynthetic electron transfer chain. This is achieved through the upregulations of superoxide dismutase (SOD), ascorbate peroxidase (APX), and catalase (CAT) activities, which regulate reactive oxygen species (ROS) production under salt stress. Under salinity stress, the foliar treatment of ZnO-NPs (20 mg L-1) improved stomatal density, Hill's response, relative water content (RWC), and reduced root ion leakage. The foliar application of Se-NPs (10 and 20 mg L-1) improved strawberry plants developing in a range of salty soils (0,25,50, and 75 Mm NaCl) because of the pigment's ability to shelter photosynthetic pigments, which leads to greater growth and production metrics. Following foliar treatment with Se-NPs, strawberry's resistance to salt was increased due to a decrease in lipid peroxidation and an increase in the activity of antioxidant enzymes such as peroxidase and superoxide dismutase.

4.3 Regulation of iconic homeostasis for protection

One of the unforeseen effects of salt stress is a disturbance in the plant life cycle brought on by toxicity or an iconic imbalance. According to research, a main phenomenon following salt exposure is expected to be imbalances in nutrition, since the salt treatment can impede the availability, absorption, and transport of nutrients in plants. Due to depolarization of the root plasma membrane and the stimulation of outward-rectifying K channels in the guard cells brought on by the salt, the plant has higher levels of Na+ and a lower concentration of K^+.

High sodium (Na^+) concentration caused *Lycopersicum* seedlings to have reduced calcium (Ca^{2+}) and magnesium (Mg^{2+}) I their roots and shoots as well as a greater Na^+/K^+ ratio. Plant metabolism is disturbed when there is an excessive amount of Na^+ in the soil because it replaces potassium in the enzymes. Furthermore, the plant exercises iconic stress and an imbalance in mineral homeostasis as a result of reduced K^+, Ca^{2+}, and Mg^{2+} uptake brought on by a higher level of Na^+ buildup. In contrast to the stress-free condition, researchers observed that during salt stress, O. Sativa seedlings ion homeostasis was hindered by an increase in Na^+ and the Na+/K+ ratio. In this research, it was discovered that the amount of Na+ in the shoots was higher than that in the roots. In addition to K_{+}, Ca_{2}^{+}, and Mg_{2}^{+}, salt stress also reduced the amount of zinc (Zn) in the plants' roots and stems. Nanoparticles (NPs) can significantly influence plant mineral nutrition by affecting nutrient absorption, translocation, and final allocation. It is commonly known that NP-mediated antioxidant activity may raise the K^+/Na^+ ratio, which enhances the osmotic potential of the plant and encourages growth in salinity-stressed environments. Applying nanosilicon foliarly at concentrations of 0.5 and 1 nm to Glycine Max L. can reduce the effects of salt stress by increasing antioxidants activities and non-enzymatic chemicals that raise K^+ and lower Na^+ concentrations, lipid peroxidation, and the generation of reactive oxygen species. By improving APX, GPX, SOD, CAT, and phenols- all upregulated by 104%, 140%, 26%, 8%, 93%, and 16% in leaf tissue, respectively- and improving the Na^+/K^+ ratio, foliar administration of Cu^- NPs (250mg L-1) decreased salt stress (50nM NaCl) in tomato crops.

Nanoceria, or cerium oxide NPs, possess the remarkable capacity to lower ROS levels, particularly OH, which do not have enzymatic scavenging routes. When 50mg L^{-1} of cerium-NPs are applied to *Arabidopsis thaliana* leaves, they can scavenge OH^-, preventing its buildup for a longer period of time. It achieves this by activating the plasma membrane's K^+ selective (GORK) channels, which are in charge of delivering potassium into the cell. Because of its functions in enzymatic processes that contributes to photosynthesis, and improves photosystem efficiency, iron (Fe) is thought to be essential for the development and the growth of plants. The majority if iron in a variety of soils is found as complexes of ferric iron with organic substances and as insoluble ferric ($Fe3^+$) hydroxides. Conversely, ferrous ($Fe2^+$) iron is found in soluble inorganic iron at very low concentrations (0.01-1 nM). For the transfer of iron from root transporters to the other pant tissue sections, the concentration of iron is insufficient. Plant utilizes two brief ways to absorb iron from the soil: chelation and phytosiderophores. Soil pH and ROS levels rise during iconic stress, although soil Fe content falls. To make the plants more resistant to abiotic stressors, iron nanoparticles are employed. Applying iron nanoparticles enhances plants nutrition absorption by starting a chain reaction at the cellular level. Being exposed to Fe-NPs boosted the development, quantity of chlorophyll, and antioxidant enzymes in wheat plants, reducing their susceptibility to salt. These nanoparticles (NPs) can reduce cellular salt accumulation, thereby shielding plants from iron stress.

5. NPs Mediated Antioxidant Enzymatic Response for Mitigation of ROS under Salinity

5.1 NPs mediated Antioxidant response for ROS homeostasis

Antioxidants are a defense mechanism used by plants against oxidative damage. Ascorbate (AsA), GSH, phenolic compounds, flavonoids, alkaloids, -tocopherol, and non-protein amino acids are examples of non-enzymatic antioxidants. Enzymatic antioxidants include SOD, APX, CAT, DHAR, MDHAR, GPX, GR, GST, and POD/POX activity (Figure 3). In order to protect themselves from ROS-induced injury, plants use SOD as their main antioxidant activating system. ROS-induced damage is caused by the conversion of O_2 into H_2O_2, which is then accumulated by CAT, APX, GPX, PPO, PRX, and TRX enzymes, or it can be catalyzed in the Asada-Halliwell pathway (AsA-GSH cycle). Crops involve four enzymatic antioxidants (APX, DHAR, MDHAR, and GR) in the AsA cycle, which helps to maintain redox equilibrium by lowering ROS generation, besides to the non-enzymatic antioxidants AsA and GSH.

Numerous studies have demonstrated that applying NPs increases the level of antioxidant enzymes. Additionally, it has been proposed that the salt threshold, the duration of salinity exposure, and the phases of plant growth affect how these antioxidants work. A few NPs have the properties of antioxidant enzymes that enable them to help plants fight oxidative stress, based on several experiments. For instance, NPs composed of cobalt, iron, and cesium have activity profiles similar to those of the enzyme CAT, and NPs composed of cesium, manganese, copper, and iron have activity profiles comparable to those of the enzyme POD.

In a research, cerium oxide nanoparticles enhanced ROS-NSCC activity, which aids in the scavenging of ROS in *Arabidopsis thaliana* plants treated with cerium. Furthermore, consideration was given to the possibility that silver nanoparticles might interact with protein Sulphur groups, namely on antioxidant enzymes, and obstruct antioxidant action. The use of NPs as radical scavengers, for their redox potential, or as carriers for antioxidant chemicals are some of the novel research avenues in the field of oxidative stress studies. Ag-NPs are often made using plant extracts, although the production technique may affect the antioxidant qualities of Ag-NPs. These phytochemicals are valuable because they shed light on Ag-NPs' nanoforms and antioxidant capabilities.

Brassica oleracea leaves with a high scavenging percentage (60-80%) are utilized in the production of Ag NPs. These NPs' significant amount of phenolics and flavonoids produced on the surface acting as capping agents may be the cause of their antioxidant action. Rises in antioxidant enzyme levels, such as APX, GPX, and SOD, as well as chlorophyll were seen in tomato leaves treated with Cu-NPs. Researchers investigated *Pennisetum glaucum* L. under salinity stress (0,120, and 150 Mm NaC1). Their results showed that the presence of Ag-NPs considerably improved the growth features of the plants treated with them. This was due to the downregulation of the sodium-to-potassium ratio and the overexpression of antioxidant enzymes including SOD, CAT, and GPX.

Similar to this, during salinity stress, foliar administration of ZnO-NPs (10mg L^{-1}) derived from *Phoenix dactylifera* leaflet extract boosted the function of ROS-detoxifying enzymes (SOD and CAT) in *V. unguiculata* and *A. esculentus*. Furthermore, ZnO-NPs (17mg L^{-1}) derived from Carcia papaya extract were assessed for their influence on the antioxidant metabolism in *Carthamus tinctoriuos* during salt stress. The results showed that ZnO-NPs enhanced the activity of antioxidant enzymes (POD, SOD, CAT, APX), which in turn reduced the concentrations of Na+ (74% in leaves and 60% in roots), MDA (leaves 54%), and proline (61% and 63% in roots).

These NPs countered the harm done by lipid peroxidation and hydrogen peroxide radicals by boosting the activity of enzymes like SOD and POX that shield cells from stress. The removal of free radicals that lessened the effects of salt stress on plants can be facilitated by antioxidant enzymes and anthocyanins. According to a recent study, different NPs detoxified ROS, which lessened the salt effect on plants summarized in Figure 3 and Table 1.

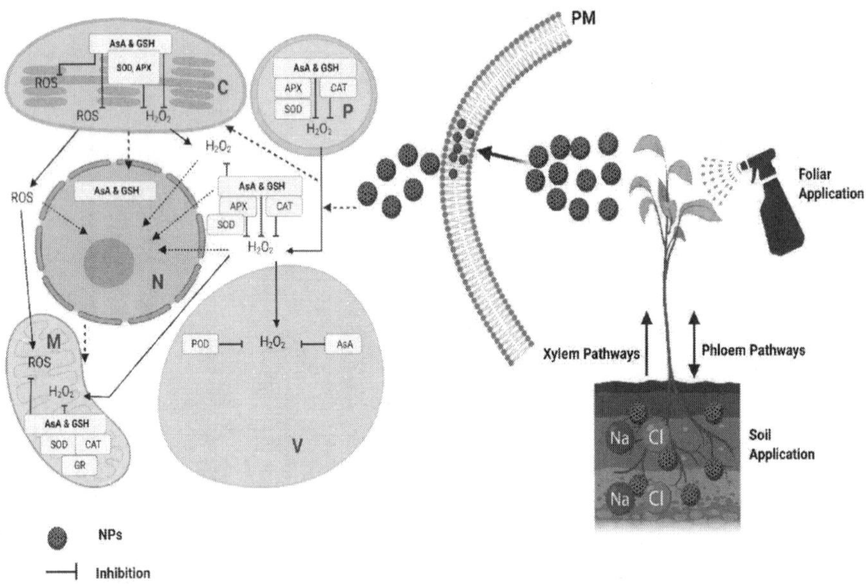

Figure 3. Nanoparticle-mediated reduction of reactive oxygen species (ROS) in various plant cell organelles exposed to salinity stress. Increase subcellular levels of reactive oxygen species (ROS), focusing on hydrogen peroxide (H_2O_2) and its detoxification by non-enzymatic and enzymatic antioxidants in plants. Possible signaling pathway: dotted arrows. AsA, ascorbate; APX, ascorbate peroxidase; C, chloroplast; CAT, catalase; GSH, glutathione; M, mitochondria; N, nucleus; P, peroxisome; POD, guaiacol-type peroxidase; V, vacuole. The diagrammed routes are draw on the findings of prior research.

5.2 A Subcellular level ROS homeostasis and protection in plant

5.2.1 Chloroplast

Ascorbate and glutathione are prominent nonenzymatic antioxidants of chloroplasts, whereas SOD and the enzymes of the Foyer- Halliwell-Asada cycle or an ascorbate–glutathione pathway are important enzymatic antioxidants (Figure 4). Diverse antioxidant combination plays a critical role in maintaining the equilibrium of ROS concentration within "functionally useful" limits, which are necessary for signaling several plant processes and stress responses. One of the main systems that control ROS equilibrium in chloroplasts, the water cycle, is necessary for salt tolerance. Excited photosystems, in particular the thylakoid membrane-bound main electron acceptor of PSI, may absorb electrons from the ground state of O_2• produced at the time of photolysis of water in chloroplasts owing to the Mehler reaction where O_2• is produced as a consequence. The acceptor component of the electron transfer process might potentially be PSII's generation of O_2• from electron leakage to O_2. Cu/Zn SOD, an enzyme attached to the thylakoid membrane, converts O_2 to H_2O_2. Ascorbate peroxidase, another enzyme linked with the thylakoid membrane, then reduces this H_2O_2 back to water. The ascorbate- glutathione cycle, often referred to as the Foyer-Halliwell-Asada pathway, is responsible for scavenging ROS by employing NADPH in chloroplasts. By providing NADP, the final electron acceptor of PSI, this route also aids in relaxing the excess reduction of photosystems, which in turn reduces the likelihood of more ROS formation .H_2O_2 is generated when O_2 is dismutated by SOD. The stromal APX then employs ascorbate (AsA) as an electron donor for neutralizing this H2O2.The enzymes DHAR or MDHAR (dehydroascorbate reductase or monodehydroascorbate reductase) are responsible for recycling oxidized ascorbate. GR recycle GHS that have been used by the farmer when NADPH is utilized as an electron donor. As opposed to sensitive species, chloroplasts have been shown to enhance the activity of antioxidant enzymes in response to environmental stressors. One example is the halophytic wild *Lycopersicon pennellii*, which, when exposed to salt, has greater levels of SOD, APX, and MDHAR activity in its chloroplasts that the common tomato.

Additionally, it has been observed that GPX and thioredoxin/peroxiredoxin (Tr/Prx) absorbed the extra H_2O_2 that salinity induced in the chloroplasts. Antioxidants like as carotenoids and tocopherols found in chlorophyll; membranes counteract $^1O_2$ reduced by PSII. During the xanthophylls cycle and direct quenching, carotenoids remove $^1O_2$ (NPQ). It is believed that a small number of C4 plants, namely those of the NADP-malic enzyme (NADPME) subtype, are unable to produce $^1O_2$ due to the absence of PSII in their bundle sheath chloroplasts. Several studies have shown that treating plants with foliar nanoparticles (NPs) significantly increases their chlorophyll levels. This allows the plants to from more complexes for light harvesting, which increases the quantity of light energy absorbed and improve photosynthesis. Using the nano- Fe complex under salt stress (100 and 200m M NaCl) lessens damage to the chloroplasts and prevents the pistachio plant from degrading.

The pigment responsible for photosynthesis are enhanced by the application of carbon NPs (100, 50, and 25mg L^{-1}) and ZnO-NPs (50mg L^{-1}) and NPs (50gm L^{-1}) because they increase the activity of APX and SOD antioxidant enzymatic activity during salt stress. APX is primarily founded in chloroplasts and efficiently converts H_2O_2 (released by SOD) into water. When *A. esculentus* L. is treated to ZnO-NPs, it shows decreased proline and total soluble sugars and enhanced photosynthetic pigments, CAT, and SOD activities. When ZN-NPs were applied to *Brassica napus* L. plants exposed to salt stress, their detrimental impacts were lessened by increasing their ionic regulation, osmolyte production, and antioxidant activities.

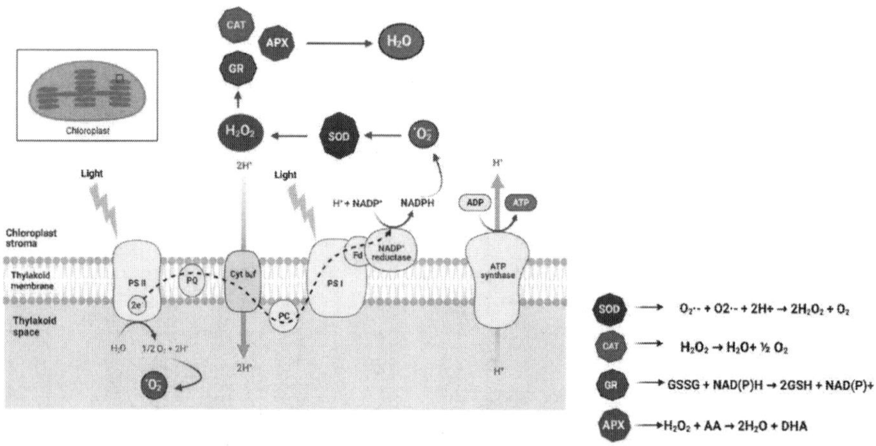

Figure 4. A graphical representation of the process occurring in the chloroplast during salinity stress, in which ROS are produced because of CO_2 inhibition, and low water levels are caused by stress. The electron transport chain (ETC) of the photosystem (PS) is the primary source of ROS and mode of action of antioxidant enzymes is chloroplast under salinity stress. Application of various NPS can enhance the activity of antioxidant enzymes that reduced the accumulation of ROS for better adaptation of plants under saline conditions.

5.2.2 Mitochondria

Mitochondria are another possible site in plants where O_2, H_2O_2, and OH• are produced. The mitochondria generate less ROS than the chloroplast (Figure 5). The way that animals and plants produce ROS in their mitochondria is different. Photorespiration occurs in the mitochondria of plants, which are surrounded by solute composed of carbohydrates. ROS are produced in several sites inside mitochondria. This organelle's NADPH dehydrogenase [complex] converts oxygen to molecular oxygen. Excess ubiquinone pool (UQ) reduction can be converted to reactive oxygen species (ROS) via the electron transfer complex found in mitochondria complex l and III. An electron is sent to cytochrome C whenever UQ is totally reduced. This results in the creation of O_2• and a very fragile radical complex. There is a retrograde electron transfer from complex II to I when complex NAD is inadequate. Because of its distinct redox potential, when is based on the straightforward transition between the $Ce4^{+}$ and Ce^{3+} state of oxidation, nano-Ce is ideally suited to counteract this impact and functions as a ROS scavenger? Furthermore, at lower Ce^{4+}/Ce^{4+} ratios, a CeO^{-}NPs mimics the catalase function; at higher $Ce^{3+}/Ce4^{+}$ ratios, it shows the elimination action and mimics the superoxide dismutase function. Through ATP generation, the redox mechanism within mitochondria and all forms of stress impact ROS, resulting in a reduction in the ubiquitin (UQ) pool. The amount of ROS within mitochondria is regulated by the combined action of many antioxidant enzymes. Several NPs that can shield the mitochondria from ROS through salinity stress were covered in Table 1.

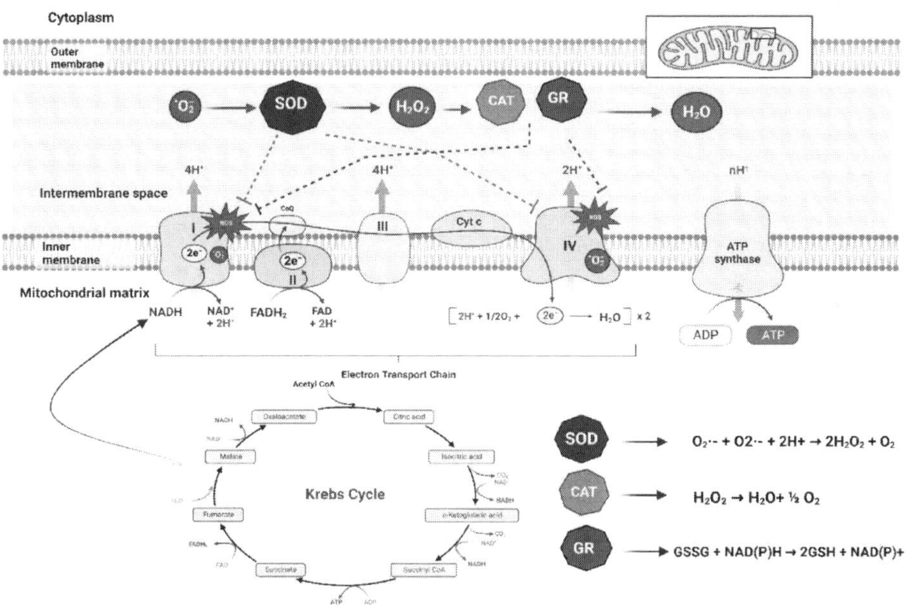

Figure 5. Electron transport chain (ETC) schematic shows the electron flow through ROS generation and oxidative phosphorylation during formation of ATP. Enzymes like catalase (CAT), glutathione peroxides (GPx), superoxide dismutase (SOD), and others make up the antioxidant defense system, which works to neutralize ROS. Application NPs enhanced the performance of enzymes involved in antioxidant processes. NPSS can enhance the rate of detoxification of ROS from mitochondria cell and help in improvement of plant survival under saline conditions.

Table1. Effect of NPs in upregulation of different antioxidant enzymes to mitigation salinity stress.

NPs	Plants	Scientific Name	Application Concentration	NPs Characters	Mode of Synthesis	Salinity Level	Antioxidant Enzymes Upregulation	Mode of action
ZnO-NPs Si-NPs	Okra	*Abelmoschus esculentus* L.	ZnO-NPs 10mg/L	hexagonal and spherical in shape and at size 16 to 35 nm	Chemical grade	0, 10, 25, 50, 75 and 100% Sea water	SOD and CAT antioxidant enzymes	Increased photosynthetic pigment contents and lowered the accumulation of proline and total soluble sugar
	Canola Barley	*Brassica napus* L.	ZnO-NPs 10mg/L	Size 30nm	Chemical grade	3000, 6000, 9000, and 12,000 mg/L	SOD, CAT and POD antioxidant enzymes	Electrolyte leakage, concentrations of MDA and H_2O_2 decreased, and key nutrient levels rise
		Hordeum vulgare L.	ZnO-NPs 100 mg/kg, 100 mg/L	size 20–30 nm	Chemical grade	50 and 100 mM	SOD, CAT POD and APX antioxidant enzymes	Increased biomass, yieldattributes photosynthesis, nutrient contents, decreased Na^+, MDA and ROS accumulation
	Wheat	*Trilicumaes-tivum*L.	0, 20, 50 and 80 mg/L	NA	Chemical grade	6.3 dS m^{-1}	CAT, POD and ascorbic acid	Increased chlorophyll contents, plant height, shoot lengths, root fresh, dry weights

NPs	Plants	Scientific Name	Application Concentration	NPs Characters	Mode of Synthesis	Salinity Level	Antioxidant Enzymes Upregulation	Mode of action
	Sorghum	*Sorghum bicolor* L.	5 and 10 mg/L	Hexagonal wurtzite, size 27.5 nm	Biogenic synthesized	400 mM NaCl	SOD, CAT and APX	Decreased Na^+/K^+ ratio, MDA, proline, soluble sugars
	Wheat	*Trilicumaes-tivum*L.	100mg/Kg	NA	Chemical grade	100 mM	SOD, POD and CAT	Mitigated salinity stress and improve seed germination, photosynthetic pigments
	Soybean	*Glycine max* L.	0.5 and 1 mM	20-30nm	Chemical grade	5 and 10 dS m^{-1}	CAT, POX, APX, SOD	Increased K^+ and decreased Na^+, lipid peroxidation, and reactive oxygen species production
	Quinoa	*Chenopodium quinoa* Willd	0, 20 mg/kg	NA	Chemical grade	12.4 dS m^{-1}	CAT, SOD, POD	Improved physiological and yield attribute and reduced salinity the phytotoxicity in quinoa
	Tomato	*Solanumlyco-persicum*L.	100 mg/L	NA	Chemical grade	250 mM	CAT, APX, SOD, GSH and AsA	Improved photosynthesis, mineral concentration, chlorophyll index, and growth attributes of tomato plants

NPs	Plants	Scientific Name	Application Concentration	NPs Characters	Mode of Synthesis	Salinity Level	Antioxidant Enzymes Upregulation	Mode of action
	Pea	*Pisum sativum* L.	500 mL	NA	Chemical grade	0, 100, 150, 200, and 250 mM	SOD, CAT, POD	Increased plant height, fresh and dry weight, total yield, and antioxidant defense systems
Ag-NPs	Wheat	*Trilicumaes-tivum* L.	0,2,5 and 10 mM	15-29nm	Chemical grade	25 and 150 mM NaCl	POD	Increased the soluble sugars and proline content
	Pearl millet	*Pennisetum glaucum* L.	0, 10, 20 and 30 mM	50 and 100nm	Chemical grade	0, 120 and 150 mM NaCl	CAT, SOD, POD, GPX, GR	Sodium (Na+) and Na^+/K^+ ratio was decreased, and potassium (K^+) increased total phenolic and flavonoid
	Roselle	*Hibiscus sabdariffa* L.	0, 15, 30, 45, 60, 75, and 90 mg/L	Spherical shape with sizes between 10 and 50 nm	Biogenic	0, 15, 30, 45, 60, 90, 120, 150, and 180 mM	SOD, POD, CAT, APX	Increase in the amounts of proline, soluble sugars and carbohydrates for osmoprotection, flavonoid, anthocyanin contents and upregulated chalcone synthase (CHS), flavanone 3 hydroxylase (F3H), and anthocyanidin synthase (ANS) genes experssion

NPs	Plants	Scientific Name	Application Concentration	NPs Characters	Mode of Synthesis	Salinity Level	Antioxidant Enzymes Upregulation	Mode of action
	Forssk	*Maeruaoblon-gifolia*L.	0, 10, 20, and 30 mg/L	Spherical shape with sizes between 9 and 30 nm	Biogenic	0, 50, 100, and 200 mM NaCl	CAT and SOD	Reduced oxidative stress by reducing proline, soluble sugars, H_2O_2, MDA, and total phenolic and flavonoid contents
Cu-NPs	Tomato	*Solanumlyco-persicum*L.	250 mg/L	Spherical shape with sizes between 20 and 50 nm	Chemical grade	50 mM NaCl	APX, GPX, SOD, and CAT	Improved the Na^+/K^+ ratio while also increasing the accumulation of vitamin C, lycopene, total phenols, and flavonoids (in both leaves and fruits).

5.2.3 Peroxisome

Glycolate is produced more when O_2 and CO_2 concentrations inside the cell reduced during a drought. Glycolate oxidase in the peroxisome subsequently converts the glycolate to H_2O_2. Like the mitochondria and chloroplast, the peroxisome creates O_2• during regular metabolism. Peroxisomes regulate their ROS levels by carefully balancing the process of generation and scavenging. O_2 is created at two different sites in the Peroxisomal matrix when O_2• functions as an electron acceptor of NADH and Cyt b, and xanthine oxidase converts xanthine and hypoxanthine into uric acid and O_2. Three metabolic pathways in the peroxisome produce H_2O_2: radical disproportionation, -Oxidation, and the flavin oxidase pathway. H_2O_2 is scavenged by GSH inside peroxisomes by APX and ascorbic acid (AsA). CATs cleanse the body of H_2O_2. and lipids in Peroxisomes undergo oxidation due to a decrease in AsA-GSH Concentration. Furthermore, it has been demonstrated that POX, an enzyme that degrades polyamines, controls stress-responsive genes to promote the production and removal of reactive oxygen species in the peroxisome.

5.3 NPs Mediated Non-enzymatic Antioxidant based ROS Homeostasis

A few of the most harmful ROS, such as $^1O_2$ and •OH, cannot be eliminated by enzymatic antioxidant system. As a result, plants may only remove ROS via non-enzymatic method (Figure 3). ROS often result in lipid peroxidation and a sharp rise in MDA concentrations. Alpha-tocopherol (Vitamin E), phenolics, GSH, suitable solutes, carotenoids, flavonoids, and proline are examples of non-enzymatic substances that cooperate to counteract the unchecked creation of ROS that happens under stressful situations. GPX uses the common thiol tripeptide GSH to accelerate the breakdown of H_2O_2. Becomes involved in the process of scavenging H_2O_2 and/or degrading H_2O_2 and lipid peroxides by producing a conjugate, which is mediated by glutathione S-transferase (GST) and GPX. In the AsA-GSH pathway, ascorbic acid often known as vitamin C, serves as a co-factor of POXs and an electron donor for APX. Tocopherols and xanthophylls, which are both replenished by AsA, contribute to the quenching of this excitation energy. Enzymatic and non-enzymatic antioxidants acquire vital avenues for tightly controlling reactive oxygen species (ROS) in plant cells. These pathways are responsible for efficiently reducing oxidative damage brought on by abiotic stress. Numerous studies have demonstrated that the addition of NPs increase the effectiveness of non-antioxidants, that may lessen oxidative stress in situations where salt is a stressor (Table 1).

According to research, applying Fe-NPs (30, 60, and 90 mg L^{-1}) topically can boost the concentration of non-enzymatic antioxidants such as phenol, flavonoids, and anthocyanins, which aid in mitigating the effects of salt stress at varying levels (50-100 mM) on the *Dracocephalum moldavica* L. plant. Under salt stress, plants of *Ocimumba silicum* L. were observed having benefited from elevated levels of non-enzymatic antioxidants such as phenol when treated with carbon NPs (25, 50, and 100 mg L^{-1}).When exposed to salinity stress , titanium dioxide NPs at concentrations of 40,60, and 80 L^{-1} improved the activity of phenolic non-antioxidant enzymes, which in turn improves seedling vigor index, potassium ion (K^{+}) concentration ,lengths of shoot and root, fresh and dry weight of the seedling , germination energy, and germination percentage by lowering MDA contents.

Se NPs 10 and 50 mg L^{-1}, Si NPs 200 and 100mg L^{-1}, and Cu NPs 100 and 500mg L^{-1} increased carotene, glutathione, yellow carotenoids, flavonoids, and phenols in the salinity stressed Capsicum annuum L. pant. Another research indicated that the non-enzymatic system is more crucial than the enzymatic system in mitigating the detrimental effects of stress on this plant. The plant's enzymatic defense system starts to assist in neutralizing the free radicals causing the damage when the non-enzymatic defense system is weakened .These findings demonstrate how various NPs (both enzymatic and non-enzymatic elements) may affect plants' antioxidant defense system, hence reducing oxidative stress. Regardless of these results, nothing is known about the interactions between NPs and the antioxidant enzymes system in plant .To understand the process underpinning NPs capacity to mitigate salt stress responses, more extensive omics research is required. More investigation is needed to comprehend how NPs start this response. Since many these conclusions are based on controlled or hydroponic research, more field experiments are required.

6. NPs Based Molecular Regulation for Antioxidant Defense to plant under Salinity

Plant biological processes are determined by their molecular activities. The two primary steps in the intricate and strictly controlled process by which genes are translated into proteins in each cell are transcription and translation. Translation and transcription both contribute to gene expression. In order to regulate the transcription of genes (mRNA) in response to salinity, plant employ a variety of techniques. Regulation genes during the central dogma-from the onset of transcription to TFs' involvement in mRNA and protein synthesis is a novel approach to salt tolerance. Plants are stressed by salinity in two ways: first, it decrease the water potential of the soil, which decreases water uptake and leads to osmotic stress; second, it also leads to an ion buildup that causes ionic stress.

Salinity can be regulated at the physiological and biochemical levels with NPs, but its physio-biochemical function cannot be assessed without understanding its molecular functions. Nevertheless, not much has been done to examine how NPs interact with gene expression. Understanding the molecular interactions between NPs and salinity stress is essential for determining how effective NPs are against the stressor. This is because salinity stress modifies the expression of multiple genes that are involved in different cellular processes and their byproducts, which in turn impact plant growth. Integration of multiple 'omics' approaches, including proteomics, metabolomics, genomes, and transcriptomics, is a new and widely used tool for dealing with salinity tolerance and adaptation in plants. Research utilizing 'omics' technologies on the both control and salinity -stressed plants has provided tangible evidence of how traits, genes, proteins, and metabolites are adjusted to enable plants to develop the necessary tolerance to salt stress.

6.1 NPs and antioxidative Genomics and transcriptomics Response

Salinity causes two types of stress in plants: iconic stress, which is brought on by and accumulation of ions in the soil, and osmotic stress, which is caused by a decreased in a water uptake. Plants sequester and exclude Na+ in order to lessen the negative effects of salinity when under iconic stress. Numerous ion transporters associated with plants stress tolerance have been identified during the previous 20 years. These transporters, the high-affinity K^+ transporter (HKT) and the Na^+/H^+ antiporter (NHX, on the vacuolar membrane), are crucial for the uptake, Long-distance transport, and redistributing of Na^+ and K^+, respectively.

The Ca^{2+} ATPase (on the plasma membrane and in vacuoles), NSCC, Ca^{2+}/H^+ antiporter (CAX), Vacuolar H^+ phosphorylase (VP), H^+-PPase, and the plasma membrane H^+ ATPase pump are also involved in maintaining $Na^+/K^+/Ca^{2+}$ homeostasis under salt stress (Figure 6) Na^+ or Na^+ K^+ transporters called HKTs are essential for the plants to be able to withstand salt. It can lower the shoot's sodium content and maintain a steady potassium level in living things. In the heterologous expression system, HKT proteins are divided into two groups: those that transport Na^+ and those that transport K^+. HKTI mediates Na^+ uptake by functioning as a Na^+ transporter when external K^+ is low, and HKT_2 functions as a K^+- Na^+ cotransporter. The proton gradient created by H^+-ATPase and H^+ pyrophosphatase drives NHXs, which consequently promotes Na^+ transport across the vacuolar membrane. NHXI in *Arabidopsis thaliana* is the first NHX gene found to date. It has been proposed the overexpression of VP plus NHX or VP alone will strengthen the proton gradient produce by H^+ ATPase and H^+-PPase, hence improving the efficiency with which the NHK protein transports Na^+ into the vacuole. Researchers have found a connection between the activation of calcium-and salt-sensitive (SOS) signaling genes and pathway and plants' ability to adapt to salt and maintain cellular homeostasis. Certain genes, such as the calcium sensor SOS3/CBL4, are activated upon binding Ca^{2+} and subsequently from a protein complex with the kinase SOS2/CIPK24 (Figure5).

The SOSI (Na^+ antiporter) gens are ultimately activated by the protein complex that forms. It is the SOSI gene that causes the salt outflow in cells. With the help of Ca^{2+}, the Na^+/H^+EXCHANGER 1(NHX1) antiporter sequesters Na^+ into the vacuole. Likewise, rapeseed (Brassica napus L.) can be treated with multi-walled carbon nanotubes (MWCNTs) to increase the expression of antioxidant genes APX and SOD, as well as the transcripts that correspond to these genes, BnAPX, BNCU/ZNSOD, and BAMNSOD. This can lead to the re- establishment of ion and redox imbalances by reducing ROS and the Na^+/K^+ ratio. This advantageous outcome was likewise linked to the MWCNTs' production of an alteration in the transcripts corresponding to the K^+ transporter 1 (KT1) and Na^+/H^+ exchanger 1 (NHX1) Proteins. Se-NPs balance the cytosolic Na^+/K^+ ratio under salt stress in wheat by upregulating the expression of the genes NHX1, CAX1, SOS1, HKTI, H^+- ATPase, and AQPs. Additionally, this regulation is in sync with the overexpression of genes that encode antioxidant enzymes, including GR, SOD, MDAR, CAT, and APX, which aid in ROS detoxification and MDA reduction. In a similar vein, the foliar application of Se-NPs increased the expression of the phenylalanine ammonia-lyase gene, hence increasing the synthesis of phenolic compounds, non-antioxidant enzymes, and decreasing the levels of MDA and H_2O_2 in lemon plants grown in salinity. AgNPs derived from *V. negundu* plant extract were used to excessively express anthocyanin regulatory genes in *B. oleracea*, including PAP1, ANS, phenylalanine ammonia lyase (PAL). This led to an accumulation of anthocyanin, which was linked to an increase in CAT, APX, And SOD activity. According to different research, applying Fe-NPs and ZnO-NPs to tomato (*Solanum lycopersicon* L.) plants can enhance the expression of genes related to the antioxidant enzymes Cu, Zn-SOD, APX, And CAT2, which contribute in H_2O_2 detoxification and the maintenance of the Na^+/K^+ Balance. Implementing ZnO-NPs exogenously to rapeseed grown during salinity stress resulted in reduced ion leakage and boosted Hill reaction. This in turn affected the expression of genes related to stress response, such as AP2-related protein (ARP), and downregulated myc, p38 mitogen- activated protein kinase (MPK4), and SKRF2. These finding show that ZNO-NPs assist in lessening the negative effects of salt.

By regulating osmotic potential, osmosensors play a critical role in mitigating the effect of salt (Figure 6). The genes decreased hyperosmolality-include (Ca^{2+}) increase (OSCA1) and histidine kinase (AHK1) comprise the putative osmosensor of the Arabidopsis plant. Thus, osmoregulation results from the upstream regulation of osmosensing-related genes, water raises the potential and produces the synthesis of osmolytes. Numerous kinase families control the expression of genes that respond to stress. The main function of osmolytes is to maintain osmotic equilibrium and cellular structure by providing a constant flow of a water. Plants store proline, a necessary osmolyte that acts as a scavenger of reactive oxygen species (ROS) in response to salinity stress. Proline building is associated with the activity of the enzyme Pyrroline-5-carboxylic reductase (P5CR), which converts Pyrroline-5-carboxylic to protein. Proline keeps cell's cytoplasmic water content constant, safeguarding the structural integrity of macromolecules like enzymes. Proline regulates the AsA-GSH cycle, which shields cellular membranes from ROS-assisted oxidative stress, and functions as a precursor to other stress-response molecules. Intracellular proline increases the ability to withstand salt stress and act as a reserve of organic nitrogen for when the body recovers from the stress. Proline, an amino acid, increases the synthesis of proteins that respond to salt. Plant adaptation is facilitated by its ability to inhibit reactive oxygen species (ROS) and shield the photosynthetic system due to its antioxidant capabilities. The P5CS gene, which is activated in saline environment and raises proline concentrations, regulate proline biosynthesis. Several studies demonstrate that applying certain NPs can also upregulate the P5CS gene, which help raise proline levels in plant cells to improve resistance to osmotic stress.

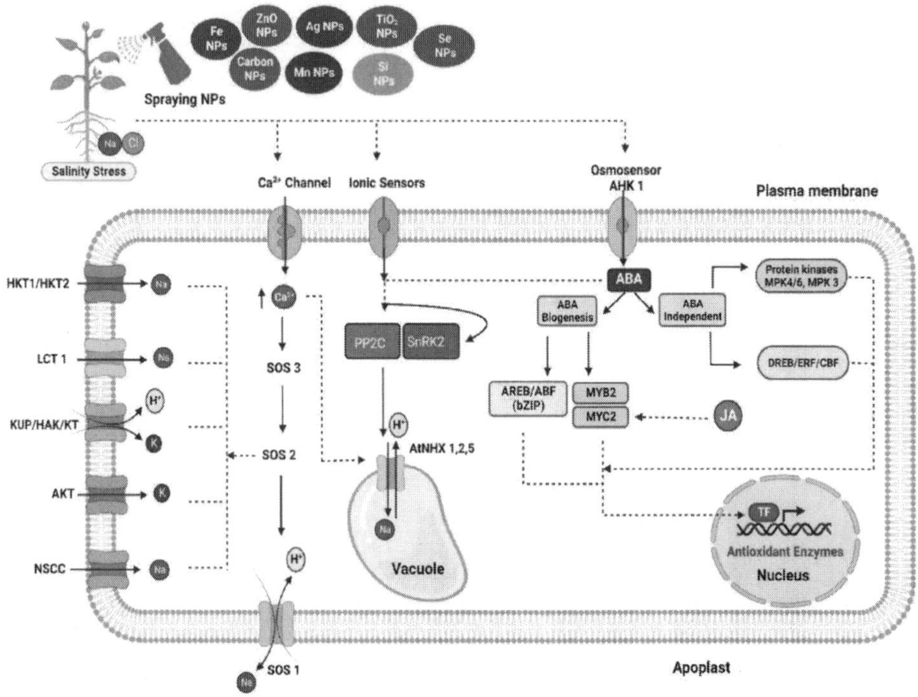

Figure 6. Overlapping of signaling pathways and NPs mediated antioxidative defense system regulation to mitigation salinity stress.

Plant tolerance to salt is associated with the activity of antioxidant genes that scavenge reactive oxygen species (ROS) (Figure 5). Transgenic rice that overexpresses OsMSRA4.1, OsVTE1, and OsECS has high salt resistance. The accumulation of ascorbate at elevated levels through enhanced seed germination and seedling growth is responsible for transgenic Arabidopsis's greater resistance to salt; this suggests that DHAR is essential for preserving the ascorbate reservoir within the cell. Elevated salt resistance has been seen in transgenic Arabidopsis plants expressing the cytosolic peroxidase gene OaAPXB or OsAPXa. Because OsAPXa-overexpressed plant is less resistant than those with high expression of OsAPXb, it is interesting that OsAPXB confers higher resistance to salt stress than OsAPXa by balancing H_2O_2 metabolism. Salt stress leads to the upregulation of antioxidative genes, including OsAPx8, OsMIOX, cCuZn-SOD2, OsGR2,3, OsSRO1c, and OsAPX7.

Transcription factor (TFs) is known to control and alter a broad variety of gene expression in response to variations in salinity. Through their interactions with promoter-specific cis-elements, the TFs drive the expression of genes. The Arabidopsis plant regulates about 428 transcription factors (TFs) in response to salt stress. In Aquilaria sinensis, salt causes 512 transcription factors (TFs) to be activated and 264 to be downregulated. Stress in plants causes the expression of several transcription factors' (TF) genes, such as myeloblastosis (MYB), WRKY, Apetala-2/enhancer-binding protein (AP2/ERF), homeodomain TF (HOX), basic helix-loop-helix (bHLH), and NAC. Many members of the AP2/ERF TF family, which is implicated in the response to salt stress, are produced by plants. Four transcriptional subfamilies make up this family: RAV, DREB, AP2, and ERF. Plant salt tolerance is mediated by both AP2 and ERF, which activate genes locate downstream of promoters. In Aquilaria, salt induced the upregulation of around 73 AP2/ERF families. A plant's resistance to salt can be increased by certain transcription factors (TFs), such as CBF2 (DREBIA/CRT/DRE-binding factor 3) and CBF2 (DREBIC/CRT/DRE-binding factor 2). The expression of 84 MYB TF, including MYB1, MYB4, MYB75, MYB39, MYB78, and MYB21, increased in response to salt stress in plants. In exposure to salt, the plant transcription factors WRKY17, WRKY25, and WRKY33 were increased. WRKY29, WRKY40, and WRKY75 were among the other TFs that were activated in result of salt stress. DNA binding motifs in TF genes such as HOX, bHLH, and NAC are present in salt-sensitive plants. According to earlier research, applying NPs under salt stress also appears to control these TFs, which aids in preventing the homeostasis of ROS.

6.2 NPs Mediated Antioxidant based Proteomics Response

Proteomics research is a useful technique for examining how plants adjust to salinity and is frequently used to learn more about the molecular processes behind plants' reactions to stress brought on by salt (Figure 7). As a plant adjusts to salinity, proteins involved in energy-producing metabolism of carbohydrates, defense mechanisms, secondary metabolic processes, cell integrity, photosynthesis, enzyme hunters of reactive oxygen species, and biosynthetic enzymes that produce osmolytes are all increased. When under stress, a lot of plants release a lot more molecular chaperone, such HSP70 and HSP90. Under salt stress, there was an increase of the proteins and enzymes involved in the glycolytic and tricarboxylic acid (TCA) cycle. Protein enzymes that participate in glycolysis include 6-phosphofructokinase 3 (PFK), pyruvate kinase isozyme A (PK), and pyruvate decarboxylase 2 (PDC); TCA cycle enzymes include PDC, malate dehydrogenase (MDH), succinate dehydrogenase (SDH), isocitrate dehydrogenase (IDH), and aconitate. As a result of the overexpression of these proteinaceous enzymes, the plant was able to enhance a variety of metabolic activities in response to salt stress, restoring normal glycolysis and the Krebs cycle while still providing sufficient energy. Antioxidant systems are well-known proteinaceous enzymatic systems that have a direct or indirect role in salt stress mitigation. Under salinity, the function of enzymes engaged in an antioxidant system such as CAT, thioredoxin peroxidase (TXN), GPx, SOD, and glutaredoxin sulfhydryl oxidoreductase increase and enhances, assisting in the prevention of H_2O_2 buildup and the scavenging of ROS. Furthermore, salinity activates all three SOD isoforms: iron-SOD (Fe-SOD), copper-zinc-SOD (Cu-Zn-SOD), and manganese SOD (Mn-SOD) in different parts of the cell. Additionally, salinity increases the activity of phytoene synthase (PSY) 2, a non-enzymatic antioxidant that neutralizes reactive nitrogen species (RNS) and ROS. Today, different research connected to NPs have proven that the administration of these NPs can increase the concentration and activity of antioxidant proteinaceous enzymes during salt stress. In previous wheat research, iron oxide nanoparticles increased antioxidant function by boosting the activity of enzymes involved in salt stress. The RuBisCO enzyme contributes greatly to the Calvin cycle by fixing CO_2. Enzymatic analysis of the small subunit of RuBisCO demonstrated that TiO_2-NPs applied topically greatly increased this enzyme by upregulating the activities of antioxidant enzymes under salt stress. The foliar application of Zn-NPs (0.03%) might increase the levels of antioxidant enzymes CAT, POD, and SOD that protect the photosynthetic components of okra (Abelmoschus esculentus L.). A separate investigation found that applying Fe-NPs and ZnO-NPs can upregulate SOD, APX, and GR genes, increasing the production of antioxidant enzymes in four tomato cultivars under different NaC1 stress conditions. This boosts the relative growth rat, dry matter percentage, osmotic potential, proline content, callus development, and shoot formation of tomatoes during salt stress.

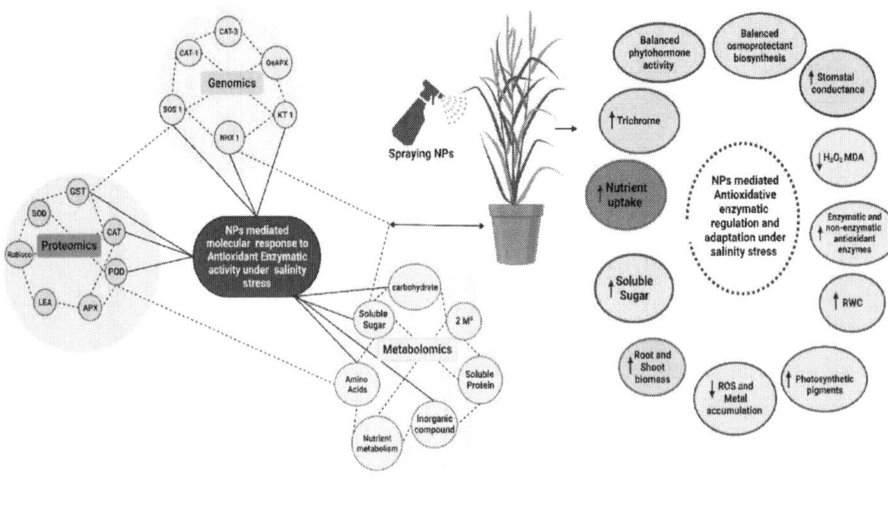

Figure 7. Diagrammatic representation of molecular level mode of action of NPs to mitigation the salinity stress effect on plants and their regulation.

Salinity also increases the rate of action of phytoene synthase, a non-enzymatic antioxidant that may scavenge ROS and reactive nitrogen species (RNS). Similarly, a 300 mg L^{-1} concentration of ZnO-NPs can alter phytoene synthase expression in Arabidopsis. Salt exclusion is further aided by a spike in transporters and protein channels, including aquaporins, K^+ transporters, chloride channels, SOS1, NSCCs, CNGCs, HKT1, NHX, V-ATPases, VSAC, and VPPases. The administration of different NPs upregulates these transporters and protein channels, which are also related with antioxidant enzymatic activity to help plants survive under saline circumstances.

6.3 NPs Mediated Antioxidant based Metabolomics Response

Metabolomics is an effective tool for studying how a plant's metabolism responds to environmental change and abiotic stress. It provides detailed insights into how plants cope with salt stress, allowing for a thorough examination of unique patterns associated with stress tolerance (Figure 7). The resulting metabolome plays a key role in processes such as maintaining osmotic balance, stabilizing membranes, scavenging reactive oxygen species (ROS), and defending against oxidative stress. A wide range of compounds, including amino acids, dipeptides, flavonoids, alkaloids, inorganic and organic acids, phytohormones, carbohydrates, sugars, and terpenes, constitute the metabolome. Metabolites, which are products of cellular regulatory mechanisms, are released by organisms in response to external stimuli. Plants release over 200,000 metabolites, which can be classified into two groups: primary and secondary metabolites, essential for their growth and development. While primary metabolites have a conserved structure across all plants, the chemical composition of plant secondary compounds is species-specific and varies. Molecular events monitor the physiological changes induced by engineered nanoparticles (NPs) in xenobiotic plants. These events reflect the involvement of metabolites in various biological pathways, including glycolysis, gluconeogenesis, the citric acid cycle, the biosynthesis of secondary metabolites and amino acids, and the metabolism of nitrogen and fatty acids. Plants produce metabolites as signalling molecules in response to various abiotic stresses, and they also alter soil chemistry and metabolic pathway to increase nutrient bioavailability. Some secondary metabolites, such as flavonoids, phenolics, anthocyanins, and carotenoids, act as non-antioxidant enzymes, helping to scavenge ROS produced in a saline environment. Several studies have shown that the use of various NPs can increase the content of flavonoids, phenolics, anthocyanins, and carotenoids. These substances are non-antioxidant enzymes that remove ROS by scavenging radicals like $^1O_2$ and •OH. Research found that copper, silver, and gold nanoparticles enhanced the quantities of flavonoids, phenolics, antioxidant enzymes, anthocyanins, and carotenoids serve as evidence indicating that the generation of free radicals is associated with an increase in the accumulation of these compounds. According to various findings, the presence of NPs induces the plant to create more antioxidant chemicals. When wheat was exposed to with Ag-NPs, the concentrations of indole-3-butyric acid, 1-naphthalene acetic acid, and 6-benzlaminopurine increased, while the concentration of abscisic acid (ABA decreased), implying that Ag-NPs influenced sprouting and grain production during salinity. Tomatoes planted with Si NPs maintained GSH (glutathione reductase) and chlorophyll concentrations, increased vitamin C levels in fruits, and improved PAL (phenylalanine ammonia-lyase) efficiency compared to untreated plants growing in salinity. Iron is an important inorganic element for all living species, involving plants. It is involved in a variety of important cellular functions, including chlorophyll production, respiration, and photosynthesis. Furthermore, iron is essential for the manufacture of numerous critical proteins involved in cellular respiration, photosynthesis, oxygen transport, metabolic pathway, and DNA repair, all of which have an influence on overall crop productivity. Fe also helps plants acquire stress tolerance. It has been discovered that Fe-NPs specifically control salt resistance by boosting important antioxidant enzymes. Carbon nano-horns (CNHs) sprayed on *Sophora alopecuroides* seedlings boosted their quality of total protein in roots and leaves, fresh weight of roots, cooper quantities, PS II efficiency in leaf, and soluble in leaves under salinity.

The metabolome analysis revealed that CNH-mediated salt tolerance was also linked to improvements in the tricarboxylic acid cycle and glycolysis, as well as resetting of nitrogen or carbon metabolic cycles for energy generation. CNH treatment enhanced the unsaturation of fatty acids in addition to increasing them, which helped to restore the membrane structure during salinity. In salinity settings (200 and 150 mM NaCl), CNPs (0.3%) enhance their impact on several lettuce varieties. By pre-treating the seed with polyhydroxy fullerene NPs, salinity in wheat plantlets was reduced. This was achieved by increasing antioxidant activity, P and K^+ concentrations, chlorophyll, amino acids, and soluble carbohydrates. The amount of oil (geraniol, geranial, geranyl acetate, and z-citral was increased by adding TiO_2 during salinity. Three different concentrations of titanium dioxide nanoparticles (TiO_2 NPs) (0.03%, 0.02%, and 0.01%) was applied via foliar spray on broad bean crops subjected to salt stress. It was observed that this treatment increased the levels of soluble sugar, proline, other amino acids, and enzymatic activities, thereby enhancing salt resistance in the plants.

7. NPs mediated antioxidant signaling pathways

7.1 Mitogen –activated protein kinases (MAPKs) pathway

ROS are a crucial molecule because they activate essential signal transduction protein known as mitogen-activated protein kinases (MAPKs) in response to a wide range of developmental and environmental stimuli. The mitogen-activated protein kinase (MAPK) cascade is a prototypical example of a signalling cassette, which is a conserved modular transduction system that distributes and combines input from several upstream signalling pathways to start a range of downstream reactions (Figure 6). The three kinases that comprise the core cascade are MAPK kinase, MAPK kinase, and MAPK. MSPKK is capable of phosphorylating and activating MAPK when MAPKKK is phosphorylates metabolic enzymes, other kinases engaged in signal transduction, and transcription factors.

A function of phosphatases is to attenuate signals. MKK5 regulates the expression of SODs in Arabidopsis when there is an excess of salt present. It is important to note that MKK5, which functions upstream of MPK6 and downstream of MRKK1, is also required for the expression of FSD2 and FSD3 in chloroplasts under salinity. Furthermore, MKK1 promotes the transcriptional activation of CAT1 under salt stress conditions, but not CAT2 or CAT3. MPK6 acts downstream of MKK1 to initiate CAT1 expressions, which is crucial for regulating ABA-mediates H_2O_2 generation. Genetic modification of MKK2-MPK4/MEKK1-MKK1 signaling transduction also results in the downregulation of the expression levels of many antioxidant enzymes. Mutants of Arabidopsis containing the genes encoding the parts of these pathways show a board range of expression levels for CSD1, CAT1, CAT2, GR and APX1. The expression of CAT1 and CAT3 in mekk1 and mekk1/mekk2 mutants, while the exception of the mpk4 mutant, demonstrates the complexity of MAPK signaling towards antioxidant genes and possible cress-talk across several MAPK signaling pathways. Because of its ability to trigger MKK1/2 and MKK5, followed by MPK3/6 and MPK4, MEKK1 seems to play a significant role in regulating antioxidant defense. MAPKs (MKK1/2/5; MPK3/4/6) are now thought to function as proteins that specify signals for certain antioxidant enzymes when they operate downstream of MEKK1. Overexpressing the GhMPK2 and GhMAP3K40 genes in cotton enhances its tolerance to salt, while overexpressing the ZmMKK1 and ZmMKK4 genes in maize promote salinity kinase activities, thereby enhancing plant salt tolerance. However, this aspect of the phenomenon has been less investigated.

7.2 Ca+ Signaling pathways

In plant cells, Ca^{2+} functions as a secondary messenger far more extensively than cyclic nucleotides do. While cells maintain considerable stocks of Ca^{2+} (1-2 mM) in the vacuole and endoplasmic reticulum (ER), the concentrations of Ca^{+} in the cytosol ($[Ca^{2+}]$ cyt) are maintained at relatively low levels. The concentration of Ca^{2+} in the cell walls is around 0.5-1mM. $[Ca^{2+}]$ cyt is measured using transformed cells producing fluorescent or luminescent dyes that respond to Ca^{2+}. This technique has shown that a wide range of biotic and abiotic factors can cause $[Ca^{2+}]$ cyt surge to approximately 1m. These factors include pathogen infection (as previously mentioned), nodulation factors, salinity, altered gravity pattern, less or high temperatures, a shortage of water, wind, red and blue light, the hormones ABA and GA, UV-B rays, and anaerobic atmosphere. ROS and Ca^{2+} have a reciprocal interaction in which the former regulates the cytosolic Ca^{2+} concentration and the latter needs Ca^{2+} to produce (Figure 6). Therefore, it is crucial for Ca^{2+}- dependent ROS signaling through NOXs as key signaling centers and for Ca^{2+}- dependent ROS signaling through direct regulation of Ca^{2+} channels and sensors to be mutually reinforced during plant immunological responses. The significance of Ca^{2+} signaling for antioxidant enzymes was demonstrated by experiments utilizing exogenous Ca^{2+} or genetically engineered Ca^{2+} transporters. $CaCl_2$ improves rice plantlets resistance to arsenic stress by lowering ROS concentration and raising DHAR, CAT, SOD, MDHAR, and glutathione peroxidase activity. Ca^{2+} dependent ATPases, or Ca^{2+} transporters, play a role in a variety of signaling pathways connected to stress as well as physiological events such as stomata closure, ROS balance, and programmed cell death. Plants that overexpress OsACA6 and Ca^{2+}-ATPase exhibit improved resistance to biotic stressors. This is because, after upregulation of CAT, APX, and GR activities, OsACA6 can modify ROS concentrations during salt stress. Reports suggest that Calcium-Dependent Protein Kinases (CPKs) regulate the function of catalase (CAT) in response to salinity stress. Specifically, CPK12 has been found to regulate the accumulation of reactive oxygen species (ROS) and iron equilibrium through the functions of ascorbate peroxidase (APX), superoxide dismutase (SOD), and CAT, thereby influencing salt stress tolerance in Arabidopsis. Additionally, CPKs have effects on other components of the ROS-degrading machinery. Consequently, OsCPK12 improves rice's ability to withstand salt by reducing the amount of reactive oxygen species (ROS) due to elevated expression of OsAPX8, OsAPX2, and decrease expression of OsBOH1, a kind of NOX. However, OsCPK12 negatively affects rice's ability to withstand blast disease. According to researchers, a complex regulatory network including several Ca^{2+}- sensing proteins work together to transmit Ca^{2+} signals to target substrates. The relevance of Ca^{2+} signaling pathways in responses caused by nanoceria and salt stress has been better clarified.

7. Limitation, Risk and Challenges of NPs Application in Agriculture

Because of their incredibly tiny size, NPs are particularly helpful in mitigating the harmful effects of ROS at the molecular, enzymatic, biochemical, and non-enzymatic levels when there is a salt stress. On the other hand, NPs can be hazardous to the environment and destroy plant life if they are sprayed in sufficient amounts. Breathing NPs that have been applied to plant leaves or soil can have an impact on the deep lung tissue. Because NPs are so tiny, they can cause oxidative stress, lipid peroxidation, genotoxicity, protein fibrillation, and inflammatory reactions when they build up in organs such as the liver, kidneys, and lungs. Because of these pollutants, harmful impacts on the environment, researchers mostly concentrated on silver NPs, ZnO NPs, nano-zerovalent NPs, and TiO_2 NPs. Implementing NPs in an agroecosystem presents a number of challenges, including toxicity, appropriate dosage, bioavailability, and an inadequate regulatory framework. When utilized improperly, NPs may be hazardous to the environment and have both beneficial and detrimental impacts on plant development, including poor plant physiology. When NPs are sprayed in sufficient amounts, they create reactive oxygen species, which harm plants through oxidative stress. It has been observed that plants respond physiologically, morphologically, proteomically, and morphologically to high concentrations of NP. The biological and geological processes that occur in soil control the bioavailability and ecotoxicity of these chemicals. Numerous scientists have spoken about what influences the toxicity and activity of NPs in soil. Relevant factors include plant species, NP content, NP shape, NP size range, NP surface charge, and NP type (carbon-based, metallic, non-metallic, and magnetic). The growth and development of plants are significantly impacted by the concentration of NPs. There is ample evidence of the risks that higher the NP application to people, animals, and plants. The broad application of NPs for the alleviation of plant salt and/or environmental stressors may be limited as a result of the previously noted safety issues. Therefore, intervention techniques must be used to reduce the threats that nanotechnology poses to the ecosystem.

8. Conclusion

For the sake of sustainable agriculture, crop yield loss caused by salinity stress needs to be handled with suitable environmentally friendly alternatives (Figure 1). Modern farming methods can be improved, agriculture productivity and product quality can be increased, and the world's population can be fed without sacrificing food supply, thanks to nanotechnology. A number of NPs are being investigated for their potential to improve agricultural productivity and plant development while lessening the harm that abiotic stress causes to plants. NPs such as TiO_2, SiO_2, ZnO, Cu, Fe, and Ag-NPs stimulate ROS generation and phytotoxicity in plants, hence activating defense mechanisms against the harmful effects of salt stress. Nanoparticles (NPs) are small enough to penetrate plant tissues easily, exerting beneficial effects on various morphological, physiological, and biochemical processes. They enhance plant growth and development and increase crop productivity, even under conditions of salinity stress. Moreover, NPs' large surface area makes a variety of particular nutrients easier to absorb and distribute. Nevertheless, there aren't many excellent researches in this area, and the application of NPs for crop improvement and sustainable agriculture is still in its infancy. Further study is required in the following areas to lessen the detrimental effects of NPs on ecosystems and crops: (a) how NPs interact with the plants and metabolic processes at the molecular and cellular levels; (b) how NPs affect different plant species and their potential toxicities; and (c) how NPs affect gene control and expression in plants.

CHAPTER 6

ROLE OF NANOTECHNOLOGY IN HEAVY METALS STRESS MANAGEMENT

Short Summary

Effective remediation techniques are required due to the serious danger that metal and metalloid pollution represent to human health and natural ecosystems. Methods based on nanoparticles (NPs) have drawn a lot of interest as effective ways to eliminate heavy metals from a variety of environmental matrices. The current review concentrated on the utilization of carbon-based nanomaterials, metal oxides, bio-based NPs, and valent iron in green manufactured NPs-mediated remediation. The mechanism of NPs interactions with heavy metals, such as adsorption, precipitation, and redox reaction, were also examined in the review. Important variables that affect the effectiveness of the remediation, such the size, content, and surface change of the NPs, are methodically investigated. Additionally, a critical assessment is conducted on the environmental destiny, transport, and potential dangers related to the application of nanoparticles. Additionally, the chapter emphasizes the many sources of metal and metalloid pollution, their effects on human health, and their translocation in plant tissues. There are suggestions for opportunities and difficulties in transferring NPs-based remediation from lab studies to practical uses. The present result will aid in guiding future investigations to support the long-term use of metal and metalloid removal.

1. Introduction

Soil contamination is a serious problem due to the long-term persistence and toxicity of metal and metalloid (HMMs) contaminants. The rapidly expanding human and technological activities are adding agrochemicals, possibly hazardous metals, and an excess of nutrients to the soil. Due to the widespread use of chemical fertilizers, farmyard manure, sewage sludge, atmospheric deposition, and the quickening pace to industrialization, HMMs are found in agricultural soils. Cadmium, arsenic, mercury, nickel, zinc, lead, copper, and chromium are among the pollutants found on more than 20 million hectares (ha) of land worldwide. According to the Agency for Toxic Substances and Disease Registry (ATSDR), four heavy metals—mercury, lead, arsenate, and cadmium—are very detrimental to plants and people. HMMs may infiltrate plant system and contaminate food chain. This poses a major risk to human health as well as food quality.

For agricultural development to occur, soil is necessary because it enable plants to absorb nutrients. Despite the fact that farming increases human well-being and is necessary for existence, it usually has a detrimental impact on the environment. The growing demand for natural resources by a growing human population presents a conflict between the benefits of arable land and the requirement for sustainable land management, a topic that has been extensively discussed in the scientific literate. Land pollution threatens people's livelihood, sustainable development, and standard of living. Soil preservation is of utmost importance due to the growing population and shrinking amount of arable land caused by technological activities. Pollution of the air, water, and soil pose a serious threat to human health and ecosystems worldwide. Numerous investigations have revealed the detrimental effects of HMMs on a wide range of species, including microorganisms, human cells, and plants. Researchers claim that HMMs have phytotoxic effects that impair the development and structure of plant cells. When cells are directly exposed to HMMs, they undergo cellular oxidative stress, which results in DNA damage and protein denaturation.

In the past few years, a variety of in-situ and ex-situ remediation technique have been used to enhance HMs in contaminated soils. Surface capping, vitrification, bioremediation, solidification, soil flushing, and electrokinetic extraction are a few of these. Physicochemical remediation techniques that are conventional include slowly releasing hazardous substances into the environment without any treatment. Furthermore, bioremediation employs biological processes, like as microorganisms and plants, to lessen the toxic effects of HMMs in contaminated environments. However, these approaches have a lot of disadvantages. In real- world field settings, for instance, they function poorly, are costly to run, take a long time to finish, do not offer the best conditions for microorganisms, and enhance the bioavailability of pollutants both spatially and temporally. Their application is determined by the following factors: public acceptability, operational cost, time required, geography of the polluted site, soil pollutants characteristics, effectiveness, and readiness for these traditional remediation procedures to be implemented. New adsorbents that are affordable, nontoxic, recyclable, and ecologically friendly must be created in order to immobilize HMMs on a large scale. A novel and possibly helpful strategy known as "nanoremediation" has emerged in response to these challenges. It entails using engineered nanoparticles (NPs) to clean up contaminated areas.

Contaminated regions may be cleaned more effectively and economically with nanoremediation. NPs have a high surface area-to-mass ratio, distinct electrical and catalytic properties, and responsiveness. NPs help in environmental cleanup mostly through chemical reduction and catalysis. Additionally, because of their vast surface area and uniformly dispersed removal techniques. NPs' tiny particle size and dispersion capabilities make them perfect for remediating soil and water. Membranes based on NPs have also been used in water nanofiltration: these membranes effectively prevent larger particles from passing through their pores. Furthermore, the separation of smaller molecules is facilitated by the contact with the membrane. Metal oxides, carbon nanotubes, quantum dots, and biopolymers are among the nanomaterials employed for remediation of water, soil, and air. Thus, we may address every aspect of HMM remediation based on nanotechnology, from fundamental to molecular level, in this study.

2. Global Status of HMMs Problem

Ecosystems can contain both naturally occurring and artificially created HMMs. Precipitation, snowfall, volcanic eruptions, and wildfires are examples of natural sources; fertilizers, fuel combustion, mining, construction, deforestation, and other industrial activities are examples of man-made sources. Arsenic is one of the prevalent elements that surround humans. The element in question is the twelfth most prevalent elements that surround humans. The element in question is the twelfth most prevalent in the human body, the twenty most prevalent in the crust of the Earth, and the fourteenth most prevalent in water by itself. Australia has an arsenic (As) concentration of more than 10,000 milligrams per kilogram, while the US is home to 41% of the 1.4 million arsenic-poisoned sites worldwide, according to the US Environmental Protection Agency. Apart from surpassing the threshold of 10 parts per billion, established by the US Environmental Protection Agency (EPA) for arsenic concentration, a number of shallow reservoirs and bore wells in Pakistan are contaminated with arsenic. There have been reports of groundwater pollution with arsenic in a number of nations worldwide. The toxicity of the groundwater in South Asian nations is a reason for concern. Numerous locations have reported seeing this occurrence, including the Bengal Basin in India, Bangladesh, and Pakistan; the Terai Belt in Nepal; the Mekong Basin in Cambodia and Vietnam; and the Red River Delta in Vietnam. As a result, at least 100 million people in these countries are at risk of developing cancer and other asbestos-related illness. Compared to trivalent arsenic, pentavalent arsenic is easier to dispose of. Global output of Cr is estimated to have increased to 105.4 million metric tons since the beginning of the industrial revolution, a significant rise from the 1950s. Cr has mainly escaped attention in pollution studies due to its poor rates of uptake and transfer by plants. In natural settings, thus, phytotoxicity, and the buildup of Cr in food chains are uncommon phenomena.

Cadmium is a hazardous metal that has detrimental effects on all living things, making it a serious environmental risk. Compared to zinc, it is more mobile at pH 4.5 and 5.5. However, once the pH rises over 7.5, it becomes stagnant. Over the past 130 years, soil Cd concentrations in the UK have been rising continuously; the last 20 years have seen the most notable increase. As per a data source, the high consumption of rice cultivated in irrigation water tainted with cadmium has made environmental exposure to the metal particularly troublesome in Japan. Lead is a dangerous environmental toxin that may seriously damage a variety of body systems. A large portion of the housing stock constructed prior to 1978 has lead paint. The federal government of the United States made lead-based illegal in 1978. The EU's 2003 Restriction of Hazardous Substances Directive further restricted the use of lead.

3. Sources of HMMs

Scientists distinguished between two main categories of HMM sources: synthetic and natural. According to scientists, natural sources include sedimentary rocks, volcanic eruptions, soil formation, and rock weathering, whereas anthropogenic sources include industry, agriculture, mining, and residential effluents. However, pollution indicators may be used to adequately describe geogenic and anthropogenic soil contamination. It should be noted, nonetheless, that in many situations, source apportionment can be difficult even with the use of sophisticated research methodologies. Alloway cited a variety of HMM sources and origins, including aerosol particles sedimentation, rainfall containing HMMs, and agrochemicals. Although several other metals were discussed, the study's main focus was on HMMs like Cd, Pb, Cu, and Zn.

3.1 HMMs Natural Sources

The most common natural sources of heavy metals are thought to be sedimentary and igneous rock. The ranges of igneous and sedimentary rock's heavy metal contents, expressed in parts per million (ppm). It has been found that different types of rocks have different amounts of the same elements, and that these proportions also vary from one type of rock to another. Even for parts found in the same kind of rock and the conditions of the environment around them. Along with the deposition of sediments in rivers, the development of soil is also considered to be a major factor in the accumulation of heavy metals.

3.2 Anthropogenic HMMs Sources

Wastewater from human-induced sources, including mining, agriculture, and industry, can include heavy metals. Such events, like smelting, which releases Cu, Zn, and As pesticides, which also releases As; burning fossil fuels, which releases Hg; and vehicle exhaust, which helps release Pb, are major contributions to the rise in the concentration of heavy metals and pollution in the ecosystem. Furthermore, the regular human activities of farming, manufacturing, and other industrial processes upset the biosphere's natural rhythm.

3.3 Agricultural HMMs Sources

Since the formation of the globe, HMs have been a component of the Earth's crust. Moreover, as seen in figure 1, Soil erosion, floods, volcanic activity, sediment resuspension. Metal corrosion, geological weathering, and metal evaporation from soil and wastewater are the main natural processes that contribute to HMM contamination in agricultural soils. In recent times, there has been a remarkable increase in the levels of heavy metals due to growing urbanization and industrialization. In the poor world, HMM pollution in terrestrial areas, particularly agricultural fields, is a serious problem. The main cause of heavy metal pollution are anthropogenic activities such as foundries, mining, smelting, leather industries, sewage, leaded paints, burning crop residues, petroleum distillate spills, care, landfills, animal manure, wastewater irrigation, and runoff from various factories. Secondary sources of heavy metals (HMs) in agricultural areas include pesticides, fertilizers, herbicides, and other crop improvement products. The HMMs cannot be broken down by typical physiochemical processes, hence they will remain in the soil indefinitely since they are not biodegradable. When plants take up HMMs from the soil along with other nutrients and water, they can make their way up the food chain. The utilization of HMMs damages soil quality and renders agricultural land unusable by intensifying the effects of soil emissions and altering the characteristics of the soil, such as pH, porosity, natural chemistry, and color.

By incorporating organic matter into the soil and providing it with different nutrients that plants require to flourish, fertilizers increase plant development and productivity. Soil productivity is raised by fertilizers. Fertilizer comes in two primary varieties: inorganic (man-made) and organic (natural). Ammonium fertilizers (sulfate and nitrate) are produced as organic or biofertilizers after the anaerobic digestion (AD) process. Fertilizers that are manufactured synthetically or chemically, sometimes referred to as inorganic fertilizers, are composed of both inorganic materials and chemical compounds. The usage of chemical and organic fertilizers directly contributes to the presence of HMMs in the soil. Although phosphorus is an essential component of many fertilizer formulations, when added to soil, it also greatly increases the accumulation of heavy metals. The precipitation of metal phosphates, which occurs when water-insoluble phosphorus fertilizers are applied, is essential for immobilizing metals in soil and produces phosphate rocks. Excessive fertilizers use results in heavy metal deposition in agricultural soils, which reduces soil fertility and reduces crop yields and plant development. Once HMMs have contaminated the soil, it is very difficult to restore the environment. Fertilizer use over an extended period of time enhances the possibility for Cu, Zn, and Cd to accumulate in agricultural soil. Phosphate fertilizers, liming materials, and bio-fertilizers are the main types of inorganic fertilizers that release heavy metal into agricultural soil for plant uptake. Consequently, animals and ultimately humans eat them.

4. Impact on plant and human health

4.1 Impact of HMMs on Plant Health

HMMs impede the growth and development of plants since their density is five times greater than that of water. HMMs are widely present in plants-soil interactions, although their concentration varies geographically. While many plants organs require HMMs in low concentrations to function properly, when HMMs levels exceed the acceptable threshold, plants are at risk. The perception of HMMs as significant pollutants only intensifies with time. While certain metals, like Cd, Cr, and Pb, are not very useful to plants and can even be poisonous if present in excess for an extended length of time, others, including Cu, Mn, Fe, Zn, Co, and Ni, are required for plants growth as micronutrients (Fig 1). HMMS may enter plants through adventitious roots, where they can impair growth, interfere with essential biological functions like photosynthesis, and eventually be harmful to the health of the plant. According to a recent study, barley roots may experience growth retardation and radial cell proliferation as a result of poisoning.

These morphological changes are brought on by the buildup of reactive oxygen species (ROS), such as hydrogen peroxide, which causes cellulose deposition and cell death. The toxicity can lead to cytoskeleton damage, delayed cell division, formation of reactive oxygen species, damage to the plasma membrane, and decreased plant growth. Similarly, due to A1 toxicity, black gram (*Vigna mungo* L.) Hepper displayed indications of oxidative damage, cellular damage, and root injury. According to different research, the substitution of magnesium ions in chlorophyll and the absence of ferric ions caused by Cd absorption reduce photosynthesis in plants. Plants are toxic to high levels of Cd in soil, and this can have negative impacts on growth at both the physiological and morphological levels. A reduction in mitotic activity in meristematic cells is indicative of cadmium poisoning and may impede the growth of plants. HMMs are poisonous and seriously impede plants' ability to absorb essential nutrients from the soil. For instance, by generating insoluble complexes and obscuring the absorption site, an excess of Cr in the soil might decrease the amount of calcium, phosphorus, magnesium, and iron that plants absorb. Cr has the potential to alter a plant's nutritional balance; nevertheless, in Citrullus plants, moving Cr to various parts led to higher phosphorus and manganese concentrations and lower concentrations of iron, zinc, copper, and sulfur in the leaves.

4.2 Impact of HMMs on Human Health

Toxicological exposure to As, Pb, Cr, and Cd in humans is the most common kind of HM-related harm. The ingestion of paints containing lead, contamination of the air or water, and inadequately covered food containers are all possible sources of mercury poisoning. Numerous significant human diseases, including cancer, Parkinson's disease, Alzheimer's disease, renal pathology, respiratory difficulties, mental abnormalities in children, visual problems, depression, dementia, and neurological disorders, can be brought on by exposure to heavy metals (HMs) (Fig 1). After eating tainted food, HMMs are incredibly tenacious throughout the human body. It was shown that long-term exposure to Cd has detrimental effects on the quality of semen, spermatogenesis, sperm motility, and hormone production in males. Numerous studies have discovered negative consequences of cadmium exposure on women's reproductive health.

5. Uptake and Translocation of HMMs in plants from the Soil

The risk that HMMs pose to natural systems has risen because to rising urbanization and globalization. The health of people and animals as well as plant and agricultural production are all negatively impacted by an increase in HMM concentration in landscapes. Pb, As, and Se are examples of irrelevant HMMs, where Zn, Cu, Fe, and Mn are examples of important HMMs. Agricultural yields are lowered by these elements' increased quantities in the soil, even though crops require them for growth and development. The farming sector does not employ unnecessary HMMs like Cd, Hg, and Pb for any particular biological purpose. Because to the harmful effects of HMMs on plant growth, plants experience chlorosis and necrosis, which eventually cause a lack of chlorophyll and stomata to close, causing the plants to die. Additionally, exposure to contaminated soil, water, and agricultural spray can lead to a buildup of these chemicals in crops, which can result in health problems for people such as kidney disease and cancer.

Plants absorb both required and non-essential HMMs through the usage of protein transporters in their membranes. Numerous transporters, including as aquaporin, IRT, NRAMP, Cu-transporter, ZRT, and low-affinity divalent ions, are involved in the uptake of Ca and Cd by plants. When HMMs infiltrate plant cells, transporters of heavy metals (HMs) including P1B-type ATPase, HMT, NRAMP, and cation diffusion transporters are detected in vacuoles, Golgi bodies, and the endoplasmic reticulum. HM isoprenylated plant proteins (HIPPs), phytochelatins (PCs), metallothioneins (MTs), chemicals, ligands and compartmentalization are among the mechanisms that reduce or remove the harmful effects of HMMs.

5.1 Arsenic (As)

A mixture of naturally occurring and human-induced activities, such as mining, is increasing the quantities of the lethal metalloid arsenic (As) in soil. As is found in soil at different quantities and in a range of oxidation phases. While As^{3+}, an inorganic form, is more dangerous and soluble, both organic and inorganic chemicals in the environment are harmful to plant growth, Since, As depends on accessibility, plants mostly absorb As in an inorganic state (As^{3+}, As^{5+}) via transporters. During uptake, the plant goes through a number of physiological, structural, and optical changes. It is challenging to regulate as several critical element transporters have an impact on plant uptake. Extended exposure modifies physiology (Chlorosis), phenotypic (decreases in growth), and metabolic rate. It may have detrimental effects on plant growth, such as the production of reactive oxygen species (ROS), protein degradation, chlorosis, tissue necrosis, and the loss of essential nutrient intake, which can result in plant mortality.

5.2 Nickle (Ni)

Ni (nickel) is a mineral that plants need, but only in modest mounts (0.05 to 10 mg kg^{-1} dry weight), as large concentrations of Ni in the environment are detrimental to plant growth and development. Increased Ni concentrations in crops have been shown to impede growth, promote leaf senescence, reduce nitrogen absorption, and alter the concentration of Fe. Although the concentration of Ni in soil varies depending on the conditions, it is frequently more than 20 mg kg^{-1}. Ni^{2+} levels in contaminated soils are 20-30 times greater than in clean soils. Human activities such as fertilizing with phosphate, using pesticides, burning coal and oil, emitting pollution from industry, and mining lead to an increase in Ni^{2+} levels. The roots have the ability to actively diffuse or transport Ni throughout the vegetative system. Ni travels from the root system to the leaves and shoots after being moved into the xylem). Ni is a pollutant that plants may absorb and transport both actively and passively because of its negative effects, which include protein degradation, enzyme activity suppression, and damage to DNA.

5.3 Chromium (Cr)

Seas and the uppermost layers of the earth contain Cr. It is typical to find the transition metal Cr in the trivalent (Cr^{3+}) and hexavalent (Cr^{6+}) oxidation states. Compared to Cr^{3+}, Cr^{6+} is less volatile, which makes it more transportable but potentially more dangerous. Plants that take up Cr from the soil may experience negative effects on their physiology. There has been no validation of studies on the particular mechanisms of Cr absorption and translocation. However, compounds that are mixed with organic acids allow roots to take them up. Only a tiny portion of Cr is absorbed by crops through their roots and reaches various tissues and organs. The concentrations of Cr in several plant organs rose when Pisum sativum L. was raised in a solution containing potassium dichromate. From germination until the roots, stems, and leaves start to grow, a plant's total moisture content and physiological activity can be impacted by Cr exposure.

5.4 Cadmium (Cd)

Cadmium (Cd) poisoning affects marine life, terrestrial animals, and humans. Growing industrialization and industry density are associated with rising ambient Cd concentrations. The release of pesticides, mining activities, and human-caused chemical industries are the main sources of cadmium in the atmosphere. People, animals, and plants are at considerable risk for health products due to the buildup of mercury in the environment. When soil is polluted, lead may move. Lead poisoning damages people's bones, liver, and kidneys and reduces their ability to absorb calcium. Plants may readily absorb cd from subterranean water, and it is then carried to the emergent tip by a variety of transporters via the phloem. It is transferred from soils to roots by transporters, such as those in the rice OsNRAMP family, OsIRT1, and Arabidopsis AtIRT1. While OsLCT1 transports Cd from the stems to other tissues, AtHMA4 and AtHMA2 move zinc and copper from the root to the shoots. It is then carried to different tissues by specialized transporters called ATPases via the phloem. It physically, phenotypically, and physiologically, cadmium damages plant growth.

By decreasing biomass, reducing seedling development, and shortening root tips, Cd lowers plant production. It has been demonstrated that Cd (0.06 g/kg)-induced soil toxicity from potato crops lowers biomass and crop yield. Chlorophyll concentrations drop, reactive oxygen species (ROS) rise, and pant membranes and cellular macromolecules are harmed by Cd. Cd causes chlorosis and necrosis of the leaves in plants due to blocked mineral uptake. Cd obstructs the essential aspects of absorption and transportation needed for plant development. It was discovered that cauliflower plants exposed to Cd had shorter cultivars and signs of insufficient leaves. This feature has been seen in soybeans, radishes, and lettuce. The most recent study indicates that Cd poses a hazard to global health because it negatively affects plant development, oxidative damage, and ROS formation. A large body of scientific research, including data on chlorophyll content, growth, seedling germination, enzyme activity, nutrient intake, ROS, plant transpiration ratios, and crop production, supports the detrimental effects of Cd on crops.

6. Conventional and NPs Based Techniques for HMMs Remediation

6.1 Conventional Techniques

6.1.1 Physicochemical Method

Physical and chemical techniques like ion exchange, filtration, solvent extraction, evaporative recovery, electrochemical treatment, electrodialysis, electrokinetics, landfilling, chemical oxidation, chemical leaching, chemical reduction, and mechanical metal separation could all be used to help with this remediation. These procedures have several downsides, including the creation of toxic waste products, significant solvent requirements, and incomplete removal of metal. They tend to affect the soil in addition to being bed for the ecology. In addition, they take a lot of money, time, and energy.

6.1.2 Bioremediation Method

There are techniques for both in-situ and ex-situ bioremediation. In situ bioremediation techniques eliminate pollutants from the location without trying to remove soil. These techniques do away with the necessity to remove contaminated samples from the environment by using the microbial system's metabolic ability to eradicate contaminants form the surrounding environment. According to researchers, bioremediation involves soil from its original position, treating it, and then replacing it. Compared to in situ cleanup techniques, *ex situ* remediation is more costly.

6.2 NPs Based HMMs Remediation

Bioremediation, both in situ and ex situ, has benefits and drawbacks. Phytoremediation is one of the most economical, sustainable, and successful forms of bioremediation. The phrase "Phytoremediation", which comes from the Greek words "phyton" for plants and "remediare" for "to remediate," refers to a method wherein specific plants work with soil microbes to transform toxins into forms that are safe and frequently lucrative. The discovery of a variety of wild plants—many of which are indigenous to naturally mineralized soils and accumulate significant amounts of metals in their leaves—inspired the idea of using plants to remove dangerous metals from polluted soils. The plant kingdom is resilient to xenobiotic pollution in that it can withstand relatively high amounts of xenobiotic chemicals without experiencing adverse effects. Another example of this resilience is its ability to quickly absorb toxins and convert them into less dangerous metabolites.

Plants that absorb significant concentrations of heavy metals from their environment are known as hyperaccumulator plants, or plants that are hyper-enriched in heavy metals. The use of this technique to clean up locations contaminated with harmful metals and chemicals is growing in popularity. For phytoremediation to be successful, contaminants must be present in plant root zones. The main mechanisms by which plants carry out remediation include phytoextraction, phytodegradation, phytostabilization, phytovolatilization, and rhizofiltration. The inexpensive and ecologically benign process of "nanoremediation" may be used to cleanse soil and other environmental HMM contaminants. It has been demonstrated that this novel remediation approach is effective in getting rid of heavy metals via absorption, decreasing the harmful valence to a stable metallic state, and speeding up the process. The technical procedures used in nanoremediation to apply NMs to remove or immobilize HMMs from contaminated soils include adsorption, heterogeneous catalysis, photodegradation, deployment of electrical fields (electronanoremediation), and involvement of microorganisms (nanobioremediation). A range of NPs, including metallic, metallic oxide, carbonaceous, polymeric, and nanocomposites, have been successfully used to remove heavy metals.

For example, it was reported that the heavy metal Cd and the hazardous organic solvent dichlorobenzene were removed from the soil by the carbon nanotube (CNT) nano-sponges. In general, the effectiveness of removing Cd and other contaminants from soil is increased by iron and sulfur-containing nanosponges. The capacity of CNTs to immobilize HMs ions (Ni^{2+}, Zn^{2+}, Pb^{2+}, and Cu^{2+}) in soil remediation was examined by researchers in a similar manner. The results of this study show that even a small amount of CNT addition may significantly increase adsorption capacity and increase HM immobilization in contaminated soils. A different study found that treating soil with $Fe_3(PO4)_2$ NPs effectively immobilized Cd by 70%. This investigation also demonstrated that a major mechanism for reducing the bioaccumulation of lead in soil is the production of cadmium phosphate.

7. Nanotechnological Approaches for Remediation of HMMs

7.1 Mode of Action of NPs for Mitigation of HMMs: Insights into Mode of Action

The management of heavy metal molecules (HMMs) and nanoparticles (NPs) by plants relies significantly on their interactions with the environment, plant physiology, and the stability of these substances. The behavior of NPs is influenced by their properties. Additionally, the morphological and physiochemical traits of NPs also contribute to their absorption. Strategies such as decreasing the bioavailable concentration of HMMs in soil, enhancing plant defense mechanisms, improving physiological functions, and regulating the expression of HMM transport genes are effective in helping plants combat HM stress. Plant cells facilitate NP entry through both plasma membrane channels and direct membrane penetration. These NPs serve a signaling function by inducing the expression of genes involved in the plant's stress response.

Furthermore, NPs have the capability to modify or sequester HMMs within the soil, thereby reducing their availability and movement. For instance, a study demonstrated that the mobility of cadmium was reduced when Fe_3O_4 NPs were applied observed high surface reactivity, electrostatic attraction, large surface area, and presence of capping molecules in FeO NPs facilitated efficient absorption of Cd from the soil. The competition for the same transporter pathway within cells results in a reduction of metal ion concentrations in plants, as Cd^{2+} and FeO NPs vie for transport. Additionally, scientists conducted an investigation and discovered that the electrostatic attraction, reactivity, and large surface area of Cu NPs hindered the translocation of Cd from the soil to the aerial parts of wheat plants. The presence of capping molecules on biological Cu NPs may have aided in the immobilization of Cd in the soil, with proteins coating the Cu NPs potentially contributing to Cd immobilization. This competition may lead to an increased demand for Cu-bound potentially contributing to Cd immobilization. This competition may lead to an increased demand for Cu-bound nutrients, consequently triggering a growth spurt in wheat. To facilitate the conversion of cadmium into more stable forms, recent research has highlighted the crucial role that mercapto Si NPs can play. A detailed examination of how HMs toxicity induces an increase in intracellular ROS levels will be provided later. Concerning HMMs, the mechanisms underlying ROS generation are distinct. An elevation in intracellular ROS production serves as an indicator of HM toxicity. However, cadmium (Cd), classified as a redox-inactive HMM, may only indirectly generate ROS by enzyme inactivation and increased lipoxygenase (LOX) production in plant tissues, unlike copper (Cu), which can directly do so. The plant's antioxidant defense mechanisms are activated in response to HMM stress. Nonetheless, excessive ROS levels can disrupt cellular processes, potentially accelerating senescence when combined with ethylene and jasmonic acid (JA). Intracellular ROS accumulation is countered by antioxidant enzymes such as peroxidase (POD), superoxide dismutase (SOD), ascorbate peroxidase (APX), catalase (CAT), glutathione peroxidase (GPX), and glutathione reductase (GR). Besides enzymes, low-molecular-weight non-enzymatic metabolites have been shown to effectively counteract ROS. The body's mechanisms for clearing ROS are activated in response to adversity. To mitigate oxidative stress, plants synthesize shikimate-phenylpropanoid and alanine, while metabolizing galactose, ascorbate, and aspartic acid. CeO_2 NPs, Mn_3O_4 NPs, and Fe_3O_4 NPs, activated by antioxidant enzymes, have been found to reduce crop losses. Researchers successfully stimulated the antioxidant defense system and prevented Cd accumulation in rice growth under hydroponic conditions by using a foliar application of CeO_2 NPs (200 mg/L). Researchers observed a negligible impact on rice seedling biomass with TiO_2 NPs when grown hydroponically. The addition of TiO_2 NPs alleviated the negative effects of Cd stress on rice seedlings, leading to improvements in root length, plant height hormone levels, antioxidant enzyme activity, and other physiological parameters. Plants respond to stress induced by heavy metal molecules (HMMs) by producing specific amino acids, proteins, genes, and signaling molecules. Recent studies have demonstrated an increase in the production of heat shock proteins (HSPs) in response to exposure to HMs.

There is evidence suggesting that proline (Pro) plays a role in detoxifying HMMs, as higher levels of proline (Pro) were detected in the Cd hyperaccumulator plant Solanum nigrum compared to the nanoaccumulator plant Solanum melongena. Roots secrete chelating substances such as phytosiderophores, nicotianamine, and organic acids, which may impact HM uptake.

The mitogen-activated protein kinase (MAPK) cascade signaling pathways, known to be involved in various stress conditions like salinity and drought, have also been observed to respond to HMMs stress. Exposure of Lucerne (*Medicago sativa* L.) seedlings to Cu or Cd ions activates four different MAPKs: SIMK, MMK2, MMK3, and SAMK. Recent reports suggest that the application of nanoparticles (NPs) can also regulate the expression of genes related to HMMs stress. Researchers found that nanoscale zero-valent iron (nZVI) increased the levels of superoxide dismutase (SOD) and peroxidase (POD) in sunflower leaf tissue, decreased HM formation, and promoted plant development. Expression of Fe transporters responsible for both Fe and Cd absorption was dramatically down-regulated after the application of Nzvi to rice seedlings. Cd was sequestered in vacuoles due to the overexpression of the OsVIT1 and OsCAX4 genes. Additionally, the addition of Si NPs significantly increased the percentage of viable rice cells in a Cd solution, possibly through the up-regulation of Si uptake genes (OsLsi1) and the down-regulation of Cd uptake and transport genes (OsLCT1 and OsNramp5).

8. Conclusion

The current chapter highlights the significant increase in metal and metalloid pollutants in the environment, particularly in soil systems, due to various natural and anthropogenic sources. This poses alarming concerns for food safety and human health, as these elements also degrade soil health, leading to production of low-quality food. Traditional approaches for the elimination or stabilization of these pollutants in soil and preventing their accumulation in plant tissues have been extensively discussed, revealing limitations and inefficacies. In contrast, discussed nanotechnological approaches have demonstrated effectiveness and eco-friendliness, showing promising outcomes in eliminating metal and metalloid pollutants. In summary, nanoparticle-based remediation emerges as a promising solution for addressing environmental contamination. The review emphasizes crucial mechanisms, recent advancements, and the importance of considering environmental impacts. While recognizing the potential benefits, a balanced approach is necessary, taking into account both efficacy and environmental considerations. Future research should prioritize scalable and cost-effective technologies, fostering interdisciplinary collaboration for practical implementation. This review lays the groundwork for the sustainable and responsible use of nanoparticles in remediating metal and metalloid pollutants.

CHAPTER 7

GENOME EDITING TOOL AND TECHNIQUES FOR CROP IMPROVEMENT WITH COMBINATION APPROACHES OF NANOTECHNOLOGY

Short Summary

For a considerable period, the enhancement of new plant traits, particularly stress tolerance to abiotic and biotic factors, has heavily relied on labour-intensive breeding methods. Genome editing (GE) has emerged as a prominent set of molecular techniques enabling precise modifications at specific locations within the genome to enhance desired traits swiftly. This approach not only facilitates the exploration and fundamental understanding of plant molecular mechanisms but also enables the attainment of diverse objectives aimed at enhancing crop quality and yield. Through GE tools, researchers have been able to identify stress-responsive genes, elucidate their functions, comprehend associated signalling pathways, and ultimately utilize this knowledge to enhance the stress tolerance of existing crops. Novel GE technologies such as meganucleases (MN), Zinc Finger Nucleases (ZFNs), Transcription Activator-Like Effector Nucleases (TALENs), Clustered Regularly Interspaced Short Palindromic Repeats/ Cas9 (CRISPR/Cas9), Base Editing (BE), and Prime Editing (PE) have proven themselves as efficient and reliable tools for crop improvement. This review outlines the development of first and second-generation GE tools and their contributions to the genetic enhancements of plants. Furthermore, it discusses the advantages and applications of CRISPR/Cas9, BE, and PE compared to other GE techniques. The chapter thoroughly examines the main challenges associated with integrating second-generation GE tools like CRISPR/Cas9, BE, and PE with other crop improvement strategies and explores their prospects.

1. Introduction

In recent years, plant biology research has focused on enhancing different agronomic traits such as yield, abiotic and biotic stress tolerance, and increasing the nutritional value of food to address global food security concerns. This research has significantly contributed to understanding various biological mechanisms and elucidating the roles of genetic and epigenetic factors. Traditional breeding methods have been utilized by breeders to develop improved varieties of several crops, but these methods are challenging and time-consuming due to the lengthy process of germplasm selection.

Genome editing (GE) has emerged as a more precise alternative to traditional breeding techniques and conventional genetic engineering strategies. It enables accurate modifications to DNA by targeting specific nucleotides within the genome of living organisms. The recent advancements in GE technologies have overcome the limitations of conventional breeding programs, ushering in a new era of crop improvement.

Numerous issues have been brought to light, such as the need for effective and targeted genome control of crops that are difficult to propagate by conventional breeding techniques. In this direction, a number of molecular and genetic strategies and processes have been toughly studied in order to acquire knowledge that has allowed researchers to replicate in vitro experiments and develop novel techniques for manipulating DNA through the creation of numerous vectors and delivery tools. The ability to precisely replace a dominant allele with a substitution and integrate a foreign gene into a specific region of the genome has been made possible by recent GE methods.

From 1994 to 2019, many GE techniques were developed. These include base editing (BEs), prime editing (PE), zinc finger nucleases (ZFNs), transcription activator like effector nucleases (TALENs), clustered regularly interspaced short palindromic repeats/Cas9 (CRISPR/Cas9), and engineered endonucleases/meganucleases (E/MNs) (Fig.1). These techniques are essential for plant research because they allow future crops to be modified. The discovery of the "zinc finger", a protein domain that functions as site-specific nucleases that cut precisely specified areas of DNA when paired with FoKI endonuclease domains, in 1996 sparked interest in these methods. A triplet of nucleotides that make up a chimeric protein with a modular assembly is identified by the "zinc finger" domain. Later on, this technique was used to modify cultured cells for both model and non-model plants. Novel GE techniques, like TALENs and CRISPR/Cas, were produced via persistent research and development. Site-specific nucleases allowed for high GE efficiency across eukaryotic species, which has revolutionized biological research and its applications in a variety of crops.

While creating TALENs necessitates the re-engineering of new proteins for every target, creating CRISPR is rather straightforward. Following the introduction of CRISPR further genome editing (GE) tools such as base editing (BE) and prime editing (PE) were developed. These techniques were more focused on specific targets than CRISPR/Cas9. By employing these editing tools to manipulate the genome effectively, the intricate issues associated with creating transgenic and mutant plants might be resolved. This study covers the development chronology (CRISPR, BE, PE, GE) tools, as well as their application, safety evaluation, and regulation of GECs.

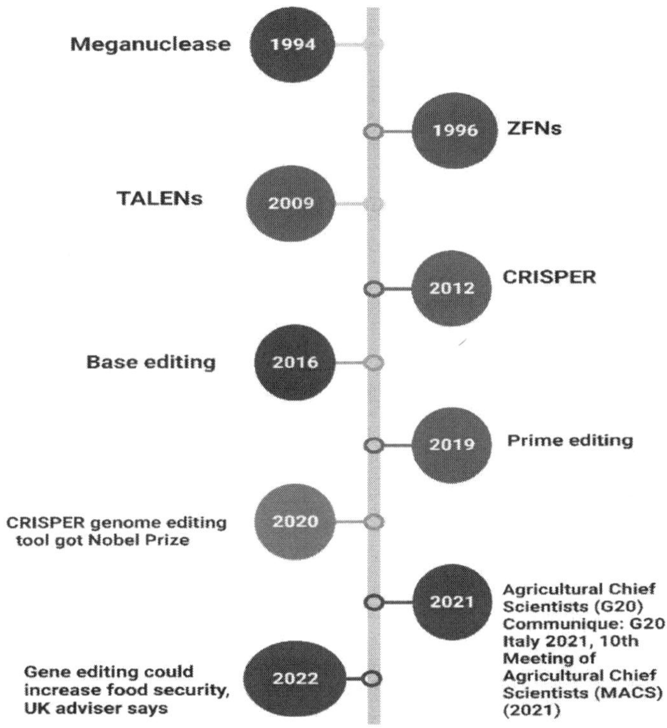

Figure 1. Timeline of first- generation MNs, ZFNs, TALENs, and second-generation CRISPR, BEs, PE, GE tools from 1994 to 2022. The CRISPR GE tool got Nobel Prize in 2020 in Chemistry, developed by two female scientists Emmanulle Charpentier and Jeniffer A.Doudna. Agricultural Chief Scientists (G20) Communique: G20 Italy 2021, 10th Meeting of Agricultural Chief Scientists (MACS) was organised in Italy held a special discussion panel on the potential benefits of genome editing for food production, environmental preservation, and climate change adaptation ("G20 Italy 2021, 10th Meeting of Agricultural Chief Scientists (MACS) Communique")

2. GE in Crop Improvements

Numerous developments in molecular biology have made it possible to precisely modify the genome at certain locations. The two domains of the engineered nucleases are the non-specific nucleases domain and the sequence-specific DNA binding domain. Sequence-specific and non-specific DNA binding domains are merged. Target DNA nicks may be produced by precise and site-specific cleavage, which can be filled in at the cleaved location by fused nucleases at a particular site using non-homologous end joining (NHEJ) and homology-directed repair (HDR). NHEJ is widely used, however, its effectiveness is limited by drawbacks such as allowing off-targets and producing insertions and deletions of varying sizes during repair. However, because HDR uses homologous DNA to repair double-strand breaks, it is more accurate and reliable than NHEJ. Here, genomic sequences are manipulated by sequence-specific nucleases (SSN) that introduce or remove certain nucleotides at the designated locus. Additionally, it significantly raises homologous recombination's effectiveness, allowing for precise gene replacement operations.

Genetic engineering has advanced significantly in recent years thanks to the development of MNs, ZFNs, TALENs, CRISPR/Cas9, and PE. These GE tools have the potential to alter the genome through gene knock-in, knock-out, and mutagenesis. Plant genome engineering has advanced significantly since 1996, when homologous recombination was first shown in plant tissues using SSNs, discovered ZFNs. Plants underwent the first effective SSN-based mutagenesis following the inactivation of genes using ZFNs. As a result, ZFNs are widely employed in genetic engineering to produce double-strand breaks (DSB) that are particular to certain sites. Genetic modification by Ens was promoted by the addition of TALENs to the SSN toolset for encoded genome editing. The first CRISPR/Cas9-based GE event in eukaryotes occurred recently as a result of the addition of CRISPR/Cas9, an emerging GE nuclease, to the toolbox. The aforementioned methods have the potential to produce changes by site-directed substitutions, insertions, deletions, and replacements at specific genomic locations.

First-generation genome manipulation technologies, such as TALENs, ZFNs, and MNs require a long target specificity protocol and require a lot of time to process. Furthermore, ZFNs, TALENs, and MNs cannot be constructed without protein engineering. As a result, MNs, ZFNs, and TALENs have not been used extensively in plants. But compared to first generation editing technologies tools like CRISPR/Cas9, BE, and PE are more reliable, less costly, and easier to build.

2.1 Meganucleases (MNs)

According to research, meganucleases are a novel class of nucleases that are often used in plant genetic engineering. Homing endonucleases are a class of enzymes that cause double-strand breaks in multiple genomes. Given that

MNs can recognise target DNA sequences of between 12 and 40 bp, they make up one of the most effective delivery systems among all the vectors that make up plant RNA viruses. Aside from having the most maximal genomes, MNs are less harmful to cells than other GE techniques like ZFNs. Although the MNs are ideal for genetic engineering because of their large detection site, their nature is restricted and they cannot adequately cover loci that may be of interest.

In addition to their many benefits, MNs have a number of disadvantages, including the need for an initial protein engineering stage in order to produce a conventional meganuclease and the laborious and costly process of assembling sequencing-specific enzymes for all feasible sequences when compared to other SSNs. However, the main factor preventing MNs from being reconstructed is the non-modular characteristics of certain proteins, which restricts the usage of MNs to MNs that are naturally prevalent in plants. It has been shown that MNs have a significant impact on crop improvement in *A. thaliana*, *G. hirstum*, and *Z. mays*; yet MN modification is difficult and time-consuming.

2.2 Zinc Finger Nucleases (ZNFs)

The discovery of zinc finger nucleases dates back to 1996, however, its practical application didn't start until around 2002. It has frequently been used to change genes in several plants, potentially resulting in genetic modifications in *G. max*, *Z. mays*, *N. tabacum*, and *A. thaliana*. It was used to induce mutagenesis and targeted cleavage in Drosophila. The ZFNs are chimeric proteins that contain two domains: the N-terminal Cys2-His2 zinc finger (DNA binding) and the C-terminal FokI restriction enzyme cleavage domain (Fig.2A). ZFN enzymatic activity requires the dimerization of the FokI domain. It results in double-strand breaks at certain loci of the target site. Three to four zinc fingers make up DNA binding domains, and each finger uses binding to recognize a DNA sequence that is three base pairs long.

Numerous studies have reposted successful ZFN-based GE in plants. For example, expression of a ZFP-TF for KASII gene activation reduced palmitic acid and total saturated fatty acid content and increased total C18 in B. napus seeds; GsZFP1 overexpression improved Medicago sativa's resistance to salt and drought; and the development of artificial zinc finger nuclease (AZFNs) effectively cleaved the target sequence and prevented the growth of various begomoviruses in *N. tabacum*. In hexaploidy *T. aestivum*, resistance to imidazolinone herbicides has been established by ZFN-mediated GE of acetohydroxyacid synthase.

Nevertheless, it has a number of drawbacks, including labor-intensiveness, a narrow range of specifically designated target domains, and poor target specificity, which makes it technically more difficult and increases the likelihood of non-targeted editing.

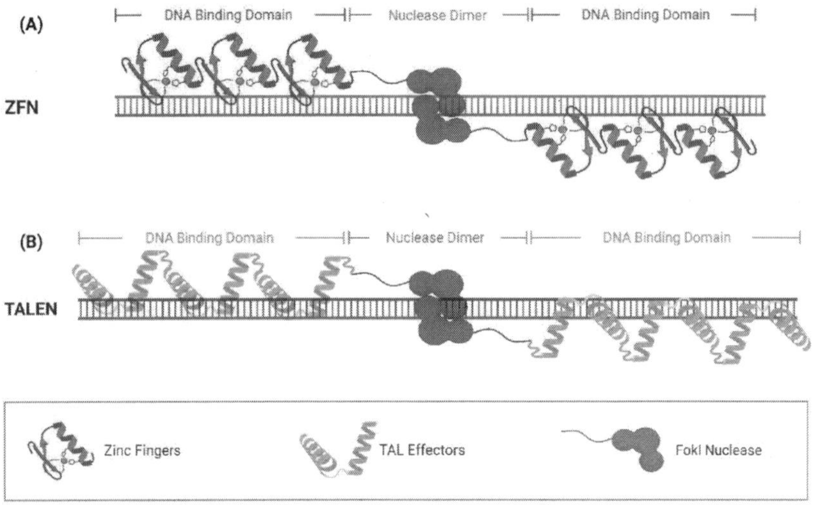

Figure 2. Diagrammatic representation of (A) ZFN base GE. The Zinc finger (DNA binding) domain leads to activation of FokI nucleases and subsequently cut double-stranded DNA (target site). (B) In TALENs base, GE techniques use TALENs proteins composed of two parts: the first part is TALE that targets the protein to a specific DNA sequence and the second part is a nuclease (N) that cuts DNA. FokI is a nuclease that is commonly used in TALENs. The outcome of breaking of target DNA and repair may create small insertions or deletions and through this process target DNA will be able to acquire new functions or lose old ones.

2.3 Transcription Activator-like Effector Nucleases (TALENs)

First discovered in plant pathogenic bacteria (*Xanthomonas campestris*), the TALENs were found. It is shown that chimeric proteins resembling ZFNs (Fig. 2B) are also known to include the FokI endonuclease and the DNA-binding domain of TAL effectors. TALENs can attach to any target DNA sequence and cut the material at specific locations when they team up with a nuclease. Because TALENs may only target a single nucleotide at the target location, they are incredibly accurate. These are easily designed, have a high success rate because of cleavage activity, and have an infinite targeting range, making them useful for genetic engineering in plants.

A number of studies have demonstrated the efficacy of TALENs-based GE in plants, including the generation of disease-resistant *O. sativa* through mutation of its S gene, OsSWEET14/Os11N3, the targeting of an OsBADH2 gene for fragrance improvement, the production of G. max oil with increased oleic and reduced linolenic acid by targeting FAD2-1b, FAD3a genes, resistance to *Xanthomonasoryzae pv*. Oryzae in *O sativa*, the removal of cytoplasmic male sterility (CMS) associated genes (orf79 and orf125) in O. sativa to treat CMS, and decreased levels of steroidal glycoalkaloids (SGAs) in S. tuberosum. Thus, to improve agricultural plant productivity, quality, and nutritional content, TALENs may be highly useful GE technology (Table 1).

2.4 Clustered Regularly Interspaced Short Palindromic/Cas9 (CRISPR-Cas9)

The CRISPR-Cas9 system is one of the most extensively used and quickly researched genetically engineered technologies (GE) in science today. In bacteria and archaea, there is an innate, site-specific nuclease system that serves as an adaptive immune system to protect against viral infections such as bacteriophage virus. First-generation GE using CRISPR/Cas9 was carried out in 2013 on *O. sativa, Arabidopsis, N.benthamiana, Sorghum bicolor, and T. aestivum.* ZFNs and TALENs, two more GE techniques that were widely employed before CRISPR, were utilized to produce novel crop characteristics. However, there are certain drawbacks to both of these methods, such as the need to create a unique nuclease set for every genomic target. The scientific community has embraced the CRISPR system quickly because of its ability to overcome these restrictions.

2.4.1 The Mechanism of CRISPR/Cas9

The two fundamental components of the synthetic CRISPR system are a guide RNA and a CRISPR-associated endonuclease, or Cas protein. A gRNA is a synthetic short RNA composed of the scaffold sequence needed to bind Cas protein. It has a length of 100 nucleotides and around 20 nucleotide sequences at the 5′ end, which are crucial for identifying the target DNA. Additionally, it has a protospacer adjacent motif (PAM) sequence, mostly consisting of the nucleotide G (guanine) and any other nucleotide N. PAM acts as a binding signal for Cas9, and gRNA contains a loop-like structure at its 3′ end that interacts with Cas9 to produce a compound structure by anchoring with the target sequence via the guide sequence. As seen in figure 3, this complex binds to the target dsDNA and forms DSB at the binding site, which is subsequently repaired by the DNA repair technique HDR and NHEJ.

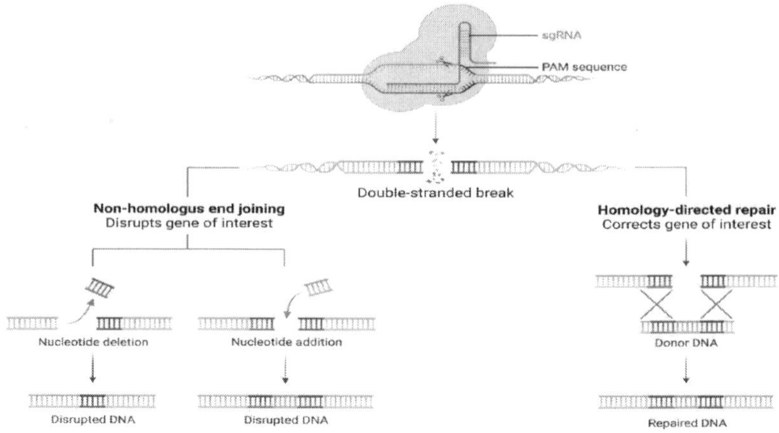

Figure 3. A brief overview of the CRISPR/Cas9 mediated-genome modifications. This GS system depends on RNA guided nuclease, Cas9 to introduce doble-stranded breaks in target DNA. A single guide RNA (gRNA), whose 20 nucleotides match the target DNA is essentially required for the cleavage of DNA in a sequence-dependent manner. Cas9 cleavage generates DSBs, which are able to repair through NHEJ or the HR pathway. Through this repaired mechanism target DNA can lose (Gene KNOCK-OUT) or gain (Gene KNOCK-IN) their function.

1. **Normal modification Using Non-Homologous End Joining (NHEJ)**: Double-strand DNA is broken by the non-homologous end-joining (NHEJ) repair process. It is called non-homologous because it can repair itself without the aid of a homologous template by simply ligating the ends together. In contrast to HDR, the non-homologous end-joining repair pathway is active and straightforward. At the DBS site, it frequently results in little nucleotide insertions and deletions known as indels. Because NHEJ-based DSB repair creates a variety of mutations, its arbitrariness is crucial. Small indels in the target DNA caused by insertions, deletions, or frameshift mutations in amino acids during NHEJ base repairing cause the premature maturation of stop codons in the target DNA's Open Reading Frame (ORF), which results in "gene knock-out", or loss of function within the targeted gene (Fig 3). Gene knock-out is an authenticating technique that allows one to investigate the role and function of a particular gene in the target organism.
2. **Precise Modification Using Homology Directed Repair (HDR):** For DSB with a higher fidelity rate procedure, homology-directed (HDR) is an incredibly effective and desired repair mechanism. When applying the HDR process for GE, a template for DNA repair is needed, and Cas9 nickase and gRNA (s) are used to insert the desired sequence into the

target cell. Anticipated modified and additional homologous sequences for the targets upstream and downstream are included in the inserted template.

2.4.2 Role of CRISPR/Cas9 method of genome editing in crop improvement

More than 20 crop species have used the CRISPR/Cas9 method of genome editing to date for a variety of agronomic traits, such as abiotic and biotic stress management with yield enhancement. This technology modifies plant breeding by covering the method in favor of a fresh viewpoint for crop improvement. Numerous studies have been published that announced the use of CRISPR/Cas9 to target specific genes involved in these agronomically significant crop attributes, improving abiotic stress tolerance, nutritional quality, and yield quality (Table1).

Table 1. Utilization of first and second-generation GE methods in various crops to upgrade abiotic, biotic stress tolerance, nutritional values and produce quality.

Plant Species	Targeted Gene(s) and Genome	Results	GE Techniques
Zea mays	MS26	Male sterile	EMNs
Gossypiumhirsutam	EPSPS	Herbicide resistance	EMNs
Glycine max	DCL	Herbicide resistance	ZFNs
Z. mays	PAT IPK1	Herbicide resistance Herbicide tolerance	ZFNs
Oryza sativa	OsQQR	Safety assessment of herbicide resistance	ZFNs
Malusdomestica	mutated uidA gene	Expressing GUS regenerated and developed transgenic plant	ZFNs
Triticumaestivum	TaMLO TaMLO-A1, TaMLO-B1, TaMLO-D1	Fungal resistance	TALENs
Solanumtuberosum	ALS	Herbicide resistance	TALENs
O. sativa	OsBADH2	Fragrant rice	TALENs
G. max	FAD2-1A, FAD2-1B	Low PUFA fats	TALENs
Glebioniscarinatum	DMC1	Male and female sterility	TALENs
O.sativa	Mitochondrial genome Os11N3 OsBADH2	Cytoplasmic male sterility Bacterial blight resistance High fragrance content	TALENs

Plant Species	Targeted Gene(s) and Genome	Results	GE Techniques
Z. mays	zMLpk	Lowering the phytic acid amount	CRISPR/ Cas9 and TALENs
O. sativa	OsERF922	Rice blast resistance	CRISPR/ Cas9
	OsSWEET13	Bacterial blight resistance	CRISPR/ Cas9
	OsRAV2	Salt stress tolerance	CRISPR/ Cas9
	SBEI, SBEIIb	Higher amount of amylose	CRISPR/ Cas9
	OsMYB108	Higher lignin content	CRISPR/ Cas9
	OsSh1	Elucidates Seed Shattering Reduction during Rice Domestication	CRISPR/ Cas9
	OsCRS2	OsCRS2 is necessary for early chloroplast development in rice	CRISPR/ Cas9
	DRO1	Modified root angle of rice to enhance yield in Saline Fields	CRISPR/ Cas9
	FLN2	FLN2 is vital for seedling growth as well as for tolerance to salinity stress.	CRISPR/ Cas9
	Acetolactate synthase (ALS)	Chlorsulfuron-resistant plants	CRISPR/ Cas9
Z. mays	ARGOS8	Drought stress	CRISPR/ Cas9
	Acetolactate synthase (ALS)	Chlorsulfuron-resistant plants	CRISPR/ Cas9
	Acetolactate synthase (ALS)	Sulfonylurea herbicide resistant	CRISPR/ Cas9
T. aestivum	TaGW7	Increase Grain Shape and Weight	CRISPR/ Cas9
N. benthamiana	BSCTV genome	Virus resistance	CRISPR/ Cas9
	TYLCV genome	Virus resistance	CRISPR/ Cas9
	AGO2	Virus resistance	CRISPR/ Cas9

Plant Species	Targeted Gene(s) and Genome	Results	GE Techniques
M. domestica	Mutate DIPM-1, DIPM-2, and DIPM-4	Resistance to fire blight Disease	CRISPR/ Cas9
	Mutate PDS gene	Increased lycopene	CRISPR/ Cas9
Citrus sinensis	Mutate PDS gene	Lycopene engineering	CRISPR/ Cas9
Populustomentosa	Mutate PDS gene	Lycopene engineering	CRISPR/ Cas9
	Mutate 4 coumarate: CoA ligase (4CL) gene	Decreased lignin amount	CRISPR/ Cas9
Musa balbisiana	B genome	Resistant to banana streak virus (BSV)	CRISPR/ Cas9
Solanumlycopersicum	SlPHO1;1	Phosphate nutrition of tomato seedlings	CRISPR/ Cas9
	SlNPR1	Drought tolerance	CRISPR/ Cas9
	Acetolactate synthase (ALS)	Chlorsulfuron-resistant plants	CRISPR/ Cas9
S. tuberosum	Acetolactate synthase (ALS)	Chlorsulfuron-resistant plants	CRISPR/ Cas9
	PDS (PHYTOENE DESATURASE)	Transgene-free potato	CRISPR/ Cas9
Citrulluslanatus	Acetolactate synthase (ALS)	Tribenuron-resistant plants	CRISPR/ Cas9
	Clpsk1	Enhanced watermelon resistance to *Fusariumoxysporum.*	CRISPR/ Cas9
Cucumissativus	eIF4E	Cucumber vein yellowing virus (Ipomovirus) infection and resistance to the potyviruses Zucchini yellow mosaic virus and Papaya ring spot mosaic virus-W.	CRISPR/ Cas9
Lactuca sativa	LsGGP2	Editing the uORF of LsGGP2, which encodes a key enzyme in vitamin C biosynthesis, increased oxidation stress tolerance and ascorbate content	CRISPR/ Cas9

Plant Species	Targeted Gene(s) and Genome	Results	GE Techniques
N. tabacum	Acetolactate synthase (ALS)	Chlorsulfuron (Cs) resistance	CRISPR/Cas9
Arabidopsis thaliana	Acetolactate synthase (ALS)	Imidazolinone herbicide resistance	CRISPR/Cas9
O. sativa	SBEIIb NRT1.1B ALS SPL14 Pi-d2	High amylose High nitrogen use efficiency Herbicide resistance High yield Blast resistance	CBEs
Lycopersiconesculentum	ALS	Herbicide resistance	CBEs
S. tuberosum	ALS	Herbicide resistance	CBEs
C. lanatus	ALS	Herbicide resistance	CBEs
A. thaliana	ALS	Herbicide resistance	CBEs
O. sativa	ACC	Herbicide resistance	ABEs
	ALS NRT1.1B SPL14, SPL16 SPL18, SPL17	Herbicide resistance High nitrogen use efficiency High yield	ABEs
A. thaliana	ALS	Herbicide resistance	ABEs
B. napus	ALS	Herbicide resistance	ABEs

3. Enhancing Crop Biotic Stress Tolerance

For many crops, including *O. sativa, T. aestivum,* and *G. hirsutum,* illnesses caused by pathogens and pests are a major problem because they affect crop production and seed quality, and in some situations, they can result in total yield loss. In many crops, the response mechanism of plants is quite intricate, and several genes are linked to this process. Although transgenic techniques effectively over-express particular genes to produce crops more resistant to a variety of viruses, only modest progress has been made since no one gene significantly influences how plants respond to illnesses. However, it seems that plants with undesirable gene knockouts are more resistant to infections. With this goal in mind, genome-edited mutants that may acquire plant tolerance to a variety of plant diseases have been created using CRISPR/Cas9. CRISPR/Cas9 methods have been applied in several recent studies to improve plant resistance to bacterial and viral infections.

CRISPR/Cas9 has emerged as the most popular SSN-based technology for gene-specific GE crop production in recent years. For instance, altering a

CRISPR/Cas9 SSN directed at the OsERF922 gene has increased the resilience of O. sativa crops against rice blast disease. In a different instance, it was demonstrated that CRISPR/Cas9 technology was effective in changing the expression of the sucrose transporter gene OsSWEET13, which is a PthXo2-related disease susceptibility gene and necessary for the development of rice bacterial blight. In *N. benthamiana*, a model plant used in virus-based research, the lack of appropriate genetic mutations significantly reduces AGO_2's contribution to viral immunity.

Therefore, the CRISPR/Cas9 technique was utilized to knock out the AGO_2 gene in *N. benthamiana* in order to ascertain the precise function of AGO_2 in plant immune responses against viruses such turnip crinkle, turnip mosaic, and potato mosaic. The B genome of the progenitor *M. balbisiana* and its descendants is known to incorporate the endogenous Banana steak virus (eBSV), severely restricting banana breeding efforts by rendering them ineffective for crop breeding projects. In order to prevent eBSV sequences in the plant host genome from converting into infectious viral particles under water stress, the CRISPR/Cas9 technique may also be used to delete the genes that code for the hormones that plants release and which expose them to a variety of illness. It is well known that phytosulfokine (PSK), a precursor of disulfate pentapeptide, inhibits plant immune responses. By using the CRISPR/Cas9 technology to knock out the Clpsk1 gene, which codes for the PSK, *Citrullus lanatus* plants became more resistant to Fusarium wilt disease.

4. Enhancing crop abiotic stress tolerance

The CRISPR/Cas9 system is being actively employed to increase crop resilience to abiotic stressors. According to recent research, the RAV TF family, which is located in the AP2/ERF domain, plays a part in adaptation to abiotic stress. The functional characterization of O. sativa RAV gene, OsRAV2, towards salt stress was made possible using the CRSPR/Cas9 method. Using the method for deletions and targeted alterations in the gene's promoter region, the function of the GT-1 element- which is situated 664 nucleotides downstream from the likely translation start position- in controlling the salt response of OsRAV2 was discovered. Using the CRISPR/Cas9 editing method, the role of heterotrimeric G protein mutants shown increased salt tolerance and yield components has been identified. While other G protein mutants showed increased salt tolerance, the dep1 and gs3 G protein mutants had acceptable agronomic characteristics. Similarly, O. *sativa's* tolerance to alkaline stress has been enhanced by ppa6 gene knockout. By monitoring the rise in grain output during periods of reduced water availability, the improvement in drought tolerance of crops was assessed. Furthermore, larger yields are produced by *Z.mays* decreased susceptibility to ethylene. Increased plant tolerance to drought stress results from increased expression of ARGOS genes, which are thought to be an inhibitor of ethylene production. These lowers yield loss.

It has been demonstrated that the ZmARGOS8 gene of Z. mays is a negative regulator of ethylene signaling. By introducing the GOS2 promotor from Z. mays into the untranslated region of the gene using the CRISPR/Cas9 system, new variations of ARGOS8 have been developed. The increased expression of the gene has boosted Z. mays's grain yield even under stressful situations. This guarantees that the CRISPR/Cas9 system may be used to change individual endogenous genes to create distinctive variations that could produce particular stress tolerance and aid in breeding initiatives. The CRISPR-based CRISPR-bzr-1- and BZR1 overexpressing lines demonstrated how the Feronia (FER) genes regulate tomatoes' thermotolerance, with Brassinazole resistant 1 (BZR1) playing a role. Furthermore, the CRISPR-Cas9-generated CBF1 mutants had severe cold damage symptoms, including increased electrolyte leakage, suggesting that C-repeat binding factor 1 (CBF1) protects plants from cold injury.

5. Nutritional quality improvement

Next-generation genetically engineered technology paved the path for a revolutionary prospectus that enhances nutritional quality. Numerous investigations have demonstrated the role of starch branching enzyme (SBE) in controlling the structural characteristics of starch that may provide various health advantages. Targeted SBEIIb mutants in *O. sativa* using CRISPR/Cas9 were able to create longer chains of amylopectin, enhancing the plant's resistance to starch and amylose and producing a more nutritious, high-amylose variety of the plant. Sufficient amounts of phosphorous in the soil are important for maximal growth and development of *S. lycopersicum*, since they are a limiting factor in its output. In *A. thaliana*, the homologs of PHOSPHATE 1 (PHO1) play a crucial role in the uptake and transport of phosphate. Its role in inorganic phosphorous transport in *S. lycopersicum* seedlings was confirmed when its homolog, S1PHO1;1, was mutated in *S. lycopersicum* using CRISPR/Cas9. This resulted in severe phenotypic signs of phosphorous deficit in plants. It has been demonstrated that Target-AID (Target activation-induced cytidine deaminase) BE technology is a potent method for precise base editing. Conversely, in the case of the tomato and O. sativa, only one gene target was initially known. Furthermore, several genes linked to the accumulation of carotenoids in tomatoes may be targeted by the combination of the CRISP/Cas9 system and the Target-AID method. Target-AID technique based on CRISPR/Cas9 was used to replace the nucleotide cytidine with thymine in three target genes of *S. lycopersicum*: S1DDB1, S1DET1, and S1CYC-B. This modification produced significant changes in carotenoid accumulation and demonstrated the technology's ability to edit many genes at once.

6. Improvement of Economical Plants Traits

Gene editing (GE) based on CRISPR/Cas9 has been widely used to improve commercially significant qualities in crops and has the potential to significantly change the way modern agriculture progresses in a number of areas. With the use of this method, it is possible to effectively eliminate genes that negatively regulate plant features linked to yield. The technique was utilized to eliminate the genes tgw6, gw5, and gw2, which are detrimental regulators of grain weight. The resultant mutants showed increased grain weight and size in O. sativa. In developing O. sativa grains, the vacuolar invertase2 (vin2) gene regulates the sucrose transport, which in turn affects the size and weight of the seeds. The reduction in grain weight and seed size seen when the CRISPR/Cas9 method was used to knock out the vin2 gene verified this. Similarly, higher grain weight in T. aestivum was discovered by the CRISPR/Cas9 system-mediated mutation of GASR7. By using the CRISPR/Cas9 system to change the cis-regulatory element CLV-WUS, *S. lycopersicum* was able to produce larger fruit in addition to increased grain weight. Under low nitrogen and drought circumstances, O. sativa showed higher grain yields when the 30-end of the OsLOGL 50s coding sequence (CDS) was edited using CRISPR/Cas9. The results were consistent at different sites.

Since lignin increases the pace at which biomass heats up and makes it a dominating source of aromatic compounds, enriching biomass with lignin through breeding techniques may be a useful tactic to enhance the biorefinery process. Lignin-rich *O. sativa* is produced when the transcriptional repressor OsMYB108 gene is targeted for mutation using CRISPR/Cas9-based GE. With a higher selling price than conventional non-fragrant *O. sativa*, fragrant rice, particularly Thai Jasmin and Indian Basmati varieties, has won praise from all across the world. Using TALEN technology, scientists were able to eliminate the OsBADH2 gene and create a dysfunctional badh2 allele. This allele encodes Betaine Aldehyde Dehydrogenase (BADH2), which in turn creates 2-acetyl-1-pyrroline (2AP), which is the true cause of the fragrance in the rice. Increased grain size and weight resulted from CRISPR/Cas9 editing of the TaGW7 gene in the wheat B and D genomes, indicating that TaGW7 is involved in the development of yield component features in T. aestivum. According to the findings under discussion, the CRISPR/Cas9 GE technology is a suitable approach for precise breeding aimed at enhancing various agricultural attributes. Even if it was effective in every way, the future's favored operational domains may be disease resistance, enhanced economic characteristics, nutritional qualities, and – most importantly – yield.

7. CRISPR 2.0

1. BEs and PE

As was previously mentioned, targeted DSB in DNA is caused by genome editing using CRISPR/Cas9. NHEJ and HDR might be used to fix these breaks. On the other hand, NHEJ is a more effective process that results in random insertion and deletion, disrupting the target gene at the location. On the other hand, HDR-mediated insertion is inefficient and needs the right template. Furthermore, it has been demonstrated that HDR-mediated repair is incredibly ineffectual, particularly in non-dividing cells, and that Cas9-mediated DSB may result in undesirable indel mutations. To get over these problems, base- and prime editing allows a specific tool.

2. DNA- based Editing

A site-specific DNA binding Cas enzyme and a single-stranded DNA modifying enzyme for changing the targeted nucleotide are the two main components of base editing in DNA. These are two groups of DNA base editors: those that edit adenine or cytosine bases. Transition mutation (C T, T C, A G, and G A) is handled by available CRISPR/Cas9 base editors. Researchers have published the designs of two unique base-editors that can be achieved with a single DNA base-editor at this time.

3. Cytosine Base Editor (CBE)

First generation cytosine base editor (CEB1) was created by researchers. Apolipoprotein BMRNA Editing Enzyme Catalytic Subunit 1 (APOBEC1), a rat-derived cytosine deaminase, was joined to the amino terminus of catalytically dead Cas9 to generate dCas9. CBE1 converts cytosine to uracil, which the cell's replication machinery interprets as thymine and causes the base change from C-G to T-A. Because of CBE1's limitations and poor editing efficiency, many basic editors have been developed to improve editing with it. One of them, CBE2, was produced by joining the C-terminus of BE1 with an inhibitor of uracil DNA glycosylase (UGI), therefore preventing UDG (uracil DNA glycosylase) from acting. Editing efficiency in human cells was tripled when BE2 inhibited the base excision repair. In order to produce a base editor employing Cas9 nickase, BE3, which was made by reinstalling histidine at position 840 (H840, HNH catalytic domain) in dCas9, significantly boosted editing efficacy (nCas9). CBE was further improved to reduce indel creation during base-editing, boost editing efficiency, and reduce the editing window. In order to do this, a fourth-generation cytosine base-editor (CBE4) was created by fusing an additional copy of UGI to the N terminus of nCas9 using an enhanced 27 bp linker.

4. Adenine Base Editor (ABE)

The spectrum of the disease-causing mutations that the cytosine base-editor (CBE) can fix is limited by C-G to T-A mutation. It is noteworthy that methylated cytosines are highly vulnerable to high rates of spontaneous cytosine determination. An ABE can be used to convert an A-T base pair back into a G-C base pair in order to reverse this process, introducing the A to G conversions, a novel class of adenine base-editors. GE's ABE mechanism is similar to CBE. CBE and ABE-mediated DNA editing operate via comparable mechanisms. The ABE-dCas9 fusion's deoxyadenosine deaminases domain binds to a target DNA sequence and catalyzes the conversion of adenine to inosine. During DNA replication, guanine is translated from isosine. Consequently, the initial A-T base pair may be replaced with a G-C base pair at the target location. Two herbicides, sulfonylureas and imidazolinones, block acetolactate synthase (ALS), an enzyme that is essential for the production of the branched-chain amino acids isoleucine, leucine, and valine. Herbicide-resistant crops against sulfonylurea and imidazolinones herbicides are produced in rice, wheat, tomato, potato, and watermelon by BE substitution of a particular amino acid in the ALS enzyme. The modification of acetyl-coenzyme-A carboxylase gene utilizing ABE base change can contribute in the growth of O. sativa plants that are herbicide-resistant. Additionally, BEs aids in the growth of virus-resistant plants. Organic variance when eIF4E proteins were independently inserted into the eIF4E1 gene of A. thaliana, they demonstrated resistance against the clover yellow vein virus (ClYVV) without impairing plant development. These proteins have previously given resistance to several crops against potyviruses and BEs in the eIF4E viral-resistance allele of *Pisum sativum.*

5. Prime Editing

The "search-and-replace" technique is the foundation of the prime editing (PE) genome editing tool, which was initially described by David Liu's group ("Genome-Wide Study of Off-Target Effects of Prime Editing in Rice Plants Demonstrates High Specificity"). The inability to produce precise base alterations is the primary disadvantage of using CBE and ABEs, aside from the four transition mutants. Recently, prime editing has been used to get around these limitations. Unlike base editing using CRISPR, prime editing does not need DSBs. A designed reverse transcriptase for prime editing was created by combining Cas9 nickase with a prime-editing guide RNA (pegRNA) (Fig. 4). Notably while being very different from regular sgRNAs, the pegRNA plays a crucial role in the system's functioning. The main advantage of using perRNA is that it provides complementary target sites to direct nCAs9 to its target sequence together with an additional sequence that spells the desired sequence modifications. To disclose the non-complementary strand, the 5'end of the perRNA binds to the DNA's primer binding site (PBS). By creating a nick in the PAM-containing unbound DNA strand, Cas9 links the nCAs9-linked reverse transcriptase (RT) primer. Furthermore, the RT uses the inside of the pegRNA as a template to lengthen the nicked PAM-strand. As a result, two redundant PAN DNA flaps are generated: the modified 3' flaps that were reverse transcribed from the pegRNA, and the original, unedited 5' flap. It depends on the equilibrium mechanism, wherein the flap that hybridizes with the DNA strand without PAM is most likely the completely complementary 5', which is thermodynamically preferred. But the 5' flaps are selectively destroyed by the ubiquitous cellular endonucleases that are present during the lagging-strand DNA synthesis. Ultimately, a resolved heteroduplex containing an unedited strand and an edited 3' flap is effectively integrated into the host genome.

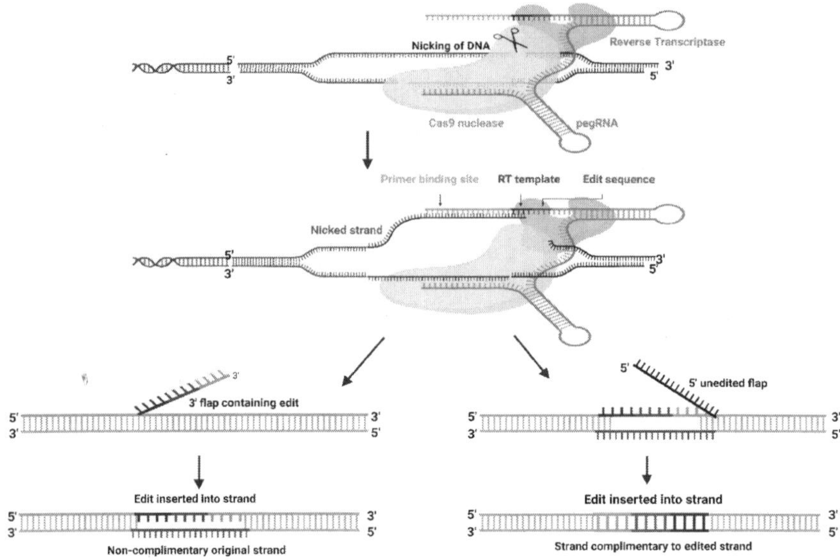

Fig 4. A brief overview of the prime-editing mediated-genome modifications.

Plant prime editor systems (PPEs) are employed to improve crops, especially *O. sativa* and *T. aestivum*. One of the greatest examples of GEC is herbicide resistance in *Oryza sativa*. However, there is still a great deal of research being done on PE for crop improvement. Scientists from several nations are examining the effectiveness and precision of PE for the production of GECs, such as *S. tuberosum*, *S. lycopersicum*, and *O. sativa*.

8. Safety Assessment of Genome-edited Crops

There are still safety issues associated with the usage of genome-edited plants. The principal concern is to undesired genetic modifications resulting from off-target mutations, guide RNA design, target gene site selection, and delivery techniques. The PAM requirement raises severe concerns regarding the selectivity of the Cas9 protein and the insufficient number of targeted sites. However, this problem has been resolved by improving the specificity by modifications to the Cas9 protein and guide RNA design. A key contributing factor to society's anxiety over genome-edited crops is a lack of understanding regarding the fundamentals and practical uses of this technology. Crops with altered genomes can be classified as transgenic or non-transgenic. Crops may be classified as non-transgenic via gene editing if the transgene contains the CRISPR/Cas9 cassette. This means that the plants have been edited and do not have transgenes with a genetic structure resembling the product mutations. It becomes imperative to educate people about the basics of genome editing in order to prevent misunderstandings.

9. Regulation of Genome-edited crops

While some nations throughout the world forbid the consumption or cultivation of genetically engineered plants, other nations do so. Policies and laws pertaining to genome editing techniques for global agricultural advancements vary throughout nations. America, Canada, Colombia, Brazil, Argentina, Chile, Australia, Israel, and Japan are the nations that do not regulate GECs. On the other hand, certain nations, such as China, India, Russia, Philippines, Bangladesh, Nigeria, Kenya, Paraguay, Uruguay, and the United Kingdom, are continuously debating the formation of regulations for the regulation of GECs; these nations fall under the category of ongoing research. Nevertheless, the GECs are fully regulated in Europe, Argentina, New Zealand, Ukraine, and nations that resemble Mexico (Fig.5). Two conditions must be met for genome-edited crops to be under directional control. A few nations steer the product's direction, while other nations standardize its characteristics.

The US Department of Agriculture said in 2018 that genome editing is the same as traditional breeding techniques. Regulation of genome-edited crops is therefore not necessary. Two examples of such crops are waxy maize which contains starch and browning-resistant mushrooms. According to researchers, these crops were authorized for commercial use in the USA without the need for any kind of control. Canadian legislation states that any genome editing method that produces new goods must be further regulated in order to account for allergies, toxicity to humans and animals, and effects on creatures other than the intended target. Two genome-edited products- non-browning apples and potatoes with no black spots—were authorized by Canada. Argentina has also established laws governing goods with altered genomes. In accordance with this method, a product will be labelled as non-transgenic if it was developed using transgenic technology and does not contain the transgene. Governments in the European Union oppose the use and consumption of crops modified by genome editing.

Because mutagenesis procedures have a track record of safety and are utilized in a variety of applications, they are expected from these criteria. The Hazardous Substances and New Organisms Legislation (HSNO) of 1996 governs the growth, import, release, and testing of genetically modified organisms (GMOs) in New Zealand. The HSNO legislation prohibits the cultivation and sale of commercial plants within the nation. A 2016 revision to the HSNO status stipulates that genome editing is subjected to the same regulations as genetically modified organisms. In India, the laws governing the use and cultivation of genetically modified organisms were established in 1989. Genetically modified food is defined by the Food Safety and Standards Authority of India (FSSAI) as food made using contemporary biotechnology. Therefore, India's GMO control policy also covers genome editing technology. Japan's Ministry of Health, Labor, and Welfare has a policy that states food produced using genome editing technology that doesn't include any transgenes isn't deemed genetically modified organisms (GMOs) and doesn't need to be regulated. Moreover, the ministry states that off-target mutation in food that has been genome-edited shouldn't raise any red flags.

Fig 5. World map representing the countries with regulated, non-regulated and ongoing research categories for GECs regulation. The three colours in figure first green shows the regulated, second shows the non-regulated and last third dark-cream colour shows under ongoing research categories of countries for GECs regulation.

10. Impact of Plant Genome Editing in the Post-COVID Pandemic Situation

Early in October 2020, Jennifer Doudna and Emmanuelle Charpentier made history by winning the Nobel Prize in Chemistry for their work discovering genome-editing technology. The United States World Food Program uses this groundbreaking finding to fight hunger and promote peace. There is a vital relationship between hunger and peace in the context of food policy, particularly in the post-COVID-19 age. The post-COVID-19 crises revolve around food and nutrition security. According to the most recent FAO (Food Agriculture Organization) repost, two billion people lack micronutrients and over 800 million people experience chronic hunger.

Food supply systems were impacted by the COVID-19 pandemic crisis, which also resulted in a rise in the number of hungry people and financial losses. In addition, another issue brought on by environmental stressors such as locust attacks in Africa and South Asia and Australian bushfires in 2020 was also caused by climate change. Globally, these variables have had a significant influence on agricultural production.

The development of genome-editing techniques, primarily through CRISPR-Cas9, has emerged as a potent scientific instrument of the modern period for the quick and accurate creation of high-quality agricultural products that benefit both procedures and consumers. The definition of genetic engineering (GE) has changed over time in accordance with technical advancements. Unlike previous methods of genome modification, current genome editing does not fundamentally involve the insertion of foreign DNA into plant species; this is a revolutionary development in the field of genome editing. Using genome editing in conjunction with cutting-edge technology like speed breeding is the ideal method for creating novel crops. Tools for altering genomes CRISPR technology and novel coops breeding techniques, such as speed breeding, have the possibility to grow crops quickly with little agricultural resources. They can also be used to solve post-pandemic problems and provide enough food to support an increasing population.

11. Future of Genome Editing

The growing prevalence of climate change and the ever-expanding population are driving up research and development as well as the costs associated with the biotechnology and agricultural sectors. These elements all affect financing in the public and commercial sectors, encourage quick advances in sequencing and genetically engineered instruments, and discourage the labelling of genetically modified organisms (GMOs).

GE's wide range of applications drove the technology's development and worldwide study. Genome editing has made it possible to produce non-transgenic plants with unique features quickly. The creation of non-transgenic plants with novel traits allayed worries about the dangers and moral dilemmas associated with genetic engineering. North America has led the world in genome editing research because of its fast expansion in the biotechnology, pharmaceutical, and agricultural sectors. Europe has trailed suit. There are more industries and consumers of cattle, crops, and poultry meat in North America.

Therefore, new technologies like as genome editing have been permitted for increased output with richer nutritional quality. Growing numbers in the agriculture and biotechnology sectors have made the Asia Pacific and Latin America's economies emergent. As a result, it has been noted that more labs in these areas have used genome editing technologies. CRISPR/Cas9 technologies are expected to expand at the fastest rate of any genome editing technology. Research estimating the market for genome editing from 2019 to 2026 states that it would develop a steady rate of 23.35 percent (Fig 6), "Genome Editing Market | Growth, Trends, and Forecast (2019-2024),"

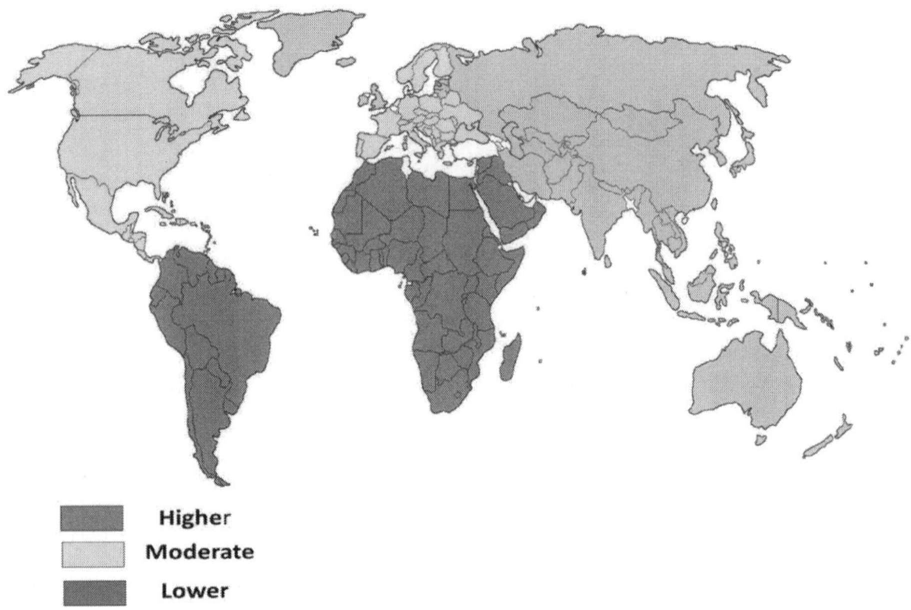

Fig 6. Global genome editing and marketing growth rate by region from 2019 to 2024 ("Genome Editing Market I Growth, trends, and Forecast (2019-2024),"). Green, yellow, and orange color shows high, moderate and low GE marketing growth rate, respectively.

12. Challenges in Genome Editing

One of the main obstacles to plant genome editing is species-dependent transformations, which call for the delivery of editing reagents into the plant cells, the selection of edited cells, and the restoration of the complete plant with the desired modifications. It is necessary for researchers to put forth fresh scientific advancements that are amenable to the various genome editing methods, such as targeted mutagenesis, gene insertions, and base substitutions. The current lack of knowledge regarding biological processes restricts the number of targets that may be used for genome editing crop development. The goal is to identify targets that are amenable to genome editing and have the potential to produce desirable agronomic features by combining several methodologies. Because certain plants have low efficiency, there are also extremely few applications and replacement alternatives for the base. Further research is needed to adapt and refine these methods for a broader range of plant species. Most government agencies classify genome-edited products as genetically modified organisms (GMOs) because they can be used to create phenotypic and genotypic variations in plants using GE techniques that are not distinguishable from those produced by natural or conventional mutagenesis processes. The public's trust may be strengthened, and the regulatory outlook must be influenced, by encouraging the open exchange and transparency of information on genome editing technologies, which in turn impacts the applicability of GE in agriculture.

13. Conclusion

These days, GE has developed into a flexible method for changing an organism's genotype and phenotype. Targeted and directed breeding now seems promising thanks to this new breeding technique. It enables the simultaneous alteration of numerous elite cultivars' genetic loci, improving a number of crops that are significant to agronomy and increasing food security worldwide. It follows that this technique will likely hasten the advancement of molecular breeding, which includes improving nutritional quality and resistance to biotic and abiotic challenges as well as strengthening economically significant features like yield-related characters. Conventional breeding techniques, however, are restricted to specific species. Transgenic food received a lot of attention as well, thus using GE with prudence is necessary. It is important to recognize the perquisites for both the beneficial and detrimental consequences of GE tools, such as CRISPR, TALEN, and ZFN have the potential to be cytotoxic. Experts express some valid bioethical issues. In the future, GE tools may be weaponry. To prevent all of these problems, GE and transgenic crops and food should be standardized for marketing management and registration. To prevent negative impacts, it is necessary to quantify, record, and publicly disclose the inevitable off-target effects, which will occur to varying degrees. Nonetheless, it seems certain that a biotechnological revolution is about to occur. The management and release of genetically engineered crops require public perspectives, professional bioethical arguments, and a controlled framework.

CHAPTER 8

NANO-OMICS: A SHORT STORY FOR SUSTAINABLE APPROACHES FOR CROP PRODUCTION

Short Summary

Nano-Omics: A Short Story for Sustainable Approaches for Crop Production explores the integration of nanotechnology and omics sciences to enhance sustainable agriculture. The narrative delves into how nanotechnology, through nano-fertilizers and nano-pesticides, can improve nutrient delivery and pest control, leading to higher crop yields with minimal environmental impact. Omics technologies, encompassing genomics, proteomics, metabolomics, and more, provide deep insights into plant biology, enabling precise modifications and improvements in crop traits. The fusion of these advanced technologies offers a holistic approach to addressing the challenges of modern agriculture. By employing nano-omics, scientists can develop crops that are more resilient to climate change, diseases, and pests. The story emphasizes the importance of a multidisciplinary approach, combining the strengths of nanotechnology and omics to achieve sustainable crop production. Key benefits highlighted include reduced chemical usage, enhanced crop productivity, and improved soil health. The narrative also touches upon the potential socioeconomic impacts, suggesting that these innovations could lead to more equitable food distribution and better livelihoods for farmers. Overall, NANO-OMICS chapter presents a forward-looking vision for the future of agriculture, where cutting-edge technology and sustainability go hand in hand to secure food production for a growing global population.

1. Introduction

Molecular activities have a complex role in regulating biological processes that occur in plants. There is a very complex and well-regulated system inside every cell that converts genetic information into proteins. Translation and transcription are the two main functions of these processes. Both transcription and translation lead to the complex machinery that produces gene expression. Plants use a wide range of responses to salt in order to either boost or reduce gene transcription, or more particularly, mRNA synthesis. This novel strategy comprises controlling genes across the core dogma, starting with transcriptional imitation and going all the way up to Transcription Factors (TFs') participation in the mRNA and protein synthesis. This thorough molecular regulation offers a fresh approach to improving a plant's resistance to salt stress. Plants are subject to two different kinds of stress due to salinity. First of all, it lowers the soil's water potential, which results in less water being absorbed and osmotic stress. Second, it causes an ion buildup that leads to ionic stress. Although NPs can be used to regulate the physiological and biochemical components of salinity, a thorough knowledge of their molecular activities is essential to assess how well they work in these capacities. This is especially important since the stress caused by salt changes the expression of several genes that are involved in different cellular processes and the products that arise from those activities. This affects plant development. In response to this problem, a novel and extensively used strategy for improving plant adaptation and salt tolerance is combining many 'omics' techniques, including transcription, proteomics, genomes, and metabolomics. Studies comparing salinity-stressed and control plants have yielded important insights into the regulatory characteristics, proteins, genes, and metabolomes that are essential for plants to adopt NP-based salt stress resistance (Figure 1). These insights have been obtained through the application of 'omics' technology. This all-encompassing strategy makes it possible to comprehend how NPs interact with molecular mechanisms on a deeper level, which illuminates how effective they are against salt stress.

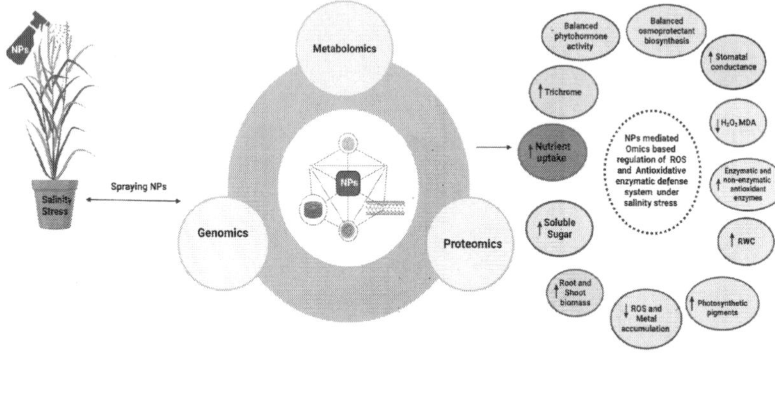

Figure 1. NPs-based Nano-Omics Response to plants under Salinity stress.

2. Navigating Plant Stress Responses: Insights from Nano-Omics-Based Genomics and Transcriptions

Plants are subject to two different types of stress caused by salt: ionic stress, which is caused by a buildup of ions in the soil, and osmotic stress, ethic is caused by reduction in water intake owing to salinity. Plants use efficient systems to reject and sequester sodium ions (Na^+) in order to counteract ionic stress. Much work has been made in the last several decades in discovering the plant ion transporters that are essential for salt resistance. Among them, the Na^+/H^+ antiporter (NHX) and the high-affinity K^+ transporter (HKT) on the vacuolar membrane is essential for controlling the absorption, transport, and redistribution of Na^+ and K^+. The balance of Na^+, K^+, and Ca^{2+} is also maintained under salt stress by a number of additional transporters, including Ca^{2+}-ATPase, NSCC Ca^{2+}/H^+ antiporter (CAX), Vacuolar H^+ phosphorylase (VP), H^+-PPase, and plasma membrane H^+-ATPase. Interestingly, HKTs, which function as Na^+ or Na^+-K^+ transporters and lower sodium levels in shoots while preserving potassium stability, are intimately associated with plant salt tolerance. HKT proteins fall into two categories: those intimately associated with plant salt tolerance. HKT proteins fall into two categories: those that facilitate K^+-Na^+ co-transport based on the availability of exogenous potassium and those that transport Na^+.

H^+ATPase and H^+-pyrophosphatase contribute to the proton gradient that drives Na^+ transport by NHXs across the vacuolar membrane. The first NHX gene identified, AtNHX1, from *Arabidopsis thaliana*, is essential to this procedure. It has been proposed that increasing the concentrations of NHX and VP, or VP alone, can improve the efficiency of Na^+ shuttling into vacuoles by strengthening the proton gradient produced by H^+-ATPse and H^+-PPase. Salt and calcium-sensitive (SOS) signalling genes and pathways are strongly related to plant salt tolerance and cellular homeostasis. Here, genes like calcium sensors SOS3/CBL4 are activated when they bind Ca^{2+}. They then interact with the protein kinase SOS2/CIPK24 to generate a preteen complex that in turn activates the Na^+ antiporter gene SOS1, which is in charge of cellular salt efflux (Figure 1). With the aid of Ca^{2+}, the Na^+/H^+ EXCHANGER1 (NHX1) antiporter promotes Na^+ transport into vesicles. Moreover, the use of NPs has shown promise in upregulating the expression of antioxidant genes like SOD and APX, which would restore ionic and redox equilibrium by lowering ROS and preserving a normal Na^+/K^+ ratio. These beneficial effects are linked to the changed transcript levels of important genes that NPs activate, such as NHX1 and K^+ transporter1 (KTX1). By upregulating a number of genes, including NHX1, CAX1, SOS1, HKT1, H^+-ATPase, and AQPs, selenium NPs (Se-NPs) have also demonstrated potential in supporting a balanced cytosolic Na+/K+ ratio during salt stress. Applying Se-NPs topically to lemon plants increased the expression of genes linked to phenylalanine ammonia-lyase, which in turn increased the synthesis of phenolic compounds and non-antioxidant enzymes. This, in turn, decreased the levels of hydrogen peroxide (H_2O_2) and malondialdehyde (MDA) in the plant under salinity. Ag-NPs, which were extracted from the *V. negundo* plant extract, caused anthocyanin regulators to be expressed in Brassica oleracea, which resulted in an accumulation of anthocyanins and elevated levels of SOD, CAT, and APX activities. Furthermore, it is found that the administration of Fe-NPs to tomato plants increased the expression of genes for antioxidant enzymes, including Cu, Zn-SOD, APX, and CAT2. These genes support H_2O_2 detoxification and preserve the Na^+/K^+ balance. Further reducing the negative impacts of salinity, ZnO-NPs also had a favourable influence on stress response gene expression. Plants use osmosensors, which aid in regulating osmotic potential, to combat osmotic stress (Figure 2). Potential osmosensors in Arabidopsis include genes such as histidine kinase (AHK1) and increased expression of the OSCA1 gene, which reduces hyperosmolality-induced (Ca^{2+}) levels. The production of osmolytes and a rise in water potential are the results of osmoregulation, which starts with the upstream regulation of osmosensing-related genes. Because they hold onto water, osmolytes like proline are essential for preserving cellular structure and osmotic equilibrium. Additionally, proline scavenges reactive oxygen species (ROS) and shields the cell membrane. The buildup of proline under salt stress requires the enzyme pyrroline-5-carboxylic reductase (P5CR). Proline is a precursor to other stress-response compounds in addition to helping with osmoregulation. It enhances the synthesis of stress-responsive protein in plants and functions as an organic nitrogen store during stress recovery. Proline biosynthesis is controlled by the P5CS gene, which is increased in saline environments. Notably, it has also been demonstrated that applying different NPs upregulates the P5CS gene, raising proline levels and improving plant resistance to osmotic stress.

A plant's ability to withstand salt stress depends on the activation of antioxidative genes, which scavenge reactive oxygen species. OsMSRA4.1, OsVTE1, and OsECS overexpressed transgenic rice had increased salt resistance. In a similar vein, transgenic Arabidopsis plants that exhibit higher ascorbate levels due to OsAPXb or OsAPXa overexpression have enhanced salt tolerance. OsAPXb is very useful for granting salt tolerance and regulating H_2O_2 metabolism. In response to salt stress, several antioxidant genes are increased, including OsMIOX, OsAPx8, OsGR2,3, cCuZn-SOD2, OsSRO1c, and OsAPX7. Transcription factors (TFs) are important regulators of a broad variety of gene expression in response to salinity variations. To increase gene expression, TFs interact with cis-elements in gene promoters. For example, during salt stress, Arabidopsis plants show a control of about 428 TFs. Stress triggers the expression of several TF families, including WRKY, MYB, bHLH, HOX, AP2/ERF, and NAC. By trigging downstream genes, the AP2/ERF TF family—which is further subdivided into DREB, ERF, RAV, and AP2 subfamilies—contributes significantly to salt tolerance. Two well-known examples are CBF2 and CBF3, which improve plant's ability to withstand salt. In addition, MYB TF genes WRKY TFs such as WRKY17, WRKY25, WRKY33, WRKY 29, WRKY40, and

WRKY75 are also upregulated in response to salt stress. Plant salt sensitivity is influenced by several TFs, including members of the HOX, bHLH, and NAC families (Figure5). To summarize, plants utilize a complex molecular response that includes antioxidant genes, transcription factors, osmosensors, ion transporters, and osmoregulation. By strengthening these molecular pathways, the application of NP has the potential to increase plant resistance to salt stress.

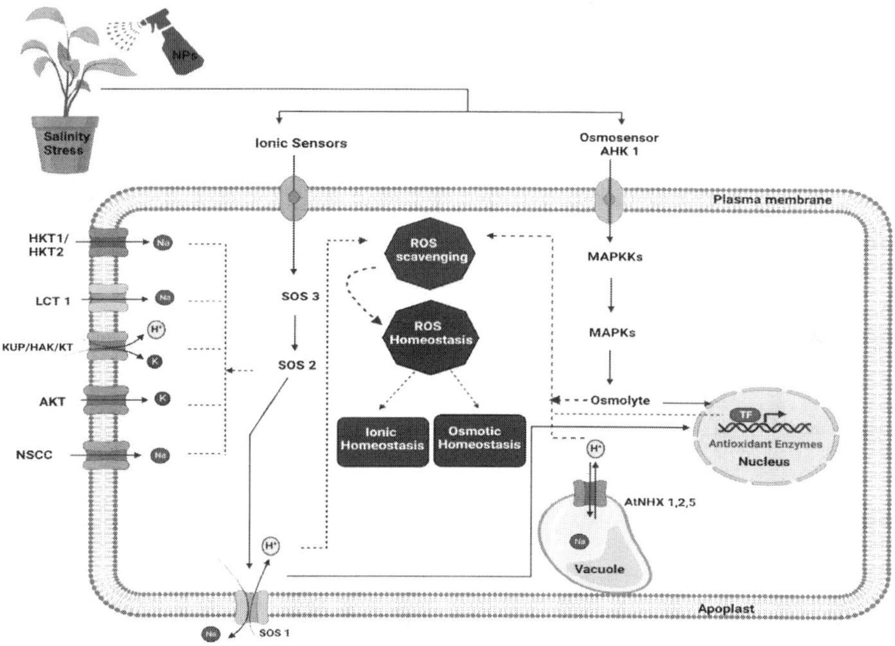

Figure 2. Molecular mechanism of plant cells during the application of Nano-biosensors (ionic sensors and osmosensors) to reduce the production of ROS.

8.2 Unraveling Plant Responses to Salinity: Insights from Nano-Omics-Based Proteomics

Proteomics has become an effective technique for studying how plants adjust to salinity, offering insightful views into the molecular processes controlling plant's reactions to stress brought on by the salt. Plants intentionally increase the amounts of certain proteins in response to salinity problems, which enhance their metabolic, defense, cell integrity, photosynthesis, reactive oxygen species (ROS) scavenging, and osmolyte manufacturing activities. The significant rise in molecular chaperones such as HSP70 and HSP90, which assist in protein folding and stability, is a noteworthy solution to salt stress. Furthermore, there is an upregulation of glycolytic and tricarboxylic acid cycle (TCA) proteins and enzymes during salt stress. Aconitate, succinate dehydrogenase (SDN), malate dehydrogenase (MDH), isocitrate dehydrogenase (IDH), pyruvate kinase isozyme A (PK), pyruvate decarboxylase 2 (PDC), and 6-phosphofructokinase

3 (PFK) are among the TCA cycle enzymes that exhibit increased expression. The plant can increase a number of metabolic processes as a result of this upregulation, guaranteeing a consistent flow of energy. The activity of antioxidant systems, which include enzymes like glutathione peroxidase (GPx), superoxide dismutase (SOD), catalase (CAT), thioredoxin peroxidase (TXN), and glutaredoxin sulfhydryl oxidoreductase (GPx), is also increased by salinity stress. These enzymes are essential for reducing the accumulation of hydrogen peroxide (H_2O_2) and reactive oxygen species (ROS) brought on by salinity. All three isoforms of SOD—manganese SOD (Mn-SOD), copper-zinc-SOD (Cu-Zn-SOD), and iron-SOD (Fe-SOD)—are activated by salinity in different parts of the cell. Furthermore, salinity conditions increase the activity of non-enzymatic antioxidants such as phytoene synthase 2 (PSY2), which neutralizes reactive oxygen and nitrogen species (ROS and RNS). Notably, the use of NPs in a number of investigations has shown that they can improve, it was shown that iron oxide nanoparticles improved antioxidant performance in wheat by upregulating enzymes linked to salt stress. Additionally, by upregulating the small subunit of RuBisCO, a crucial enzyme in the Calvin cycle, and enhancing the activities of antioxidant enzymes under salt stress, topical administration of TiO_2-NPs significantly improved photosynthesis. It has been demonstrated that foliar spraying Zn-NPs increases the levels of antioxidant enzymes, such as CAT, POD, and SOD, in okra (Abelmoschus esculentus L.), protecting photosynthetic components. Applying Fe-NPs had comparable results, upregulating the genes for APX, GR, and SOD in tomato plants stressed by salt, which improved the plant's ability to grow, retain water, and respond to stress. The administration of NPs can affect the production of non-enzymatic antioxidants that respond favourably to salinity, such as phytoene synthase. For example, ZnO-NPs were reposted to alter the expression of phytoene synthase in Arabidopsis at a concentration of 300mg/L.

Plants improve their salt stress exclusion mechanism by activating protein channels and transporters in addition to their antioxidant responses. According to researchers, they comprise VDAC, V-ATPases, NSCCs, chloride channels, HKT1, aquaporins, SOS1, K^+ transporters, CNGCs, and NHX. Numerous studies have shown that adding different NPs can increase the activity of antioxidant enzymes and upregulate these transporters and protein channels, all of which help plants survive in salty environments.

3. Unlocking Plant Responses to Salinity: Insights from Nano-Omics-Based Metabolomics

A powerful scientific method of examining how plants adjust their metabolism to abiotic stress and environmental changes in metabolomics. It explores the complexities of plant responses to stress caused by salt, illuminating several metabolic pathways linked to stress tolerance. Plants respond in a series of ways to salt stress, and metabolomics provides a window into this dynamic process. The resultant metabolome plays a critical role in several vital functions, such as preserving osmotic equilibrium, stabilizing cell membranes, scavenging reactive oxygen species, and enhancing antioxidant defenses. A wide range of substances are included in the metabolome, including alkaloids, amino acids, dipeptides, flavonoids, organic and inorganic acids, phytohormones, carbohydrates, sugars, and terpenes. When an organism is exposed to external stimuli, it releases metabolites, which are the byproducts of cellular control. More than 200,000 metabolites found in plants may be divided into two groups: primary and secondary metabolites. Each category has a specific function in the growth and development of the plant. While secondary metabolites vary depending on the species and have a variety of functions in plant physiology, primary metabolites are structurally conserved and serve essential functions in all plants. The modifications in physiological processes brought about by synthetic nanoparticles are monitored by molecular events, which signify shifts in essential biological processes like the Kerbs cycle, glycolysis, synthesis of primary and secondary metabolites, nitrogen and fatty acid metabolism, and amino acid production. In response to abiotic stresses, plants change the chemistry of the soil, release metabolites that act as signaling molecules, and shift metabolic pathways to increase the availability of nutrients. Significantly, in saline settings, secondary metabolites such as flavonoids, phenolics, anthocyanins, and carotenoids function as non-antioxidant enzymes and help scavenge reactive oxygen species (ROS). Numerous studies show that adding various NPs can raise the concentration of these substances. As non-antioxidant enzymes, these metabolites are essential for the detoxification of reactive oxygen species (ROS) because they scavenge radicals such as hydroxyl radicals (•OH) and singlet oxygen ($^1O_2$). For instance, it was discovered that *Momordica charantia* and *Prunella vulgaris* exhibit higher amounts of flavonoids, phenolics, and protein in response to copper, silver, and gold nanoparticles, indicating their function in mitigating the production of free radicles.

Moreover, metabolomic research suggests that plants create more antioxidant chemicals when NPs are present. For example, it has been demonstrated that Fe-NPs improve salt tolerance by activating important antioxidant enzymes. Interestingly, *Sophora alopecuroides* seedlings treated with Carbon Nano Horns (CNHs) showed significant increases in copper levels, fresh root weight, total protein content, photosystem II (PSII) efficiency in leaves, and soluble sugar content after salt stress. Improvements in the tricarboxylic acid cycle, glycolysis, and metabolic pathway resetting were also found by metabolomic research, and these processes all contributed to the enhanced generation of energy. Furthermore, by raising the unsaturation of fatty acids, the use of CNHs helped to restore the normalcy of membrane structure under salt stress. The administration of C-NPs (0.3%) under salinity conditions produced similar results, improving a range of metrics in distinct varieties of lettuce. Additionally, via increasing antioxidant activity, P and K^+ levels, chlorophyll content, amino acid content, and soluble sugars, pretreatment of wheat seedlings with polyhydroxy fullerene NPs boosted salt tolerance. Furthermore, the treatment of TiO_2 NPs (at different doses) strengthened the resistance to salt in broad bean crops by raising the amounts of soluble sugars, proline, and other amino acids as well as by proving enzymatic activities.

All things considered, metabolomics provides a wealth of information on the complex metabolic reactions of plants to salt stress and the possibility of NPs to improve plant stress tolerance by modifying metabolite profiles.

CHAPTER 9

CONCERNS OF NANOTECHNOLOGY FOR AGRICULTURE, ENVIRONMENT, PLANT AND HUMAN HEALTH

Short Summary

Nanotechnology holds great promise for agriculture, yet it also brings several concerns for the environment, plant, and human health. One major issue is the potential toxicity of nanoparticles. While they can improve pest control and nutrient delivery, their long-term impact on soil health and microbial communities is uncertain. These particles might accumulate in the soil, disrupting beneficial organisms and altering soil chemistry, leading to unintended ecological consequences. For plants, nanoparticles can be both beneficial and harmful. They may enhance growth and resistance to stress, but excessive or improper use might lead to phytotoxicity, stunting growth or causing cellular damage. Understanding the optimal types and quantities of nanoparticles is crucial to avoid negative effects on crop health and yield. Human health concerns are significant, given that nanoparticles can enter the food chain. There is limited knowledge about the long-term effects of consuming foods treated with nanomaterials. Potential risks include toxicity, bioaccumulation, and unforeseen health issues. Ensuring thorough research and regulation is essential to mitigate these risks. Environmental concerns also encompass water bodies, where runoff containing nanoparticles could affect aquatic ecosystems. The persistence and bioaccumulation of these particles in marine life pose additional risks, potentially impacting entire food webs. To address these concerns, comprehensive risk assessments, robust regulatory frameworks, and ongoing research are necessary. Developing safe and sustainable nanotechnology applications in agriculture requires a balanced approach, considering both the benefits and potential hazards. This ensures the protection of environmental integrity, plant health, and human safety while harnessing the advantages of nanotechnology.

1. Introduction

Nanoparticle—particles with sizes ranging from 1 to 100nm—have different physical and chemical properties from their bulk counterparts. Because of their size-dependent physical and chemical properties, nanoparticles (NPs) can be formed in a way that is similar to naturally occurring supermolecule ensembles. Based on their size, shape, chemical composition, surface, phase state, and other characteristics, the many kinds of nanosized materials are displayed in Table 1. The assessment of possible environmental risks and the identification of toxicity to biological systems are significantly influenced by the physical and chemical characteristics of produced nanomaterials. In-depth investigation of the physiochemical properties of NPs is necessary in order to perform the toxicological test on developed NPs. Because of recent advancements in nanotechnology, a large range of nanosized materials with many applications in different fields are now widely available, spamming over 1300 nanomaterials across several sectors. They were used in food, cosmetics, furniture, clothing, and bottle coatings. Opacifiers, microelectronic devices, SIM cards for mobile phones, pharmaceuticals, and more. Results from nanotechnology during the past 10 years indicate that NPs have a wide range of applications, including fuel additives, water treatment, the oil sector, and catalysis. Furthermore, nanofluids are used as cutting-edge new generation materials because of their well-known thermal properties and capacity for heat transmission. Prior to expanding the usage of nanoparticles (NPs) in agriculture, it is vital to take into account factors like known impacts, monitored density, potential toxicity, and degrees of overdosing. Every ecosystem depends on plants, and NPs have the ability to interact with the environment. Therefore, it is reasonable to assume that NPs will interact with plants and that these interactions will impact the fate and movement of NPs in the environment through a variety of processes, such as plant biomass accumulation and income. NPs in the environment through a variety of processes, such as plant biomass accumulation and income. NPs can bind to plant roots and cause harm to plants in addition to their physical and chemical toxicity. The idea of positive interactions between soil bacteria and plants cannot be completely ruled out. These data clearly show that nanoparticles can penetrate the living tissues of plants; nevertheless, this has consequences for their build-up in the food chain and their use as intelligent delivery systems in plants.

We are unable to evaluate NPs' effect if we are unaware of how they circulate throughout plants. One crucial topic to address is whether plants can absorb and transfer intact NPs to other plant tissues. Researchers working in this field found that NPs may enter the plant tissues by a number of different entry points, including wounds, root connections, aboveground organs and tissues (including cuticles, trichomes, stomata, stigma, and hydathodes), and root tissues. Only a few studies have reported on the "direct" uptake, translocation, and localization of NPs in plants using different insoluble NPs, such as carbon nanotubes , fullerenes (c70), quantum dots, gold nanoparticles, titanium dioxide nanoparticles (TiO_2) NPs), Iron (II, III) oxide (Fe_3O_4) NPs, and virus-based NPs. Stable nanoparticles that dissolve in water have the ability to build up in aquatic environments and modify the behaviour of aquatic organisms in a concentration-dependent manner. The majority of studies have looked at the effects of NPs on fish, amphibians, freshwater creatures, algae, plankton, and earthworms and manufacture highly reactive particles known as free radicals. The imbalance between the production of ROS and the detoxification and repair of metabolic damage is known as oxidative stress. Oxidative stress is caused by metal-based nanoparticles, carbon nanotubes, and fullerenes, all of which harm live cells. Free radicals and peroxides, which harm proteins, lipids, and DNA, are produced by oxidative stress. In addition to indirectly inducing oxidative stress and inflammatory reactions, carbon nanotubes can interact directly with DNA. NPs pass through cell membranes to enter cells. NPs come into contact with biomolecules (lipids, carbs, and proteins) within the body. Target organelles in oxidative stress caused by nanoparticles are crucial, including mitochondria. Explaining NP transportation and interaction inside cells is made easier by cellular absorption. It has been investigated how hazardous NPs are to plants. Additional studies show that plants treated with nanoparticles either benefited or did not. Although, the exact mechanisms underlying nanotoxicity remain unknown, the surface area, size, shape, and chemical makeup of nanoparticles are probably all involved. Nanoparticles on the root surface need to be taken into consideration since they can change the chemistry of the root surface and impact nutrition absorption. Nanoparticle toxicity is caused by stress stimuli arising from the particle surface, size, or shape, as well as chemical toxicity resulting from chemical composition, such as the release of (toxic) ions. The benefits and drawbacks of nanoparticles (NPs) and how they affect plants will be covered in this study. There will also be a conversation regarding the possible introduction of modified nanoparticles into the food chain.

2. Cons: Toxicity Challenges of NPs in Plants

2.1 Toxicity Concerns in Plants

2.1.1 Seed Germination

One common negative effect of metal-based nanoparticles is that they decrease seed germination. It is found that zinc NPs on ryegrass and zinc oxide NPs on maize decreased seed germination at 2000mg L^{-1}, perhaps due to the toxicity of dissolved zinc cations. The detrimental effects of Ag NPs and zero-valent iron NPs, which have an average particle size of 1 to 20nm, on the germination of ryegrass, barley, and flaxseeds were examined. The concentrations of Ag and zero-valent ion NPs that were evaluated ranged from 0 to 100mg L^{-1}, and 0 to 5000mg L^{-1}, respectively. At 250 mg L^{-1}, zero-valent iron NPs were observed to have inhibitory effects in aqueous solutions. Nucleovalent iron NPs were shown to fully germination at 1000-2000mg L^{-1} doses. Ag NPs reduced seed germination at lower concentrations, but these effects were never fully eliminated and were not size-dependent. The use of NPs to seed germination can have both detrimental and beneficial effects. In a different investigation, it was demonstrated that TiO_2 NPs at concentrations of 2 and 10mg L^{-1} improved wheat biomass and germination. Impacts on most germination parameters appear to be concentration-dependent, with low concentrations of Zn nanoparticles showing either no effect at all or a positive effect, and high quantities demonstrating an inhibitory effect. Several crops, such as *Solanum lycopersicum, Lathyrus sativus, Raphanus sativus, Sinapis alba,* and *Zea mays*. Germination may be impeded or unaffected, depending on the research, as seen in the examples of *Oryza sativa, R.sativus, S. lycopersicum,* and *L.sativa*. Plant response to nanoparticles is dependent on a number of factors, including concentration of nanoparticles, sort of NPs, and species of plant. Conversely, because of the higher concentration of Cu^{2+} in the fluid, a higher concentration of Cu nanoparticles results in less germination and more toxicity. This was seen in plants like *E. sativa* and *L. sativa*, which are very susceptible to dissolving Cu^{2+}. Plants and seeds that directly absorb copper nanoparticles may sustain damage that inhibits germination and root development. Cu's redox characteristics, however, could contribute to its toxicity since Cu^{2+} boosts the synthesis of reactive oxygen species. This ultimately results in damage to proteins, nucleic acids, and lipids as well as oxidative stress. The level of Cu oxidation in the nanoparticles determines the influence on wheat germination. It was shown that low concentrations of Cu^{2+} aided in germination, but CI^{1+} had less effect. Depending on the target crop, Fe NPs have negative effects on germination over a certain level (0.25-0.5mg L^{-1}).

2.1.2 Root Growth and Shoot Development

Two investigators found that a typical sign of environmental toxicity is the suppression of root growth. Six higher plant species, including cucumber, radish, rape, ryegrass, lettuce, and maize, were examined to see how NPs affected the elongation of their roots. All of the investigated species' root elongation almost stopped when they were subjected to suspensions of 2000mg L^{-1} Zn NPs or ZnO NPs. 50% inhibitory concentrations (IC50) for zinc NPs and zinc oxide NPs were estimated to be around 50mg L^{1} for radish, and 20mg L^{1} for rapeseed and ryegrass. Research was done on how phenanthrene (Phen) affected the roots of cabbage, carrots, soybeans, cucumbers, and maize. It was demonstrated that attenuation of particle-induced reduction of root development occurred when either 10.0%, 100.0%, or 432.4% of the monomolecular layer (MML) of Phen was loaded onto the particles. Seven higher plant species, including radish, rape, tomato, lettuce, wheat, cabbage, and cucumber, were studied to see how sensitive they were to four rare earth oxide nanoparticles (NPs) during the root elongation process. CeO_2 NPs, La_2O_3 NPs, Gd_2O_3 NPs, and Yb_2O_3 NPs. NPs had a substantial impact on root growth in many different ways. Zinc oxide nanoparticles (ZnO NPs) have been shown to induce oxidative stress in soybean (G. max L.) seedlings at 500mg L^{-1}. ZnO NPs stress markedly decreased plant growth, root stiffness, and root cell survival. Genes involved in the oxidation-reduction cascade were down-regulated after being treated with ZnO NPs. Among these genes were quinone reductase, galactose oxidase, GDSL motif lipase 5, and SKU5. Researchers assessed the impact of commercial ZnO nanoparticles (<100nm) on wheat (*T.aestivum L.*) cultivated on sand, a solid matrix. ZnO NPs' solubilization of metals in the sand was comparable to that of their bulk counterparts. Root growth was significantly (p=0.05) supressed by ZnO NPs (500mg kg1) applied to sand; however, this impact was lessened by applying the bulk amendment. Plants exposed to ZnO NPs were seen to bioaccumulate zinc in the form of zinc phosphate in their shoots. When plants were exposed to ZnO NPs, they displayed symptoms of oxidative stress such as increased activity of the enzyme's peroxidase and catalase in the roots, increased levels of lipid peroxidation and oxidised glutathione in the roots, and decreased levels of chlorophyll in the shoots. According to research, the biomass of seedlings was shown to be significantly decreased to 35% of the control group when exposed to 1000mg L^{-1} of ZnO NPs. The median inhibitory concentration of ZnO NPs was 215mg L^{-1}. Transmission electron microscopy demonstrated that ZnO NPs were found within certain root cells and has a high adherence to the root cell wall. The amount of zinc that one-month-old maize plants absorbed ranged from 69 to 409mg Kg^{-1} in their roots and from 100 to 350mg kg^{-1} in their shoots in soils contaminated with different concentrations of ZnO NPs (from 100 to 800mg kg^{-1} soil). Researchers found that ZnO NPs aggregates passed through the endodermis and entered the xylem arteries using a confocal microscope. This demonstrates that the aggregates did not utilize the apoplastic pathway, but rather the symplastic one. Still, most of the aggregates were concentrated in the area abutting the endodermis. The following study investigated the phytotoxicity and accumulation of copper oxide (CuO) nanoparticles (NPs) on *Elsholtzia splendens*, a CuO NPs exhibited a much higher Cu content than those subjected to 0.5mg L^{-1} CuO and CuO bulk particles. CuO NP-like deposits were seen in the cells of the roots and leaves. Subsequent analysis using Cu K-edge X-ray absorption near-edge structure revealed that the predominant form of the accumulated Cu species in the plant tissues was CuO NPs. These results suggest that CuO NPs can be taken up by *E. splendens* at the root level and then transferred to the shoots. The effect of laboratory-prepared FeOx NPs (likely -Fe_2O_3) on the germination of lettuce (*L. sativa*) seeds in the water medium was investigated by researchers. Over the course of NPs' concentration range (5-20mg L^{-1}), root elongation was enhanced by 12-26%, and NPs exhibited much lower toxicity than their ionic counterparts. On the other hand, root development was inhibited at 50mg L^{-1}.

2.1.3 Cellular and Biochemical Concerns

(A) Cellular Concerns

Some NPs cause harm to cells and subcellular organelles, resulting in mitochondrial malfunction and damage to cell membranes. In comparison to the control group, soybeans exposed to Al_2O_3 NPs experienced damage to the root cap and minute fissures around the terminals of their roots. Al_2O_3NPs increased the production of reactive oxygen species (ROS) in mitochondria and chloroplasts after being absorbed by soybeans, indicating that ROS may be a factor in cell damage. Researchers observed that Ag NPs may infiltrate cells and harm vacuoles and cell walls, potentially affecting other organelles in a study on *Vigna radiata* and *Brassica campestris*.

Ag and ZnO NPs were shown by researchers to decrease cell turgidity and vacuole size in *B. oleracea* and maize. TEM images demonstrated lettuce cell damage caused by PS-NPs shattering the structure of the chloroplast. It has been shown by FTIR and synchrotron computer micro-tomography that PS-NPs (Pd-doped PS-NPs) may affect the cells walls of wheat roots, changing the structure of the roots.

Tissue culture is an essential approach in several domains related to plant biology. Plant tissue culture appear to be benefit from nanoparticles, according to many researches. NPs commonly used to achieve a number of objectives, such as increasing the synthesis of secondary metabolites, causing callus, organogenesis, somatic embryogenesis, somaclonal variation, and genetic modifications. Microbial pollutants in plant tissue cultures are effectively controlled by adding NPs to the culture medium. Studies on plant tissue cultures have indicated that NPs may be hazardous despite their advantages. Plant cell genetic and metabolic responses to nanoparticles (NPs) have also been researched in vitro, but nanotoxicity has been well investigated utilizing a variety of intact plants.

In comparison to whole-plant systems, tissue cultures—which include callus, cell suspensions, and hairy roots—offer convenience, speed, and the ability to exclude microflora and translocation barriers. While in vivo studies provide challenges for molecular-level toxicity research, in vitro studies provide a controlled setting despite the complexity of natural surroundings and organismal systems. They may be strong auxiliary model systems that provide important information for studies involving complete plants. Cell wall and membrane rupture occurred when tobacco plant cells were treated with water-soluble carboxyfullerenes ($C_{70}(COOH)_2)_2^{-4}$), which inhibited cell multiplication. Additionally, a plant cell's defense mechanism against fullerenes stress was demonstrated by increased glycosyl residue on the cell wall and ROS in treated cells. In N. tabacum cell culture, Ag NPs induced oxidative stress as seen by decreased cell viability and proliferation, decreased H_2O_2 levels, increased ROS scavenging enzymes, and decreased phenolic and flavonoid content. Furthermore, the unique detrimental effect of CuO NPs synthesized both chemically and physiologically were shown in tobacco cell suspension cultures. In physiological tests, chemically produced CuO NPs were more detrimental than biologically produced ones. Examining SnO_2 and Ag/SnO_2 NPs' cytotoxic effects in tobacco cell cultures emphasizes how important structural modification is to NP toxicity. SnO_2 NPs showed no toxicity in tobacco cells, whereas Ag-doping produced oxidative stress-related toxicity.

Microscopic examination demonstrated significant NR content in tobacco cells exposed to NP stress and cell death in high-level SnO_2 NP treatments (0.5mg mL^{-1}). These findings suggested vacuoler pH acidification. In *O. sativa L.*, callogenesis and regeneration were enhanced by biosynthesized CuO NPs (1-20mg/L) derived from *Azadirachta indica* leaf extract. At 20mg L^{-1}, Basmati 385 (92%) had the greatest regeneration rates, followed by Basmati 2000 (80%), Super Basmati (52%), and Basmati 370 (32%). Low concentrations of CuO-NP (1 and 15 mg L^{-1}) led to inadequate regeneration of Basmati 370. It has been demonstrated that *Linum usitatissimum L. calli* and regenerants internalize gold and silver nanoparticles. It was proposed to vacuum-deposit metal nanolayers on top of powdered hormone in the growth media in order to transfer metal NPs to plant cells. CNT concentrations up to 50mg L^{-1} greatly accelerated the development of *Saturejakhuzestanica Jamzacalli*. Elevated concentrations (500mg L^{-1}) resulted in decreased growth, elevated H_2O_2 production, and elevated PPO, POD, and secondary metabolic activities. *Trigeonella foenum-graecum L. calli* was less toxic to uncapped CuO NPs, which resulted in an increase in the production of secondary metabolites. *Prunella vulgaris L.* developed more calluses and produced more antioxidants when exposed to Ag and Au NPs. ZnO and CuO NPs treated *Stevia rebaudiana* callus may function as an antioxidant source. According to a research, Zn and ZnO NPs improved total proline, SOD, CAT, and POD but reduced callus development in vitro banana cells. NPs had significant antibacterial antifungal qualities, but they had no detrimental effects on explant regeneration. Ag NPs exhibited more phytotoxicity than $AgNO_3$ in vitro experiments on potato cells, most likely as a result of oxidative stress. Ag NPs were more detrimental than Ag ions, as evidenced by altered antioxidant enzyme activity and decreased glutathione and ascorbate in potato shoot cultures. Ag NPs stimulated growth at 25 and 50mg L^{-1} and significantly prevented at 100 to 200 mg L^{-1} in vanilla shoot cultures. The main mechanisms by which Ag NPs produced vanilla hormesis in vitro were ROS and mineral feeding. In order to increase tropane alkaloid production and cause oxidative stress in the hairy roots of the medicinal plant *Hyoscyamus reticulatas L.*, iron oxide nanoparticles (NPs) were employed. In hairy roots of *Dracocephalum kotschyi Boiss*, the concentration of important medicinal components such as xanthomicol and rosmarinic acid was greatly enhanced by SiO_2 and Fe NPs.

(B) Biochemical Concerns

1. **ROS Production:** Like other aerobic species, plants select ROS as a single molecule to trigger cellular responses to changes in their environment. NPs have the ability to change ROS levels, disrupt cell metabolism, harm

the antioxidant system, and eventually limit plant growth once they enter plant cells. Superoxide anion, hydrogen peroxide, hydroxyl radical, and singlet oxygen are a few examples of reactive oxygen species (ROS) that are continuously and spontaneously produced by cell organelles throughout metabolism. It is important to have both low levels of ROS, which trigger a productive signal, and high levels of ROS, which oxidatively damage lipids, amino acids, and nucleic acids. One helpful measure of the health of a cell's membrane is the amount of lipid peroxidation present in the membrane. ROS-induced lipid peroxidation damages membranes, resulting in ion leakage, interruption of metabolism, and eventual cell death. The low molecular weight antioxidants and antioxidant enzymes found in plants protect cells and subcellular compartments from the harmful effects of reactive oxygen species. As a result, the majority of research on the antioxidant levels as well as antioxidant enzyme activity. Ag NPs on bean leaves caused reactive oxygen species to be produced. The results demonstrated that ROS production increased with decreasing particle sizes because smaller NPs have a bigger specific surface area and give more cytotoxicity. The presence of TiO_2 increased tocopherol content and SOD activity under bright light conditions while decreasing catalase CAT activity. Increased ROS production and particle photoactivation may be linked to the membrane lipid peroxidation that is facilitated by higher NP concentrations. In response to the oxidative stress caused by polymethyl methacrylate nanoplastics (PMMA-NPs), lettuces' antioxidant enzyme activity increased; nevertheless, this did not stop further damage from free radicals. Nevertheless, its active oxygen content was greater than that of the control. Recent studies have demonstrated that when exposed to increasing concentrations of polymeric NPs in their root systems, plants are able to strengthen defensive responses and minimize excessive buildup of ROS. One typical cause of ROS is electron leakage to O_2 during electron transport in mitochondria, plasma membranes, and chloroplasts. Equilibrium is disrupted when ROS accumulate to hazardous levels in plant cells. When the ROS level surpasses the defense system, the cells enter a state known as "oxidative stress," which leads to unending damage to the lipids, proteins, and nucleic acids in cell membranes as well as the stimulation of the production of associated defense genes in reaction to these injuries.

2. **Alteration in Antioxidant Enzymatic and Nonantioxidant Enzymatic Responses**

   a. **Alteration in Antioxidant Enzymatic Response**: Plants generate ROS in response to NP stimulation, and these molecules serve a variety of purposes, including elimination. On the other hand, too much ROS damages biomolecules and eventually kills cells. Antioxidant enzymes have the task of removing ROS. Researchers measured changes in antioxidant enzyme activity to assess the impact of Se NPs capped with L-cysteine and tannic acid on L. minor. They discovered that Se NPs significantly reduced the activity of peroxidase (POD) but had no effect on superoxide dismutase (SOD). CAT activity increased, while POD activity did not. Studies have been conducted on sodium selenite (Se^{4+}) to further evaluate the toxicity of NP Se and ionic Se. Remarkably, the ionic forms of Se increased the activities of SOD and CAT while significantly suppressing POD. In addition, because ROS burst denatured the structure of the enzyme, ions suppressed POD. Since low concentrations of Se^{4+} and Se-NPs for SOD are shown to be unable to properly elicit the defense system response for POD, it is likely that the molecular structure has changed. TiO_2, CuO, and CdSe NPs enhanced SOD activity in *S. polyrrhiza* and *L. minor* in a manner akin to that observed with ZnO and Ag NPs in *S. polyrrhiza*. Further investigation on soybean seedlings and Triticum aestivum exposed to SiO_2 NPs, TiO_2 NPs, and nano alumina verified comparable outcomes, in line with previous studies. ZnSe-NPs, $ZNC_4H_6O_4$, and $ZnSO_4$ all markedly increased CAT activity while dramatically lowering POD in L. minor. As previously mentioned, NPs and the ions they produce lead to oxidative stress. The breakdown of water and other small molecule homeostasis therefore affects mitochondria, the photosynthetic system, and other cellular processes. Zinc is a vital component of many enzymes, yet in big doses, zinc is also extremely toxic. It has been demonstrated that ZnO-NPs raise CAT activity in a variety of plants, such as *S. polurrhiza, Prosopis juliflora-velutina,* and *Pisum sativum*. There was no indication of any enzyme in wheat treated with CuO and ZnO NPs. When C. vulgaris was exposed to CuO-NPs, CdSe NPs, and elemental Cd, POD activity rose; however, this rise was transient, and POD activity decreased at high concentrations, presumably as a result of H_2O_2 breakdown. High concentrations of Zn^{2+} and CdSe NPs stopped the SOD activity of *Brassica napus*, Bt-transgenic cotton, *C. vulgaris*, and conventional cotton affected by CeO^{2-}NPs because of an excess of ROS. TiO_2 and CeO_2 NPs were observed to lower SOD activity in Karenia brevis and lettuce, respectively, by the researchers. SOD activity did, however, rise when *Phaeodactylum tricornutum* was treated with 0.52 nMCdSe/ZnS QDs. After being exposed to elemental Se and Cd2, Zea. Mays and L. minor showed increased SOD activity. Furthermore, researchers found that POD activity was decreased in onion, transgenic tobacco, and *L. minor* that were exposed to high concentrations of TiO_2, CdSe NPs, and Cd stress respectively. Following exposure to CdSe, Cd^{2+}, Zn^{2+}, and ZnSe, L. gibba showed an increase in CAT activity; *Allium cepa* showed an increase in CAT following exposure to ZnO NPs and Zn^{2+}. In the cases of *S. polyrrhiza, Cucumis sativus, Medicago sativus,*

P. vulgaris, and *Hydrilla verticillate*, increased CAT activity has been demonstrated in the presence of CdS NPs.

b. **Alteration in Nonantioxidant Enzymatic Response:** A variety of nonenzymatic compounds, including glutathione (GSH), appropriate solutes, phenolics, alpha-tocopherol (vitamin E), carotenoids, flavonoids, and proline, can mitigate the unchecked ROS generation that happens during abiotic stress. When environmental conditions like UV-B radiation, dryness, and heavy metals are present, changes in flavonoids and phenols are frequently seen. ROS can be eliminated by enzymes that are not antioxidants. L-cysteine and tannic acid-capped Se NPs, ZnSe, and CdSe NPs were observed to boost the phenol and flavonoids levels of L. minor in three different investigations. Brassica nigra and castor seedlings were similarly affected by Ag NP and ZnO NPs. This growth is linked to chelation and ROS scavenging. MDA accumulation may be utilized to examine the integrity of the membrane. Phospholipids; peroxidation of unsaturated fatty acids results in the production of MDA. Damage to cell membranes by NPs results in ROS production, which in turn causes lipid peroxidation. Research has demonstrated a direct correlation between the quantity of MDA present and the existence of ROS. Radicals produced by lipids in turn denature DNA and proteins. MDA content has been measured in a number of studies; for instance, MDA content rose when Se NPs, ZnSe NPs, or CdSe NPs were administrated to L. mirror. Previous investigation indicated that *B. rapa* seedlings and lettuce cultivated with Ag and CeO_2 NPs had higher levels of MDA.

2.1.3 Molecular Concerns

1. NPs based Cell Cycle alteration and DNA Damage

NPs have a tendency to accumulate in various food levels after being introduced into the environment. Plants are very susceptible to nanotoxicity and have the ability to absorb NPs. Plants are therefore highly valued as genetic models for environmental screening and monitoring of potential hazardous compounds. Because NPs can interact with biological macromolecules (the nucleus, cytoplasmic components, or lipids) and lower the mitotic index while elevating the micronucleus and chromosome aberration index, respectively, plants may be vulnerable to cytotoxic and genotoxic effects from NPs. Among the numerous proteins and enzymes involved in regulating the cell cycle are cyclins, checkpoint kinases, and cyclin-dependent kinases (CDKs).

Cell cycle progression is facilitated by the cyclin protein family, which binds to and activates cyclin-dependent kinases (CDKs). Among the numerous proteins that CDKs phosphorylate are those that participate in cell division, chromosomal segregation, and DNA replication. Checkpoint kinases identify errors in the cell cycle and, if required, initiate the pathways leading to DNA repair or death. Cell cycle checkpoints are DNA monitoring systems that pause cell division at crucial junctures to prevent the occurrence of genetic errors. Checkpoints have the ability to either force a cell to leave the cell cycle or die when irreversible DNA damage is discovered. Cell cycle arrest between the G1/S and G2/M phases might be gradual, cumulative consequences of PS-NH_2 NP-induced cell damage. Interestingly, neither intracellular ATP levels nor NP internalization reduced in spite of the cell cycle halt. Ag-NP exposure results in decreased cell proliferation, cell cycle arrest, and the production of reactive oxygen species (ROS) in A549, BEAS-2B, and Calu-1, but not in NCI-H358. This is consistent with the results showing that Ag-NPs stall Candida albicans' cell cycle via a ROS-mediated mechanism. Researchers have shown that exposure to Si-NP causes both apoptosis and cell cycle halt. This is brought on by cell cycle positive regulator expression. Peripheral blood mononuclear cells exposed to palladium NPs had a large rise in G0/G1 phase cells and a considerable decrease in G/S and G2/M phase cells. According to researchers, ZnO-NPs cause apoptosis and cell cycle arrest in the MCF-7 cancer cell line. According to recent studies, exposure to carbon black nanoparticles increases the risk of lung cancer by disrupting the cell cycle through systemic inflammation. Positively charged NPs are more likely to change cell cycle progression by causing DNA damage, which may be related to the action of NPs on the cell cycle. In conclusion, NPs can cause DNA damage and intracellular ROS accumulation to cause cell cycle arrest; however, a decrease in intracellular ATP levels may not occur simultaneously. Though the fundamental process is unknown, it is hypothesized that cells re-exposed to NPs may avoid the G2/M phase checkpoint and remain proliferating instead of addressing DNA damage.

Ag NP-treated wheat root tip cells showed a variety of chromosome aberrations, including spindle dysfunction, lagging chromosomes, chromosome breakage, multicore, and incorrect metaphase orientation. These findings indicate that Ag NPs are easily internalized by wheat root tip cells and have the ability to interfere with normal cell function. After being treated with Cr_2O_3 NPs for cytogenetic investigation, cells isolated from onion roots showed a much-reduced mitotic index. Different exposure levels are also associated with chromosomal abnormalities. The potential of PS-NPs to dramatically reduce the mitotic index of onion and generate various chromosomal abnormalities (such as aggression, lagging chromosomes. Etc.) spindle deformation, and nuclear abnormality was used to illustrate the genotoxicity of the particles.

Genotoxicity mechanisms may generally be divided into two categories: direct and indirect. Direct genotoxicity is caused by NPs physically interacting with DNA, whereas indirect genotoxicity might arise from a reduction in DNA repair or an increase in ROS levels. TiO_2-NPs can induce ROS and oxidative stress, which can lead to oxidative DNA damage even though they have a limited cytotoxic potential. This is consistent with recent studies that demonstrate oxidative DNA damage can be triggered by just 1g/mL of TiO_2-NPs. This result is caused by decreased GSH levels as well as increased lipid peroxidation and ROS generation. According to one research, an imbalance in the oxidative-antioxidant system causes exposure to ultra-small superparamagnetic iron oxide NPs to elevate indicators of oxidative stress and induce DNA damage in mouse hearts (Celik,) in HepG2 cells and mice, inhibition of the Nef2 pathway by TiO_2-NPs has been demonstrated to cause oxidative DNA damage.

This is consistent with other studies that demonstrate that mice exposure to titania NPs results in DNA damage and chromosomal instability. The HIF-1 /miR-210/Rad52 pathway, which is involved in DNA damage and repair, was activated by Ni-NPs exposure in vitro and in vivo. This was evident in the expression of proteins of ataxia telangiectasia. Phosphorylation levels increase. After being exposed to NI-NPs, the homologous recombination repair gene Rad52 was downregulated, miR-210 was increased, and HIF-1 accumulated in the nucleus. CuO-NPs induced oxidative DNA damage in cell death in HUVECs via copper ion-mediated p38 MAPK activation. DNA damage may also be caused by a variety of nanoparticles (NPs), such as gold (Au), hafnium oxide (HfO_2), silver (Ag), tin (Sn), indium (In), and phosphorous (PS). Specifically, the majority of NPs cause oxidative stress, which in turn causes oxidative DNA damage.

2. Epigenetics Changes

Epigenetics refer to "the study of mitotically and/or meiotically heritable variations in gene function which cannot be attributed to alterations in DNA sequence". A number of methods to modify DNA expressions without changing the sequences themselves have been found during the last few decades, including non-coding RNAs, histone modifications, and DNA methylation. These findings cast doubt on the idea that variations in DNA sequences are the only factor causing heritable characteristics. Epigenetic changes are inherited by subsequent generations and maintain their effects long after the initial stimulus exposure has finished because they preserve epigenetic memory. For fruitful reproduction to continue, epigenetic modification must be stable and reliable. However, they are constantly stressed by their surroundings, which leads to the emergence of undesirable phenotypes. Concern over the possible negative long-term effects of NPs on human health is developing because NPs disturb the epigenome's equilibrium in the environment.

a. NPs-based alteration in DNA influences gene expression, cell specialization, and the evolutionary trajectory of organisms. Two more biological processes associated with DNA methylation are transposon inactivation and transcriptional-level gene silencing. Retrotransposon activity, miRNA, DNA methyl-seq, and an MSAP variation, have been used recently to identify DNA methylation. A method called methylation-sensitive amplified. Additionally, methylation-specific PCR (MSP), gene sequencing, amplified fragment length polymorphism (AFLP), coupled restriction enzyme digestion-random amplification (CRED-RA), coupled restriction enzyme digestion-inter-primer binding site (CRED-iPBS), bisulfites method, high-performance liquid chromatography (HPLC), and gene sequencing have all been used to detect changes in DNA methylation. The methylation-sensitive enzymes HpaII and MspI are used in the CRED-iPBS method to detect changes in methylation. Research on the epigenetic changes that nanoparticles induce in plant DNA is relatively new. The aim of the study was to determine whether wheat grown in a medium containing nanoparticles had DNA epigenetic changes. Previous studies have demonstrated that ZnO nanoparticles also operate on transcription factors to change the HSFA4A gene's expression level in wheat. ZnO nanoparticle-induced epigenetic phytohormones and chlorophyll, zinc is a necessary mineral. Zinc nanoparticles have been demonstrated to improve plant growth and yield in *Cyamopsis tetragonoloba L., Gossypium hirsutum, Lycopericicum esculentum,* and *Stevia*. The research claims that plants treated with zinc nanoparticles produced plants with longer roots and shoots, more protein and chlorophyll, and greater yields. An overabundance of nanoparticles may lead to an increase in ROS production. It has been discovered that metal nanoparticles stress plants. When exposed to zinc nanoparticles, *Cyamopsis tetragonoloba L.* plants showed significant alterations in their biomass, root and shoot lengths, protein, and chlorophyll contents, and enzyme activity. Micronutrient excess or shortage may be the cause of abnormalities in the cell and a decrease in biomass, production, and quality, respectively. Electron transport, cell wall metabolism, and ethylene receptor facilitation are all critically dependent on copper. Copper is mostly used in agriculture since it is involved in a wide range of physiological activities in plants. For example, copper ions help activate metabolic pathways, generate hydroxyl radicals, and shield plant cells from oxidative damage. CuO nanoparticles can disturb the Kerbs cycle at sublethal concentrations, increasing their solubility and toxicity. CuO nanoparticle toxicity has been shown to have a detrimental impact on lettuce, mung beans, kidney beans, alfalfa, wheat, chickpeas, and many other crops' seed germination and plant growth. Epigenetics includes variations in gene function that are inherited through the generations without involving a change in DNA sequence. Histone modifications and DNA methylation are two instances of epigenetic processes. The greater polymorphism percentages seen in the current investigation after treatment with non-Fe_3O_4 NPs, as well as 2X ZnO and CuO, might be categorized as hypermethylation when compared to the typical polymorphism percentage for MspI digestion.

Conversely, the 3X concentration of all nanoparticles decreased the percentages of HpaII and MspI polymorphins, which may be explained by hypomethylation. Nanoparticles have the potential to modify DNA methylation in two primary ways: by reducing the availability of methyl donors and by changing the activity of the enzymes known as DNA methyl-transferases. Because of their pro-oxidative characteristics, nanoparticles have the potential to alter DNA methylation patterns. ROS have the ability to impact DNA methylation status and gene expression levels in addition to oxidative damage. Reactive oxygen species are known to be released as a result of nanoparticles. The harmful consequences of nanoparticles include cytotoxicity, immunotoxicity, oxidative stress, genotoxicity, and cell death. Exposure to nanoparticles causes oxidative stress, lipid peroxidation, and membrane damage, which all lead to hypomethylation and some concomitant hypermethylation. Environmental factors can also produce epigenetic changes. Since, nanoparticle shapes have been demonstrated to affect transcription. The size, shape, and method of nanoparticle synthesis, among other things, are known to affect epigenetic modifications. Numerous other factors, such as the method of administration, the dose, and the duration of the nanoparticles kept on the plant, might also have an impact on the kind and strength of the effect. Because this process is influenced by many different factors further study is required to fully understand the effects of nanoparticles on plants.

b. NPs based modification in DNA histone: As our understanding of NP-mediated alterations to DNA methylation expands, an increasing number of investigations are examining the possible effects of NPs on histone modifications. The components H2A, H2B, H3, and H4 of the octameric structures of the nuclear proteins known as histones are replicated twice. By acting as a framework for DNA to coil around and create a nucleosome, histones enable the tight packing of DNA in the nucleus. The stringency of DNA packing results from many modifications to histones' amino (N)-terminal tails, which protrude from nucleosomes and influence the transcription of nearby genes. Post-translational modifications include acetylation, methylation, phosphorylation, sumoylation, and ubiquitylation. Thus far, studies on NPs have reposted histone changes such as acetylation, methylation, and phosphorylation. DNA replication, gene transcription, and chromatin agglutination are all significantly influenced by post-translational modifications of histones, including as acetylation, methylation, phosphorylation, glycosylation, and ubiquitination. Histone methyltransferases (HMTs) catalyze one of the most significant post-transcriptional modifications: the transfer of methyl groups to the lysine residues of histones H3 and H4. Similar to DNA methylation, histone modification affects how the body responds to stress. The primary sites of histone methylation and demethylation are on the arginine and lysine residues of histones H3 and H4.

Histones that have been trimethylated (me3), demethylated (me2), or monomethylated (me_1) at particular locations all affect gene expression in different ways that might occasionally be contradictory. Examples are the histone modification markers H3K27me3 and H3K9me2, which normally suppress the expression of certain areas, and H3K4me3 and H3K36me3, which normally stimulate gene transcription. The catalytic methyl transfer site of histone lysine methyltransferase (HKMT) proteins is called the SET domain, and it is made up of about 130 amino acids. Four sets of SET domain proteins are identified: SU(VAR), Enhancer of Zeste (E[Z]), Homeotic disks (ASH1), and Trithorax Group (TrxG). The Arabidopsis plant's blooming period is regulated by SET DOMAIN GROUP 25 (SDG25) (TrxG family), CLF (E(Z) family), and SDG8 and SDG26 (ASH1 family). A reduction in H3K36me2 and H3K36me3 levels at the blooming. The KDM1/LSD1 family and the Jumonji C (JmjC) domain-containing protein family are examples of histone demethylases (HDMs), which can both oxidatively remove the methyl group from histones directly. A subset of proteins containing the JmjC domain rely on Fe (II) and -ketoglutarate (KG) as cofactors to operate on mono-, di-, and trimethylated lysines, while KDM1/LSD1 requires flavin adenine dinucleotide (FAD) as a cofactor and demonstrates demethylase activity exclusively toward mono-and demethylated lysines.

Histone H3's R2, R8, R17, and R26 residues as well as histone H4's R3 residue are the primary sites of arginine methylation. So far, three distinct types of arginine methylation have been documented: symmetric demethylation (Rsme2), asymmetric demethylation (Rme2), and monomethylation (Rme1). The evolutionarily conserved protein arginine methyltransferases (PRMTs) mediate arginine methylation, while proteins containing the JmjC domain assist arginine demethylation. Histone deacetylation amplifies this impact, while histone acetylation decreases the charge connection between histones and DNA, decreasing histone affinity for DNA and changing the accessibility of the DNA template to transcription factors (TFs). Histone acetyltransferases (HATs) and histone deacetylases (HDACs) dynamically control the levels of histone acetylation. According to researchers, both of these categories of enzymes are essential for the regulation of chromatin dynamics, which is influenced by several processes such nucleosome assembly, chromatin structure and folding, and DNA repair. HATs are classified into four subfamilies in Arabidopsis: MYST, CREB-binding protein (CBP), GNAT, and TAFII250. HDACs are classified into three subfamilies: RPD3/HDA1, HD2, and SILENT INFORMATION REGULATOR 2 (SIR2). Histone deacetylation reduces gene expression whereas histone acetylation often increases gene transcription.

3. Challenges and Future Directions

In addition to pesticides and heavy metal remediation for improving nutrient absorption and pollutant breakdown, nanoparticles provide a multifaceted approach to plant toxicity. In light of apparent potential advantages, it is crucial to research and examine the many environmental effects and potential hazards associated with the application of nanoparticles in agriculture. One of the most promising approaches to reducing soil toxicity in environmental remediation is the use of nanoparticles as a catalyst. Studies in this area help to optimize nanoparticle-based solutions for sustainable and environmentally friendly agricultural practices as the science of nanotechnology develops.

CHAPTER 10

GREEN OR BIOGENIC NANOTECHNOLOGY: A SUSTAINABLE SOLUTION FOR SUSTAINABLE AGRICULTURE AND HUMAN HEALTH

Short Summary

During salinity and drought like abiotic stress like condition created osmotic and ionic effects that hinder plants from absorbing the necessary water and encourage the hyperaccumulation of ions, which enter the transpiration stream and damage plant cells, excessive salinity annually lowers agricultural yields by a significant margin. The application of green or biogenic nanoparticles (NPs) in different sizes, shapes, and concentrations has been shown to be beneficial for several agricultural crops. These NPs help crops protect against salt stress by improving growth metrics, seed germination, changing root architecture, increasing yield, and changing biochemical profiles. Some studies have demonstrated that exposing cells to NPs can alter the expression of various genes that typically respond to salt stress, despite the fact that the underlying molecular mechanisms of this phenomenon are yet unknown. Because of this, biogenic NPs are now regarded as a practical substitute method for lessening the negative impacts of salt stress on agricultural output. Among other techniques, NPs can be created chemically, physiologically, or physically. Chemical and physical methods can be used to produce NPs with the required size and high purity, but these processes are expensive and may contain hazardous substances. Biological strategies for producing nanoparticles (NPs) are becoming more and more popular among academics because of their affordability, speed, and environmental friendliness. Here, we present an overview of current studies about producing NPs using microbes and plants. We also demonstrate how these biogenic NPs may help crops resist the damaging effects of salt stress, which could be helpful in the near future for the agricultural industry in the context of climate change.

1. Introduction

Soil salinization is a widespread problem affecting around 1100 million hectares of land, or 7% of the planet's surface. Both naturally occurring geological processes and human activity may lead to soil salinization. Due to atmospheric deposition, rising temperature, rising sea levels, and saltwater seeping into clean water reservoirs, a significant portion of the cultivated land has been adversely affected by primary salinization events. Additionally, 30% of irrigated areas are believed to be more severely affected by supplementary salinization as a result of insufficient fertilizer use, poor management practices, and increased agriculture. Significant losses in agriculture productivity are brought on by the lack of fresh water for irrigation and the ongoing degradation of agricultural fields due to salt stress, particularly in arid and semiarid regions. Soils began to get saltier hundreds or even thousands of years ago. Places like the Aral Sea Basin in Central Asia, the Yellow River Basin in China, the Murray-Darling Basin in Australia, the San Joaquin Valley in California, Gujrat, Uttar Pradesh, Maharashtra, Bengal, Rajasthan in India, and the Ararat Plain region in the Republic of Armenia have all seen alarming rates of agricultural land degradation over the past 50 years. Climate change is increasing salinity, which renders around 6% of agricultural land unsuitable due to severe stress conditions. Plant growth is impeded by a strong stressor that produces osmotic pressure and ion toxicity. This, in turn, disturbs the physio-biochemical profile, nutrient absorption, and ionic homeostasis of the plant, and modifies the expression of genes linked to plant development processes.

Plant growth and development are negatively impacted by soil salinity. The accumulation of water-soluble salts in the root zone, such as potassium (K^+), sodium (Na^+), chloride (Cl^-), and sulphate (SO_42^-), causes osmotic fluctuations and impairs the ability of plant root cells to absorb water from the soil. This process is known as soil salinization. Superoxide radicals with the presence of salt ions cause hypertonic salt stress, which damages plant cells. The two most frequent ions that induce soil salinity are Na^+ and Cl^-. Sodicity is caused by an excess of Na^+ among exchangeable cations. Hyperosmotic stresses and ionic instability cause plant tissues to accumulate reactive oxygen species (ROS), including superoxide radical (O2•), singlet oxygen ($^1O_2$), hydrogen peroxide (H_2O_2), and hydroxyl radical (OH•). Lipid oxidation brought on by an excess of reactive oxygen species (ROS) results in the generation of malondialdehyde (MDA), which disrupts physiological functions and cellular mechanisms. This has detrimental effects on membrane integrity, which in turn interferes with membrane biological functions such as cellular signalling, molecular transport, and separation if activities. Salt's nutritional inadequacies not only result in hyperosmotic stress and electrolytic disruption, but also modify the physiological water potential of leaves. Water stress is brought on in plants, and this leads to stomata closing, decreasing gas exchange and raising the risk of oxidative damage. Decreased gas exchange as a result of stomatal closure, which reduces photosynthesis. The degree to which a pant can withstand salinity is also determined by its kind; for example, halophytes may withstand surroundings that are saltier than glycophytes. Because they are able to maintain ion homeostasis by removing Na+ ions from plant cells, glycophytic crops--such as wheat (*Triticum aestivum L.,* sunflower (*Helianthus ammus L.*), barley (*Hordeum vulgare L.*), peanuts (*Arachis hopogaea L.*), rice (*Oryza sativa L.),* corn (*Zea mays L.*), soybeans (*Glycine max L.*), cotton (*Gossypium hirsutum L.*), sorghum (*Sorghum bicolor L.*), and oats (*Avena sativa L.*)—are able withstand high salt levels. Glycophytic plants contain a number of defense mechanisms against osmotic stress, such as preserving ion homeostasis and boosting the accumulation of relevant molecules like proline, which helps to stabilize the tertiary structure of proteins. Proline is a nontoxic, low-molecular-weight osmoprotectant that functions as a signaling molecule, stabilizes cellular structure, and scavenges reactive oxygen species (ROS) produced in response to salt stress. Numerous antioxidant subtypes support plants including glutathione reductase (GR), catalase (CAT), ascorbate peroxidase (APX), and superoxide dismutase (SOD). These enzymes are essential in reducing the negative effects of ROS generation.

A relatively new branch of research called nanotechnology finds applications in a wide range of industries, including biotechnology, microbiology, medicines, agriculture, and health. NPs show potential as a remedy for salinity-induced plant stress. In response to salt stress, NPs can scavenge reactive oxygen species (ROS), increase the activity of antioxidant enzymes, and decrease the Na^+/K^+ ratio in plant roots, depending on their size (often ranging from 1 to 100 nm). Nanoparticles (NPs) have the potential to have a favourable or negative impact on the plant system. Many NPs in a very high concentration can generate oxidative stress on biomolecules, which can harm or even kill plant cells.

When present in appropriate numbers or at the proper dose, NPs can play a crucial role in controlling the growth and development of plants. Plant stress adaptive ability is enhanced by NP application through the modulation of important metabolic cascades, boosting of antioxidative defense machinery, and detoxification pathways from free radicals. It has been demonstrated that some plant parts, such as seedlings, fruits, flowers, stems, bark, leaves, peels, and roots, may be created. This enables the reprogramming of various physio-biochemical processes in plants under difficult conditions. Among these are iron (Fe), copper (Cu), gold (Au), platinum (Pt), palladium (Pa), silver (Ag), zinc, (Zn), selenium (Se), and so forth. Moreover, it has been shown that certain NPs have a significant role in regulating how crops respond to salt stress. Nucleic acids (NPs) possessing great potential for relaxing various environmental constraints have been to be produced by microorganisms, including bacteria, fungi, viruses, and actinomycetes, with comparable levels of effectiveness. A method for synthesizing NPs is provided by green synthesis ways like biological means, which are affordable, environmentally friendly, and less harsh than chemical or physical processes. Moreover, biological synthesis allows for the final product's size and shape to be customized for certain applications. It is now widely known that many organisms produce inorganic chemicals either extracellularly or intracellularly. Only a few of the metals being investigated for their potential as reducing or stabilizing agents in synthetic synthesis include Au, Ag, Cu, Cd, Pt, Pa, TI, and Zn NPs. As a result, some most recent review papers concentrate on how plants and bacteria produce various NPs and how salt stress employs them. Our goal in conducting this study was to gather current data about the efficacy of biogenic NP treatments in addressing salt stress.

2. Concerns about Chemical Grade NPs: Bioaccumulation and Health Implications for Plants and Humans

2.1 Application with Bioaccumulation Toxic Effect of Chemical-Grade NP in Plants

The potential of nanotechnology to improve traditional farming methods, increase food processing and distribution efficiency, improve the quality of the agricultural environment, and open the door to the development of genetically modified organisms (GMOs) bodes well for the field of agriculture. Because of their small size and higher surface-to-mass ratio characteristics, microscopic NPs are more likely to accumulate in many species and have a larger ability to pass through biological barriers. We talk about how farming is evolving in this part in light of nanotechnology. Zinc nanoparticles (Zn NPs) and other metal particles with highly defined surface areas are often rounded or polished. Zinc nanocrystals can be used as antibacterial, antibiotic, and antifungal agents in buildings, clothes, nanofibers, polymers, and textiles. Furthermore, Zinc Oxide Nanoparticles (ZnO NPs) are widely used in man cosmetics, sunscreens, textiles, hair colorants, and other beauty and hygiene products. Among the industrial used are solar cells, optical and electrical materials, antibacterial agents, floor coatings, and so on. Foliar sprays are one of the most evident ways that ZnO NPs enter the environment. ZNO NPs are sprayed on crops to protect them against abiotic stressors, such as salt stress in addition to their usage as an abiotic stress protectant. The nutritional value of ZnO NPs may support crop growth in salt settings. According to a study, treating rice with ZnO NP can mitigate the effects of salt stress. Zinc is a cofactor for the phosphate-solubilizing enzymes phosphate and phytase, whose levels were elevated by ZnO NPs. Phosphorous absorption by V. radiata increased by 10.8% as a direct result. ZnO NPs that were biosynthesized led to a considerable rise in biochemical indicators such as chlorophyll and leaf protein levels, as well as an improvement in penology such as root volume and stem height in plants. For the foliar application study, two-week-old mung bean plants were utilized since the authors considered both the treatment method and the impact. Using an atomizer to produce droplets, 25mL of ZnO NPs suspension at a concertation of 10mg L^{-1} was sprayed over each plant. It was demonstrated that these advantageous outcomes continued even following foliar therapy.

Certain plants are able to take up nanoparticles (NPs) and store them in their tissues, where they can become harmful. These are the main mechanisms by which NPs poison plants. During NP-plant contact, increased ROS formation causes oxidative damage, NP-induced transcriptional response, and genotoxicity through NP-DNA or NP-organelle interactions. Studies on the safety of food crops, such as maize, wheat, and soy, frequently look at factors including the pace of germination and root growth and the effectiveness of nitrogen fixation. It has been demonstrated that the concentrations of NP in soil are substantially greater than in air or water. The majority of NPs that are present at higher trophic levels in the food chain have plant origins. So, after exposing *Nitzschiaclosterium* samples to TiO_2 NPs of different sizes (21,60, and 400 nm) for 96 hours, researchers looked at the results. The concentrations of 21 nm TiO_2 NPs and 60 nm TiO_2 NPs that were evaluated ranged from 0 to 100 mg/L and 360 mg/L, respectively. 400 nm BPS NPs have been studies at concentrations ranging from 0 to 500 mg/L. For NPs with sizes 21 nm, 50 nm, and 400 nm, the effective concentrations (EC50) values are 80 mg/L, and 180 mg/L, respectively. The EC50 data showed a tendency to decrease intrinsic toxicity with increasing NP size. Using the following assay, the possible genotoxicity of TiO_2 NPs has been investigated in *Allium Cepa*: TiO_2 was examined for toxicity at different concentrations (0.02, 4.0, 8.0, and 10.0 nM), and the results showed that the substance was more harmful at lower concentrations because of the precipitation of NPs. Additional damage led to chromosomal anomalies, DNA folding, and the production of ROS. AG NPs were readily taken up by barley root tip cells. Since the particles can interfere with normal cell function, researchers examined the effects of AG NPs on chromosomal aberrations and cell division. They found that treated root tip cells had chromosomal abnormalities ranging from multicore to inappropriate alignment during metaphase, chromosomal damage, metaphase plate deformation, spindle dysfunction, and lagging chromosomes. The mitotic index significantly decreased in onion root cells exposed to Cr_2O_3 NPs, according to cytogenetic analysis. Different exposure levels were also associated with chromosomal abnormalities. Prior research has demonstrated that internalizing Cr_2O_3 NPs by cells can lead to DNA damage, which in turn causes an increase in SOD synthesis as a defense against oxidative stress. The proof of the gene toxicity generated by PS-NPs was demonstrated by their ability to significantly lower the onion's mitotic rate and cause a variety of chromosomal abnormalities, including clumping and trailing chromosomes, spindle distortion, and nuclear dysfunction. Over the course of six weeks, researchers watered soil-grown Arabidopsis thaliana plants at various doses (ranging from 0 to 300 mg/L) to assess the impact of ZnO NPs. It has been demonstrated that Zn^{2+} ions cause well-known oxidative stress in plant cells, which damages the cells. Furthermore, all plants showed a reduction in gene expression, photosynthesis, and growth inhibition in response to NP concentration. Similar studies on Ag NPs in *Arabidposis thaliana* found that while the NMs had no effect on seed germination, they were deleterious to root elongation. Its toxicity mechanism may be explained by root surface absorption of NPs, which disrupt the thylakoid membrane, reducing chloroplast content and eventually inducing growth suppression. Other scientists evaluated the toxicity of soluble $AgNO_3$ and Ag NPs in relation to wetland vegetation. They were able to demonstrate the considerable effects of Ag NPs (40 mg/L) coated with polyvenylpyrrolidine (20 nm) and Arabic gum (6 nm) on plant growth and seed germination. $AgNO_3$, however, showed a very little negative effect. The study discovered a perplexing relationship between NP toxicity and size, with smaller NPs having an easier time penetrating plant cell wall. Over the course of four months, wheat *(Triticum aestivum* L.) was grown in soil with different concentrations of Ag NPs (20,200, and 2,000 mg kg^{-1}). Reduced biomass, shorter plant height, lighter grains, and lower micronutrient content (Fe, Cu, and Zn) were all signs of severe phytotoxicity at high dosages. In another study, Ag NPs were also demonstrated to have an impact on chromosome abnormalities. MWCNTs are chemicals that, according to researchers, diminish the cellular density of this species in a dose-dependent manner. Larger doses (80 mg/L) resulted in intrinsic cytoplasmic discharge and disruption of cell membranes, whereas lower dosages (20 mg/L) induced apoptosis. In cellular suspension tests, cellular defensive mechanisms were seen as cells clustered together in response to the NPs. Studies on oxide nanoparticles (NPs) have also been published; assessments conducted by researchers looked at the full cycle of metal nanoparticles (NPs), such as copper oxide (CuO) and iron oxide (Fe_2O_3) at 50 and 500 mg/kg. An analysis of peanut grains (*Arachis hypogaea*) exposed to all the nanoparticles—aside from 50mg/kg TiO_2 NPs—showed a decrease in the total amino acid content. CuO NPs dramatically decreased the biomass of young shoots because of a dose-dependent accumulation within the particles. With chemical-grade NPs, dose-responsive impacts on crop quality and production could be achievable.

2.2 Bioaccumulation of Chemical-Grade NPs in the Food Chain and Potential Toxic Effects in Humans

Understanding the biodistribution of NPs in the food supply chain is essential to understanding their toxicity and bioassessment. Because they include these compounds, a variety of contaminants have the potential to introduce NPs into the food chain. Examining the food chain greatly aids research on the absorption of NPs by high-trophic-level species (Fig.1). Numerous organism-based models are used to study the effects of NPs, such as the marine food chain, *Chlorella Vulgaris*, mammalian cells, *Epinephelus cioiodes, Eisenia Andrei, Dunaliella tertiolecta*, the algal-zooplankton food chain, the planet's food chain, and the Arctic polynya ecosystem. Three of the most significant factors that might affect NPs' migration up the food chain are their physiological characteristics, the duration of exposure, and the route of transfer. Through the process of biosorption and bioaccumulation, which are both crucial to this process, NPs may be followed up the food chain. Higher up the food chain organisms are exposed to inorganic NPs through soil, water, and food. Some of the other ways that these NPs might cause harm include alternating ROS levels, oxidation stress, and the breakdown of cells, DNA, and mRNA.

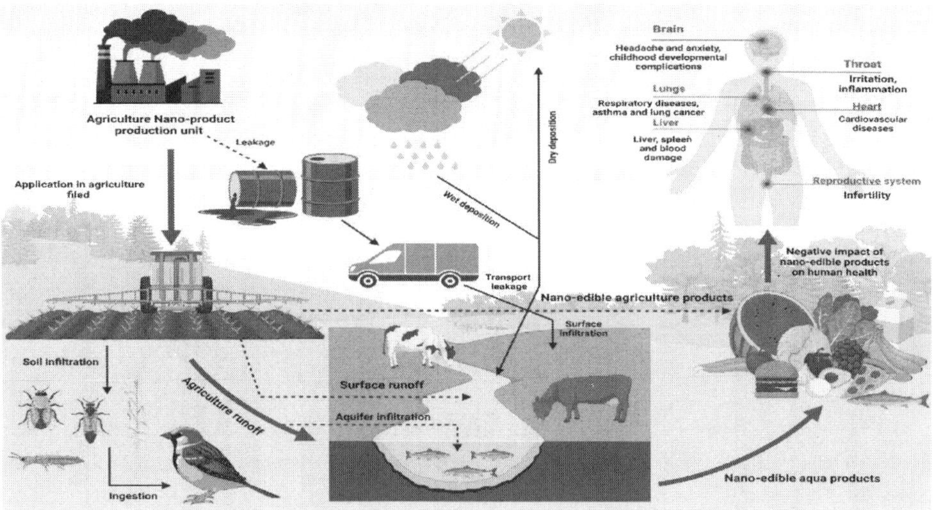

Figure 1: Flow of agronano-products in environment to food chain and their negative impact on human health.

It is impossible to corelate the potential effects of NPs on the people without first researching their toxicokinetic and biodistribution in animal models. Figure 1 examines several aspects such as cytotoxicity, genotoxicity, cancer, and reproductive toxicity. However, because of the difficulties in using experimental animal models—which need close observation over several months or even years and present ethical issues because so many animals are killed—the majority of trails and research are carried out in Virto. A common practice in animal toxicity testing models to measure alterations in organs, gene expressions, or blood parameters.

2.2.1 Toxic Effect of Chemical Grade NPs on the Human Respiratory System'

Alveoli, blood vessels, lymphatic veins, nerves, and pleura are among the tissues that make up the lungs; the respiratory tract is the set of organs that permits gases to be transferred between the human body and the surrounding air. The main factors that determine particulate matter (NPs) deposition in the respiratory system are size, shape, and surface chemistry (Fig.1). Once deposited in the lower airways, small NPs are more likely to interact with lung cells in the alveolar region. The surface chemistry of NPs has an impact on deposition as well; hydrophilic NPs are less likely to be inhaled than hydrophobic ones. Deposition is also influenced by particle type, with longer particles having a higher chance of being swallowed than spherical ones. NPs can interact with several lung cell types, such as fibroblasts, epithelial cells, and alveolar macrophages, after they reach the respiratory system.

These cells possess the capacity to both pinocytose and phagocytose NPs. NP absorption can set off a chain reaction of cellular responses that includes the generation of ROS, the release of cytokines, and the activation of inflammatory pathways. The cellular response is significantly influenced by the dosage, particle size, and surface chemical markup of particles exposed to NP. A number of detrimental effects, including fibrosis, oxidative stress, inflammation, and perhaps lung cancer, have been linked to NP exposure. The amount of exposure and the physiochemical properties of the particle have a significant impact on the negative effects of nanoparticle exposure. Titanium dioxide (TiO_2)-NPs have been shown in animal experiments to induce lung fibrosis and inflammation. The harmful respiratory effects of NP exposure are further exacerbated by preexisting respiratory conditions such as asthma or chronic obstructive pulmonary disease. Previous studies connected a number of NPs, including, SiO_2-NPs, carbon-based NPs, AG-NPs, and ZnO NPs, to harm to the respiratory system. The study demonstrated how the internalization of two distinct kinds of nanoplastic components by human lung epithelial cells resulted in cell cycle arrest and death. According to the study, SiO_2-NPs have the potential to penetrate the pulmonary capillary membrane and enter the systemic circulation, where they may result in cardiotoxicity. Ag NPs of different sizes cause respiratory discomfort and increase the production of Hsp70 protein in Drosophila. Ag NPs were also discovered to have strong harmful effects on fruit fly lungs. Numerous studies have revealed that rats given intratracheal CuO-NPs may develop tumor lesions, oxidative damage, and discomfort. NPs upset the balance of the immune system, causing immunosuppression or overactivation, which lowers the body's defense against respiratory infections.

2.2.2 Toxic Effect of Chemical Grade NPs on the Immune System

Innate and acquired are the two primary categories. Acute phase proteins, cytokines, complement, neutrophils, monocytes, and macrophages are immune system components that are members of the innate immune system, which is in charge of the body's unwavering rejection and elimination of antigenic substances that enter it. This response has a vital function in sustaining life, as evidenced by the fact that it is so widely preserved, even in the most primitive creatures. Specific immunity, or acquired immunological response, is a phenomenon that indicates the immune system of higher animals. Both B and T lymphocytes contribute to this response by exhibiting particular reactions to the antigen. When it comes to building specialized immunity, time is of the essence. Although most of these studies have concentrated on the innate immune response, research has indicated that immune cells may recognize NPs and use this identification to modify the immune response. When exposed to Cd NPs, oysters (*Crassostrea gigas*) showed less immunosuppression and haemophagocytosis. These results were supported by the discovery of transformation in mice, which resulted in immunodeficiency.

2.2.3 Carcinogenicity and Toxic Effects of Chemical Grade NPs

Cancer is a multifaced illness that impacts millions of individuals all over the world and continues to be a major killer despite substantial breakthroughs in treatment and prevention. The transformation of healthy cells into malignant tumor cells is called carcinogenesis. It's a multi-stage procedure that begins with initiation and continues through ascent and development causing precancerous lesions to form by altering the DNA or chromatin of healthy cells. It causes benign tumors to form, encouraging the growth and spread of precancerous lesions. Acquiring new mutations or epigenetic modifications that pave the way to malignancy is what we mean when we talk about progression. Numerous mechanisms have been shown to the development of cancer; these include increased cytokine and chemokine production, cellular inflammation, immunosuppression, altered cellular death, repression of cell cycle, and shortened lifespan of important cell types linked to both innate and adaptive immune responses. It is important to highlight that most of the studies on NP malignancy have been done on metal NPs, as opposed to other types of NPs. This could have something to do with NPs' molecular characteristics. For example, even in non-ionized forms, NPs have the potential to cause cancer. The carcinogenicity of NPs other than metals, such as ZnO-NPs, hexavalent Cr NPs, and Ni NPs, needs further investigation. A crucial subject for more research is the relationship between NP carcinogenicity, materials properties, material properties, and NP size at the nanoscale (Fig. 1). There aren't many studies on NPs' carcinogenicity, which could be because cancer takes a long time to develop and spread.

3. Biogenic NPs: An Eco-Friendly Approach for Sustainable Agriculture

3.1 Definition and characteristics of biogenic NPs

Green or biogenic, nanotechnology is a rapidly emerging field that offers novel ways to manufacture NPs with little environmental impact. Biological synthesis approaches enable the production of NPs utilizing natural stabilizing and reducing agents. This approach is low-cost and safe for the environment because it doesn't require energy or dangerous chemicals. In a bottom-up process known as biological synthesis, microorganisms such as bacteria, fungus, yeast, algae, and plant components including stems leaves, roots, seeds, flowers, and buds are utilized to generate NPs [55]. Both external and intercellular microbial culture filtrates are used as a reducing agent in the biosynthesis process to create

NPs. Yeast, fungus, bacteria, and actinomycetes are just a few of the many microbial kingdoms that contain metal-resistant microorganisms.

These characteristics enable microbes to create metal ions, accumulate metal reserves, and withstand exposure to metals. For example, the first bacterial Au NPs were produced using Bacillus subtilis [29]. The microorganisms were also used to make reversible-faced Ag, AU, Cu, Fe, Zn, Pt, and Se NPs. Often, the intercellular redox reaction pathway transforms metals into metal ions. The enzymes NADH and NADH-dependent nitrate reductase first bind the metal to the surface of the microbial cell, where it is reduced. The metal is then released as metal ions [30]. The function of these enzymes, which carry out electron shuttle donor actions during nanoparticle creation, has been demonstrated by *Bacillus licheniformis'* production of Ag NPs. *Fusarium oxysporum* employed anthraquinones and nitrate reductase to make Ag NPs. The microbial and fungal production of metallic nanoparticles (NPs) implies that NADPH nitrate reductase might be a crucial biofactor in the generation of metallic NPs. Even though green NPs provide an alternative to the conventional approach, it is still preferred to synthesize and produce NPs using straightforward techniques. In the fictitious universe, a wide variety of plant and microbiological life forms were available. However, from the perspective of synthesis and commercialization, these bioresources must be used. Bottlenecks in nanoparticle synthesis and use include the expense of managing microbes, the challenge of scaling up production, the risk of genetic mutation, and the challenges of mass production culture and post-production processing. Using plants as a resource to produce nanoparticles is one way that future research may proceed with more clarity. Alkaloids, flavonoids, terpenoids, polysaccharides, phenolic acids, vitamins, minerals, antioxidants, saponins, and tannins are just a few of the numerous beneficial chemical components found in plants. The elimination of these elements effectively shrunken and stabilized the NPs. The abundance of plant resources, cheap cost, easy scalability for industrial-scale production, existence of secondary metabolites, and purgative effects are other advantages of using plants as a synthesis resource. Biological entities can be optimally used to create well-characterized very stable nanoparticles.

4. Biogenic NPs Synthesis and Characterization Techniques

4.1 Approaches for the synthesis of biogenic NPs

1. Top-down approach

Top-down technique includes disassembling large materials into increasingly smaller parts to produce NPs with the lowest possible size. Such NPs (Fig. 2A) were produced using a variety of techniques, including photolithographic procedures, grinding, sputtering, and milling. Each method has advantages and disadvantages in terms of how well NPs are produced. The top-down strategy is a helpful tool that might successfully produce a large number of NPs. The methods' primary disadvantages are their high initial cost and requirement for extensive deployments. These methods are excessively expensive and inappropriate for large-scale production. The process can be used in a lab environment that is regulated. The grinding of raw materials is the main step in the process. These methods are not appropriate for soft samples. Physical vapour deposition, chemical deposition, ion implantation, electron beam lithography, and X-ray lithography are a few examples of top-down methods for creating NPs.

2. Bottom-up approach

Botom-up techniques may be used to manufacture nanoparticles. These techniques include decreasing material components to the atomic size and then going through a sequence of procedures that lead to the development of nanostructures (Fig.2A). Physical forces operating at the nanoscale integrated smaller, more stable components into bigger ones as time went on. The idea of molecular recognition, or self-assembly, is essential to this strategy. Self-assembly is the process of continuously assembling information about one's own kind with oneself. For the mass production of NPs, a number of these techniques have either recently been used or are currently in the experimental stages. Laser pyrolysis vapour deposition, sol-gel processing, chemical or electrochemical nanostructural precipitation, bioassisted synthesis, self-assembly of monomer molecules, plasma or flame spraying synthesis, and other methods are production techniques used to create NPs. The "bottom-up" approach is suggested in contrast to the "top-down" strategy, which necessitates costly and time-consuming specialized equipment and a drawn-out procedure to build NPs. The process of producing NPs via "bottom-up" green synthesis is becoming more and more popular.

4.2 Methods for the synthesis of biogenic NPs

4.2.1 Physical

Nanomaterials are produced by physical methods including deposition, sputtering, ball milling, and plasma-based techniques. The majority of these methods produce metal NPs at a relatively slow pace (Fig. 2B). For example, 50% of the materials are produced during ball milling procedures and nanomaterials. It has been discovered that there is a broad particle size dispersion since only 6-8% of the sputtered material has a size lower than 100nm. Plasma and laser ablation producers require large energy inputs. The majority of physical methods are excessively expensive due to their large variety of sizes, sluggish manufacturing speed, impurities, and high energy usage. There are two ways to approach the physical synthesis of NPs: "top-down" and "bottom-up". For example, the bulk materials in the top-down technique are reduced to nano-sized particles by high-energy ball milling. NPs of the right size and shape are not consistently produced by mechanical crushing. The bottom-up approach, as opposed to the top-down approach, could result in NPs that are more evenly distributed and uniform in size. Laser evaporation is one instance of a bottom-up strategy. Other physical techniques, like the wire explosion and inert-gas condensation processes, are also used in the synthesis of NPs.

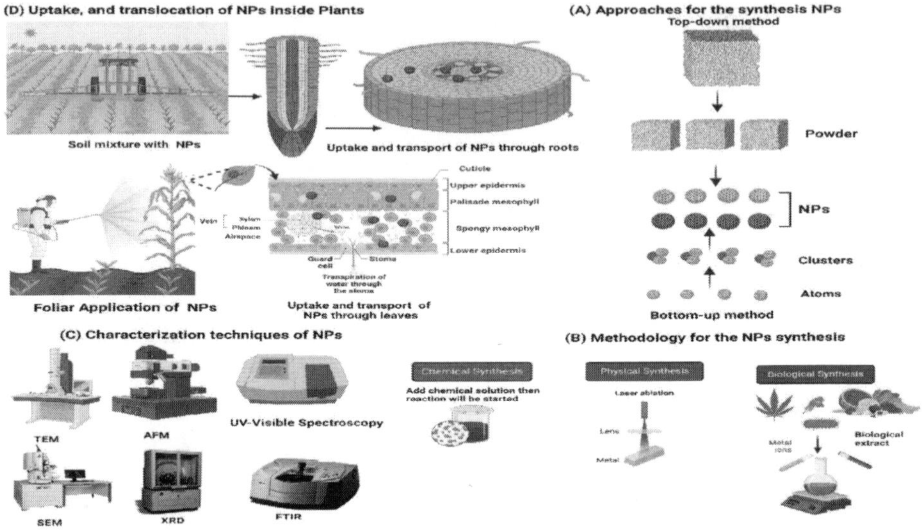

Figure 2: A figure denotes (A) bottom-up and top-down approaches with (B) chemical, physical, and biological (biogenic) methods of synthesis of NPs with different (C) techniques for their characterization and (D) soil and foliar methods of application with translocation of these NPs.

4.2.2 Chemical Method

There are several chemical methods that have been proposed to create NPs. Chemical reduction, pyrolysis, microemulsion, hydrothermal synthesis, polyol synthesis, and chemical vapour deposition are a few of these processes (Fig. 2B). Using hazardous chemicals and reagents during the synthesis process and waste product manufacturing is also bad for people and the environment. Consequently, these NPs can only be used in the biological domain. NPs are made by bottom-up chemical synthesis. Through the process of nucleation, metal cations are transformed into metallic monomers by using water-soluble cations as a precursor. Once a cluster of reduced atomic metals has formed, the particle is prevented from expanding further by capping of reducing agent. The particles attained a heat-resistant size. Atoms and smaller molecules interact to form nanomaterials. The sol-gel method, pyrolysis, chemical vapour deposition (CVD), microemulsion hydrothermal, polyol synthesis, and plasma-enhanced CVD are a few methods for chemical synthesis. Metal nanoparticles (NPs) are produced in a chemical solution during the chemical preparation process, which makes use of a range of chemical interactions and chemical compositions. Metal ions in reversed micelles are typically chemically reduced in a nonpolar solvent to produce metal NPs.

4.2.3 Biogenic Method

Biogenic nanoparticles (NPs) can originate from a variety of natural sources and may be a typical component of biological activities. Biogenic nanoparticles (NPs) are ubiquitous in diverse ecosystems and play essential roles in them. Examples of biogenic NPs that occur naturally include the following:

1. Plant Sources and Synthesis

Plant extracts are used as reducing agents in the production of nanoparticles, which has various benefits over biological procedures since labour and time-intensive phases like cell culture and maintenance are eliminated (Fif 2B, Table 1). Plant nanoparticles can be produced in an economical, humane, and ecologically responsible manner. Fruit, bark, pericarp, and root extracts are used to synthesize Ag, Au, Pt, and Ti nanoparticles in a variety of sizes and forms. Secondary metabolites such as alkaloids, glycosides, enzymes, phenolic compounds, amino acids, and flavonoids are suggested to aid in the bioreduction of metal ions.

These metabolites are found in all viable plant sources, including fruit, roots, leaves, and bark. For the production of NP, several plant leaf extracts have been used. These days, a great deal of Ag NPs with diameters ranging from 10 to 50nm are created using various leaf extracts namely those of *Piper colubrinum* leaf extracts. In a different investigation, extracts from *Youngia japonica* leaves were used to create Ag NPs that had a spherical, 20nm diameter. In the past, studies have utilized tea leaf extract to aid in the production of silver nanoparticles. For researchers, plant-based Cu NPs provide a number of benefits over synthetic ones, such as lower cytotoxicity, economic feasibility, sustainability, lifespan, strong antioxidant activity, and the potential to mitigate the effects of salt stress. Additionally, Cu NPs were extracted and synthesized from a variety of leaves, including those of *Avicennia marina, Cissus arnotiana, Moringa oleifera,* and *Allium eriophyllum*. Biogenic zinc nanoparticles are also employed in research pertaining to salt stress. When applied to brassica and faba beans at concentrations of 15 mg/L and 50 and 100 mg L^{-1}, respectively, mentha for instance, leaf extract, showed a favourable effect on the plants' growth and development. Additionally, nickel oxide nanoparticles can efficiently remove anionic pollutants from soils, they have high environmental remediation potential. Plants such as *Andrographis paniculata* and *Eucalyptus globulus* have been utilized in the biosynthesis of neodymium oxide and lanthanum oxide nanoparticles (NPs) using leaf extracts. The bioreduction activity responsible for producing nanoparticles (NPs) require the presence of plant components such as flavonoids, tannins, terpenoids, polysaccharides, alkaloids, and saponins. Furthermore, the plant components increase the stability of NPs. Using ecologically friendly methods and reducing waste is achieved by synthesizing nanoparticles (NPs) from plant biomass and extract. Plant stem extracts have been used extensively in the biological production of nanoparticles. Extracts from the stems of plants including *Momordica charantia, Salvadora persica,* and *Coleus aromaticus* have been used to synthesize Ag NPs. Stem extracts from *Leucas lavandulifolia* were utilized to create spherical Se NPs. Zinc oxide nanoparticles (ZnO NPs) derived from the stem extracts of *Mussaenda frondosa* and *Swertia chirayita* demonstrate potent antibacterial, structural, and enhanced photocatalytic properties. It has been demonstrated that these ZnO NPs inhibit the actions of -glucosidase and -amylase enzymes and increase their photocatalytic, basic and potent antibacterial activities in response.

2. Microorganisms: Sources and Synthesis

Another technique for producing NPs microbially-mediated synthesis (Table-1). Filtrates from both extracellular and intracellular bacterial cultures are reduced in this synthetic process to produce NPs. Microbes such as yeast, fungus, bacteria, and actinomycetes can survive in harsh conditions and withstand metals. These characteristics enable microbes to withstand metals accumulate metals and convert metals into appropriate ions. Additionally, the bacteria produced metallic nanoparticles (NPs) with a range of faces, such as Ag, Au, Cu, Fe, Zn, Pt and Se.

a. **Bacteria:** The use of microorganisms as a biogenic NP source has significant potential. Bacteria can also withstand extreme temperatures, high salt concentrations, extreme highs, extreme lows, and excessive alkalinity or acidity (Table 1). Microbe's ability to precipitate these chemicals from cells makes it easier for them to biologically synthesize NPs, which is useful since these substances may be collected from cells using cell filtration. To make nanomaterials, aerobically grown bacteria are allowed to reach the required optical density. The growth medium containing the bacteria is then combined with the precursor to nanoparticles. It has been incubated long enough to justify centrifugation at high rates (10,000 RPM) if there is a medium color shift. The nanomaterials get suspended in the supernatant as a result of this spin. The interactions between the bacteria and the precursors used to make them dictate the size and form of NPs. Ag nitrate and choloauric solution were interacted with by E. coli to produce Ag and Au NPs with irregular and triangular forms, respectively. The primary cause of the spherical shape was E. coli. The final size of the nanoparticle can depend on whether it is located extracellularly or intracellularly and how it interacts with bioactive substances. Compared to NPs produced extracellularly by bacteria, those produced intracellularly are often smaller. Many nanomaterials, such as zinc sulfide, lead sulfide, ferrous oxide, and selenium, have been synthesized using bacterial systems. The bacterial species *Pseudomonas proteolytica, Arthrobacter gangotriensis, Pseudomonas antarctica, Pseudomonas meridiana,* and *Arthrobacter kerguelensis* have all been associated with various types of bacteria, including Ag NPs. The periplasmic area of Bacillus sp. was used to create Ag NPs. Researchers also used Acinetobacter sp.

SW30 to manufacture Au NPs. It has been shown that using distinct strains of the same microorganisms can change the size of NPs. For instance, it was discovered that Au NPs of different sizes were produced by two different strains of *Pseudomonas aeruginosa*. Au NPs were generated in two different shapes: spherical (10-15 nm) and triangular (50-500 nm) using the *Rhodopseudomonas capsulata*. ZnO NPs have also been produced using Lactobacillus plantarum and Serratia ureilytica. Super paramagnetic Fe_2O_3 NPs with a diameter of around 29.3 nm were created by researchers utilizing the Bacillus cereus strain, which resulted in some fascinating work. *Streptomyces sp.*, $ZnSO_4$, and $MnSO_4$ were utilized to swiftly and readily manufacture Zn and Mn NPs.

b. **Fungi:** Fungi provide several advantages over bacteria when it comes to NP synthesis, including lower environmental impact, scalability, and economic feasibility in addition to producing highly effective, monodispersed NPs with desired morphologies (Table 1). The ability of fungal enzymes to synthesize is higher than that of other microorganisms' enzymes. The enzyme reductase in fungal cells is primarily responsible for the reduction process that yields metallic NPs. Because of their enormous biomass output, fungi cultures are employed for the biological generation of NPs. Fungi have a high tolerance for a variety of metals and are less challenging to work with. Fungi created a variety of proteins that help to stabilize NPs during the biogenic production process. The fungal mass's resilience to agitation and high pressure enhances scalability. By adjusting the growth circumstances, NPs with specific qualities can be produced by modifying the metabolism of fungi. Investigators by reducing aqueous $AgNO_3$, it was shown that Verticillium, a genus well-known for the destructive Verticillium Wilt, could create Ag NPs on their cell walls. Proteins present within cells can potentially contribute in the creation of NPs. *Trichthecium;s* internal enzymes produce Au nanospheres and nanorods in opposition to *Verticillium luteoalbum*. Like other ecologically benign technologies, fungus-produced nanoparticles (NPs) have a wide range of potential applications, from optoectronics to medicine. A study suggests that Au NPs can be produced using the fungi *Fusarum sp., Fusarium oxysporum, and Aureobasidium pullulans*. Furthermore, they postulated that the production of Au NPs occurs inside the fungal cells' vacuoles and that the NPs are stabilized by secondary metabolites such as protein or other biomolecules, which function as coating agents. It has also been possible to produce Ag NPs with average size of 20-50 nm using *Fusarium oxysporum*. Ag NPs were produced by researchers using *Candida glabrata* and released from the cells. Certain fungi may also produce nanoparticles (NPs) of metallic oxides; Aspergillus, for instance, is able to produce the Fe_2O_3 seen as magnetite. These nanoparticles have a variety of uses, including location sensing and magnetic resonance imaging (MRI). ZnO NPs generated by Aspergillus niger and *Aspergillus nidulans* as mediators to produce ZnO and CoO Nps, respectively. Furthermore, it was shown that *Fusarium oxysporum* may be employed to produce Pt NPs with a size range of 100-180 nm.

c. **Algae:** A variety of algae have been employed to provide ecologically friendly methods of producing metal nanoparticles. The use of algae in the synthesis of NPs has also grown in favour recently. The process of producing biogenic NPs using algae as a reducing agent or biological template is known as "algae-mediated synthesis". Diatoms, cyanobacteria, and green algae are all workable choices to finish the procedure.

Table1: The potential use of different plants and microorganisms for synthesis of biogenic NPs.

Extract Sources	Classification of Sources	Type of Biogenic NPs	Size (nm)	Shape
Sesbania aculeata	Plant	CuO	10.1	Spherical
Cuminum cyminum	Plant	TiO_2	15.17	Spherical
Cornus mas	Plant	Fe_2O_3	20-40	Spherical
Punica granatum	Plant	SNPs	20	Spherical
Alpinia galanga	Plant	AuNPs	10-30	Spherical
Withaniacoagulans	Plant	AgNPs	14	Spherical
Adiantum lunulatum	Plant	CuO	1.5–20	Quasi-spherical
Lactobacillus rhamnosus	Bacteria	Ag	10	Spherical, triangular, rod, and hexagonal
Bacillus licheniformis Dahb1	Bacteria	ZnO	100	Hexagonal
Kocuria flava	Bacteria	Cu	5-30	Spherical
Penicillium diversum	Fungi	Ag	10–15	Spherical
Aspergillus terreus	Fungi	ZnO	28–63	Spherical
Saccharomyces cerevisiae	Yeast	Ag and Au	2-20	Biosynthesis of silver NPs using Saccharomyces cerevisiae

To create biogenic NPs, an aqueous algal suspension and a metal salt solution are typically combined, and the combination is then carefully incubated. Recently, the production of biogenic ZnO NPs by *Sargassum muticum* has been explored. Furthermore, carried out a fascinating investigation that demonstrated the possibility of Au NPs being biosynthesised from *Sargassum crassifolium* in the presence of sea grass. Additionally, researchers discovered that smaller NPs were produced by larger concentrations of *Sargassum crassifolium*. Algae including *Sargassum tenerrimum, Turbinariaconoides,* and *Laminaria japonica* can also be used to create Au NPs. Cystoseiratrinodis was used to assist in the synthesis of copper oxide nanoparticles. It was discovered that the nucleation process started when calcium carbonate ($CaCO_3^{2+}$) ions were deposited on the surface of the positively charged algal cells) came into contact with them. The amount of $CaCO_3$ crystals and Ca^{2+} ions concentration both had a significant impact on the microspheres' size. Microspheres were created and utilized in the production of Ag NPs in order to promote heterogenous nucleation and maintain a consistent algal cell count. ZnO NPs may be produced from the dried powder of *S. muticum*, according to research. A zinc acetate salt solution was mixed, and the mixture was stirred for a few hours to form NPs. The generated ZnO NPs had a hexagonal shape and ranged in size from 35 to 57 nm. They also involve chemical molecules that are physiologically active, such as SO_4^{-2}, amine, hydroxyl, and carbonyl. Studies have demonstrated the use of brown seaweed algae in the synthesis of silicon dioxide (SiO_2)-zinc oxide (ZnO) nanocomposites. Green NPs made from algae were also shown to have antiviral properties. The study demonstrated how Ag and Au NPs are biosynthesized. *Oscillatoria sp. Spirulina platensis*, two blue-green algae, were used to create these NPs.

4.3 Characterization techniques for assessing their properties

Because of these unique qualities, NPs have a wide range of applications. For biological applications in particular, detailed characterization is necessary to determine their eligibility for use. To collect the information required for this, a variety of instruments and techniques are employed. Among the most often utilized techniques are dynamic light scattering (DLS), Fourier transform infrared (FT-IR) spectroscopy, atomic force microscopy (AFM), Raman spectroscopy (RS), and scanning and transmission electron microscopy (SEM/TEM). These techniques are used to examine the size, shape, structure, surface properties, and interactions of NPs with various materials, including plant tissues (Fig. 2C).

4.3.1 Size

Particle size is one of the most fundamental and important variables for characterizing NPs. A particle's size and dispersion determine whether it is nano or micro-scale. The recognized method for determining particle size and dispersion is electron microscopy (EM). SEM and TEM images may be used to quantify the size in the solid phase. Centrifugation and photon corelation spectroscopy are utilized to quantify the size of the suspended particles in a liquid.

4.3.2 Surface Area

The surface area of NPs may also be used to describe them. The ratio of a nanoparticle's surface area to volume has a significant impact on its properties and performance. BET analysis is the standard method for determining surface area. A straightforward titration may be used to examine the surface area of liquid-phase particles, but it will take a lot of human labour. NMR spectroscopy is therefore utilized. A modified SMPS and a differential mobility analyzer (DMA) may be used to calculate the surface area of NPs in a gaseous state.

4.3.3 Composition

A nanoparticle's chemical or elemental composition determines its purity and efficacy. Higher concentrations of secondary or unwanted components in the nanoparticle may lead to lower efficiency as well as the potential for contamination and secondary reactions. X-ray photoelectron spectroscopy, or XPS, is widely used to determine composition. Before the particles can be examined using techniques like mass spectrometry, atomic emission spectroscopy, and ion chromatography, they must first undergo chemical digestion. In order to collect gas or electrostatic collection is utilized.

4.3.4 Surface morphology

Because NPs come in a wide variety of forms and surface textures, it is important to take use of these features. Amorphous or crystalline surfaces can have an uneven or homogenous structure, and some of the shapes that they can make on are spheres, flat discs, cylinders, tubes, cones, and irregular shapes. To determine the surface, EM imaging techniques like as TEM and SEM are commonly employed. Particles in the liquid state are collected electrostatically or by filtering for electron microscopy imaging, whereas gaseous state particles are deposited on the surface and analyzed.

4.3.5 Surface Charge

The charge, or surface charge, of a nanoparticle influences how it interacts with the target material. One tool for assessing the stability of a solution's ability to distribute surface charges is a zeta potentiometer. DMA may be used to determine the charges of NPs in the gaseous phase.

5. Conclusion

In conclusion, green or biogenic nanotechnology emerges as a sustainable and effective solution to combat the detrimental effects of salinity and drought on agricultural crops. By utilizing biogenic nanoparticles, produced through environmentally friendly methods involving microbes and plants, researchers have observed significant improvements in plant growth, seed germination, root architecture, yield, and biochemical profiles under salt stress conditions. These nanoparticles not only enhance crop resilience but also alter the expression of genes related to salt stress responses, offering a promising approach to mitigating the negative impacts of abiotic stress on agricultural productivity. While chemical and physical methods of nanoparticle production offer precision, they come with higher costs and potential environmental risks. The adoption of biological strategies for nanoparticle synthesis provides a cost-effective, rapid, and eco-friendly alternative. As climate change continues to challenge global agriculture, the integration of biogenic nanotechnology holds great potential for ensuring sustainable crop production and safeguarding human health.

CHAPTER 11

ROLE OF NANOTECHNOOGY IN ZINC FORTIFICATION FOR COMBACTING AGAINST MALNUTRITION

Short Summery

A significant amount of the global illness burden and infant mortality rate can be attributed to deficiencies in the mineral zinc. The creation of a genotype with increased zinc bioavailability will be an effective and long-lasting solution to nutrient deficiency-related issues since biofortification is becoming more and more successful in main crops. The human bodys intricate chemistry makes it difficult for the body to absorb zinc from grains. Phytate, a significant phosphorus-storing substance found in grain and legume seeds, further reduces this complexity by interfering with zinc binding. Although it is just one component, phytate is a crucial piece in the jigsaw when attempting to determine the bioavailability of nutrients. Strong evidence for the existence of separate Zn pools may be seen in the findings of recent research on the distribution of elements and micronutrient speciation in seeds. This perspective is supported by data from research on transgenic plants that have been biofortified. Numerous investigations identify the metal chelator nicotinamide as a key component. An extremely significant observation is that Zn loading into grains has been found to rise with nicotinamide levels. Nicotinamide has the ability to significantly increase intestinal zinc absorption. According to the results of the biofortification, studies, using Nano-enabled can be a useful tactic to support plant biofortification, which can greatly increase the amount of zinc in a variety of cereal crops.

1. Introduction

A good, nourishing diet is essential to human health. Food instability, dietary imbalance, eating low-nutrient grains, a lack of dietary diversity, etc., all have a detrimental effect on human health. Dietary quality may be hampered by food and nutritional instability, which significantly raises the risk of undernutrition and obesity. Because of the increase in the production of cereal crops and cash crops in intensive cropping systems, there has been a decrease in both micronutrient intake and dietary diversification. This is particularly true in low-income countries' rural regions due to the widespread consumption of grains in the diet. Since achieving food and nutrition security is still a long-term objective for many developing nations, it is crucial to figure out how to make sure these things happen. As opposed to the target of having Zero Hunger by 2030, if the current trajectory persists, there would be more than 840 million hungry people in the world by that year. Dietary habits are significantly impacted by the economic disparity. Lower-income nations rely less on fresh vegetables and meat and more on staple foods than do high-income nations. Micronutrient malnutrition, another name for hidden hunger, is a dangerous illness brought on by insufficient consumption of certain micronutrients in the diet. Diets centred around cereals are very common, particularly in developing nations. As a result, adding micronutrients to them can enhance their ingestion. Worldwide supply of biofortified staple grains (such as wheat, rice, maize, etc.). Approximately 2 billion individuals worldwide are experiencing a zinc nutrient deficit. A zinc shortage has been linked to a number of serious health conditions, including oxidative stress, apoptosis, dysregulation of immunological response and homeostasis, and aging. Zinc supplements have been an important source of immunity-boosting nutrients during COVID-19 infections. A zinc deficit in the diet is directly related to low rhizospheric zinc availability. Because of this, it is imperative that low-income and developing countries boost the quantity of bioavailable zinc in their staple foods. Plants require zinc, one of the most crucial minerals, for healthy growth and development. It is essential for photosynthesis, cytochrome and chlorophyll synthesis, cell membrane stability, and cell transparency. Lipid synthesis, protein metabolism, carbohydrate metabolism, nucleic acid metabolism, control of gene expression, and phytohormone action all depend on it. All six groups of enzymes, including oxidoreductases, hydrolases, transferases, isomerases, lyases, and ligases, have zinc as the single essential trace element.

An eco-friendly strategy to fight human sickness and enhance people's nutrition is to biofortify crops. The exploration of nanotechnology's possibilities in agriculture is motivated by the growing need for food to feed the world's rapidly expanding population. By 2050, experts predict that there will be 9.6 billion people on the planet. Nanotechnology is bridging the gaps in crop fortification and nutrient leaching in precision agriculture. Farmers are using this technology in the nano-regime to improve the output and quality of their crops.

The application of nanotechnology to enhance crops in human cultures has garnered a lot of attention lately. There are several methods for nano-biofortifying plants, such as applying soil and foliar nutrients, priming seeds, and cultivating plants in a medium that is high in potential nutrients. Zn transport to plants is more efficiently facilitated by using Zn-based nanofertilizers than by using conventional techniques. Zn-NPs are distinct from typical Zn particles in terms of their mechanical, optical, electrical, thermal, chemical, and magnetic characteristics. They range in size from 1 nm to 100 nm. NPs that are absorbed by plants through their roots or leaves might disperse throughout the plant and have an impact on the growth and development of the plant in a positive, negative, or neutral way.

Zn NPs improve plants' capacity to make high-quality proteins, carbohydrates, and fats. Applications of ZnO-NPs have been demonstrated to improve the physiological and agronomic characteristics of crops in earlier research. ZnO-NPs at a dosage of 100 mg/L were applied as a foliar spray, which enhanced photosynthetic activity in immature Cucumis sativus plants. Applying 500 mg/kg of ZnO-NPs and bulk ZnO to Portulaca oleracea resulted in a 1000% rise in flavonoid content and a 581% increase in Zn content as compared to the control. This study aims to highlight the relationship between human health and nano-biofortification in the context of food security, as well as the possible use of agricultural products based on zinc oxide nanoparticles (ZnO-NPs) to increase the nutritional content of crops.

2. Agricultural zinc deficiency

Fertilizers are enriched with nutrients that plants require in order to improve the possibility that plants will absorb the additional nutrients. Plants can exhibit signs of nutrient deficiencies in a variety of ways, such as abnormal development of plant parts. In certain situations, though, even plants growing in soil rich in micronutrients might exhibit deficient symptoms because of the tiny pore size of their roots, which makes it difficult for the roots to absorb and transfer nutrients. We need to find ways to improve plant nutrition and, therefore, the quality of the food that plants produce if we are to feed the world's population in a sustainable manner. Using chemical fertilizers has long been a practice and has significantly enhanced agricultural yields. Nevertheless, they lead to a mineral imbalance in the soil and affect the soil's structure, nutrient level, and ecology as a whole, all of which have disastrous long-term implications. To address the issue, intelligent materials that promote a pollution-free environment and can release nutrients to certain locations when needed are necessary. Recent studies have shown that graphene is a useful substance for transporting nutrients to plants. By distributing nutrients to the plants gradually and purposefully, it can boost agricultural productivity with little negative impact on the environment. A long-term solution to hunger reduction and human health challenges is the biofortification of crops. There has been a lot of talk lately (Table 1) on how to use nanotechnology to make crops more resilient in human cultures.

Crops can be nano-biofortified using a variety of methods, including nutrient-rich medium cultivation, soil and foliar treatment, and seed priming. Numerous considerations should be taken into account in this case, including: (i) determining the most effective nanomaterials; (ii) figuring out the right dose that won't interfere with a plant's growth or physiology (iii) Examining how nanoparticles affect the uptake, accumulation, and movement of additional mineral elements.

Zinc is an essential micronutrient for plant development since it co-occurs with several other essential enzymes. A key source of worry is the global zinc shortage. West Asian agricultural productivity has suffered greatly in recent years as a result of the effects of climate change, including severe droughts, dust storms, and widespread salinization of the soil. Numerous studies have demonstrated that giving plants more zinc can help them cope with environmental stressors. The bioavailability of zinc in the edible plant portion also increases when zinc is administrated externally, either as a foliar spray on leaves or as fertilizer in the soil. Foliar fertilization with zinc enrichment (Fig. 2) is by far the most efficient and environmentally acceptable way to increase crop yields. Applying substances to leaves allows substances to reach the plant through the stomatal route or the cuticle. During the blossoming stage, rice can benefit considerably from the use of zinc foliar fertilizer. The foliar spray may not, however, work every time. The efficacy of foliar spraying may be diminished by heavy rains since the foliar spray may wash off the leaves and seep into the soil. Zinc fertilizer boosts crop yields by raising the zinc content of harvested grains. Numerous investigations on the fortification of *Oryza sativa* (rice) with zinc have led to the conclusion that, when combined with other fertilizers, adding zinc to the soil increases crop output. Zinc fertilization is less effective in soil due to the presence of fixed Zn ions, humus clay minerals, and iron and aluminium oxides. As a result, unused zinc fertilizer accumulates in farming soil, potentially damaging agricultural ecosystems. Because of this, it's critical to investigate novel avenues for zinc fertilizer that may take the place of conventional zinc inputs and exhibit excellent performance while having less detrimental effects on the environment. In 1937, reports of significant zinc deficiencies in crops were first made. That being said, the first recorded case of zinc insufficiency in humans dates back to the 1960s. After the green revolution, a greater portion of the world has suffered from acute zinc insufficiency due to newly introduced high-yielding varieties inside the system during the course of the previous five decades. When compared to previous, locally adapted (landrace) types, these new crop varieties frequently had lower yields and were Zn-inefficient. These types need extra N, P, and K (nitrogen, phosphorus, and potassium) fertilizers to reach their full production potential. In certain situations, liming the soil is also necessary to elevate the pH of the soil. In general, increased phosphorus concentrations and higher pH levels have made deficiencies in zinc worse. But such cultivars are no longer very productive. Instead, in regions where they were introduced for a reason, they now have yield plateaus. A hidden zinc deficit was discovered later. For example, Turkey was able to use the same amount of arable land to produce 10 million metric tons (Mt) more wheat grain annually in the early 1960s. The development of new wheat cultivars and increased wheat-growing intensity made this feasible. Nonetheless, zinc shortage was becoming more prevalent in some areas, such as Central Anatolia, and in the 1990s, a NATO-funded study effort identified it as the primary reason for low yields. Zn fertilizer raised wheat yields in a region where they had previously been as low as 0.25 t/ha by a ratio of 6-8. Overall, the yield of *Triticum aestivum L.* (bread wheat) rose by 43% (range 5-550%) with the addition of 10 kg of zinc per hectare.

Plants are able to absorb zinc in five distinct forms: a) the lattice structures of primary and secondary minerals; b) non-specifically absorbed cation; c) ions occluded mostly in soil carbonates and A1 oxide; d) live organisms along with other biological residues; and e) free and complex ions inside soil solution. Zinc is mostly taken by roots as the divalent cation in Zn^{2+}. However, reports of the absorption of organic Ligand-Zn complexes by roots have been made. Similar to Zn^{2+}, when plants take up the mineral in the form of divalent cations, two pathways take place in the plant, each of which is dependent on a distinct ligand released by the root. In order to release Zn^{2+} ions in the form of soluble zinc compounds, such as phosphate and hydroxides, for improved absorption by the root epidermis, the first technique involves the release of hydrogen ions (H^{+}), organic acids, and reductants. Phytosiderophores can also transport zinc into root epidermal cells by forming stable complexes with the metal beforehand. However, this process is exclusive to cereal crop roots. Phytosiderophores are organic molecules that have a reduced molecular weight and are used by plants to absorb and chelate metals.

In comparison to the average content of around 64 mg Zn kg^{-1}, soil solutions often have lower quantities of Zn (4-270$\mu g l^{-1}$). It is discovered that solubility and soil pH have a strong negative association, with very acidic soils exhibiting concentrations as high as 7,137$\mu g l^{-1}$. The "total Zn" content, pH, oxidation/ reduction levels, calcite ($CaCO_3$) and organic matter contents, concentrations of all ligands from organo-Zn complexes, rhizosphere microbial behavior, presence of other micro and macro elements (especially P) and their concentrations, and soil moisture content are among the primary variables influencing how much zinc is readily accessible to plants. High calcium carbonate concentrations, comparatively high organic matter (3%), neutral and alkaline pH, a highly accessible phosphorus profile, and have an impact on zinc levels in the soil. High levels of salt, potassium, and calcium in the soil, as well as circumstances involving extended waterlogging, such as paddy soils, are associated with the occurrence of zinc.

The majority of plants exhibit zinc concentrations between 30 and 100 grams of Zn^{2+} per gram of dry weight, indicating that the zinc equilibrium approach is not applicable to all plants. Brassicaceae are examples of zinc hyperaccumulators that may collect more than 10,000g Zn^{2+}/g dry weight without showing any signs of toxicity. In spite of this, Zn^{2+} concentrations of more than 300 mg Zn^{2+}/g dry weight are thought to be detrimental to plants. The research shown in Table 1 provides several instances of zinc's beneficial impact on plants. Thus, it is thought that an excess or deficiency of Zn^{2+} in the soil might negatively impact a number of plant metabolic processes, leading to reduced photosynthetic activity, yellowing and necrosis of the leaves, as well as restricted plant development.

3. Role of Zn in Human and Plant Health

Two processes involved in the metabolism of carbohydrates, glycogen production and gluconeogenesis, are influenced by zinc. Zinc often has little effect on a plant's ability to respire. The net photosynthetic rate can be affected by a zinc shortage by 50% to &70%, depending on the kind of plant and the severity of the nutritional deficit. Reduced activity of the enzyme carbonic anhydrase may lead to less effective photosynthesis. Compared to monocotyledon carbonic anhydrase (found in cereals), dicotyledon carbonic anhydrase is a larger molecule and includes more zinc. When zinc availability is charged, there is no direct correlation between carbonic anhydrase activity in C3 plants and photosynthetic carbon dioxide uptake or plant development. Zinc is necessary for other photosynthesis-related enzymes, such as ribulose 1,5-biphosphate carboxylase (RuBP), which is found in pearl millet, barley, rice, and beans. RuBP catalyzes the initial stage of photosynthesis and carbon dioxide fixation.

The fact that a zinc deficit reduces starch levels, enzyme activity, and the quantity of starch grains in plants suggests that zinc may be an important component is starch production. Zinc-deficient plants were found to have larger quantities of sugar and starch in their cabbage leaves and lower levels of carbohydrates in their bean roots. Although the exact cause of the reduced sucrose transfer is unknown, it might be related to zinc's function in maintaining the integrity of bio-membranes. Free amino acid levels in zinc-deficient bean leaves were 6.5 times greater than in control beans, but they decreased when zinc was given for two to three days while protein levels increased. Zinc prevents ribonuclease from destroying ribosomal RNA and is essential for the RNA polymerase enzyme's function. Meristematic tissue, which undergoes a great deal of cell division as well as nucleic acid and protein synthesis, has to have relatively high zinc concentrations since zinc is necessary for protein synthesis. Zinc's role in maintaining the integrity and functioning of DNA has the most basic impact on protein metabolism.

3.1 Role of Zinc in human's

Zinc (Zn) binds to around 10% of all human proteins, according to omics research on the human genome. This astounding finding has radically shown the significance of zinc in compounds engaged in biological functions. Zinc is necessary for the correct operation of a number of enzymes, transcription factors, and other proteins. Some protein areas that may promote zinc interaction include RING finger domains, LIM domains, and zinc-finger domains. The body's main stores of zinc are found in the skeletal muscles and bones (Fig. 1), but these tissues are limited in their capacity to store more zinc than the organism needs. Zn needs to be consumed every day for cells that depend on it to operate properly. Zinc has minimal toxicity and is normally harmless, but too much or too little of it might have detrimental effects on health. While too little zinc can lead to immune system malfunction, growth retardation, taste anomalies, eye and skin vomiting, fever, and headaches. Zinc supplementation reduces the symptoms of zinc insufficiency, highlighting the role of zinc as a trace mineral and the physiological relevance of maintaining zinc homeostasis. Healthy growth and differentiation, immunological system performance, brain functions, and protein synthesis all depend on zinc. Studies have indicated that taking zinc supplements and zinc complexes with other chemicals can have positive effects on health.

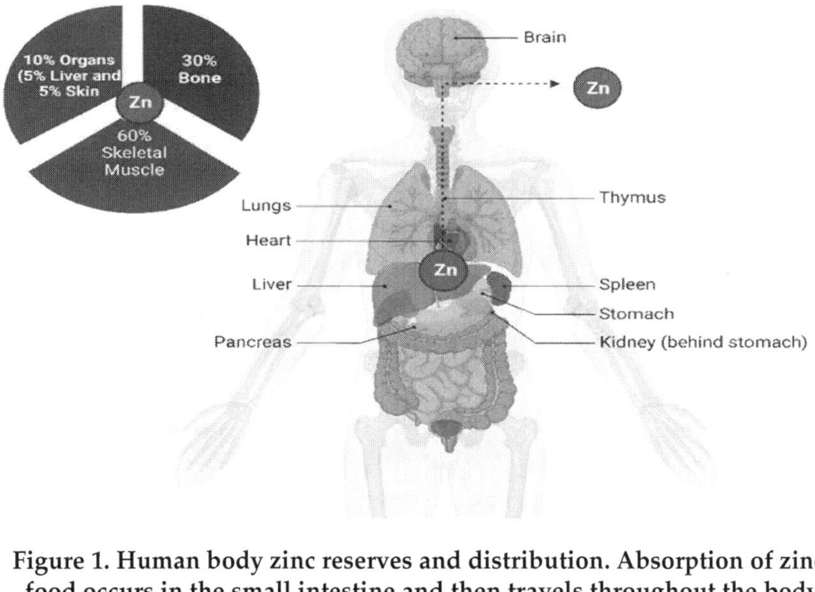

Figure 1. Human body zinc reserves and distribution. Absorption of zinc from food occurs in the small intestine and then travels throughout the body. The skeletal system, specifically the bones and muscles, serves as a major Zn reservoir tissue.

In recent years, cheatable zinc has been found to have an increasing number of physiological roles in cellular responses. Zinc is used by cells for a variety of functions, such as intracellular signal transduction and neuromodulation during synpatic transmissions. Different zinc transporters contribute to the maintenance of zinc homeostasis and are necessary for regular cellular functions.

3.2 Role of Zinc in plant health

Zinc-binding proteins make up around 6% of the bacterial proteome and 9% of the eukaryotic proteome. Zinc is found in plant species in enzymes such as carbonate dehydratase and aldehyde dehydrogenase, Zn/Cu superoxide dismutase (SOD), and Zn-finger DNA-binding proteins. Zn-Cu-SOD smothers the radicals generated by the body's regular metabolism, preventing the oxidation of membrane proteins and lipids. This vitamin stimulates the production of cytochromes, which is essential for seed development. Zinc is necessary for several physiological functions in plants, such as the synthesis of tryptophan, which is the precursor of indole acetic acid, and the input transfer process via mitogen-activated protein kinases. Auxin regulation, photosystem II renewal, mesophyll CO_2 quantity stability, and other physiological responses are among the plant-specific processes that are dependent on zinc availability. Through their interactions with functional groups or displacement of other cations from binding sites, zinc ions reduce the number of harmful effects on plant cells. Zn has to be taken up, moved, and dispersed throughout the tissues, cells, and intracellular spaces of plants in order for them to flourish. Zinc deficiency, albeit important, results in stunted development and discolored leaves. This can lower food absorption and ultimately produce a zinc deficit in human nutrition.

4. Need of nano-enabled approaches for biofortification

A significant portion of people's health is influenced by the food they eat, and plants are still the main source of nutrition for humans. Foods that are heavy in calories and frequently consumed in huge quantities are usually the ones that become staples in a community's diet. Thus, there is a strong correlation between consumers' nutritional status and staple foods, especially for those residing in rural and impoverished locations where there is a dearth of supplemental nutrients. In these areas, foods frequently lack the essential micronutrients which are widespread and have major repercussions. The nutritional needs of half of all children are not met, which increases their susceptibility to disease and lowers their cognitive potential. Deficits in iron and zinc contribute significantly to infant mortality and other global health problems. Among the suggested methods for treating nutritional deficiencies are pharmacological treatment, dietary variety, and industrial fortification. Food diversity is recommended as a long-term remedy, but those who are at risk of malnutrition cannot afford it. Foods that have been artificially fortified, with the exception of iodized salt, have not proven very beneficial. Biofortification is the process of improving a crop's nurtitional value before, during, and following harvest. It differs from other micronutrient deficiency therapies in that it is inexpensive and available to everyone. This is because these fortified crops provide a staple food for a large number of individuals. As a result, it doesn't require any extra funding and is independent of people's social customs. Notwithstanding their high vitamin and mineral content, vegetables and other non-staple foods are costly. A large portion of the food budgets of low-income individuals are allocated to basics that give them very little energy, leaving them with very little money left over for more costly meals like meat, fruits, and vegetables.

Among the biofortification procedures that are commonly employed to improve the nutritional status of crops are agronomic and breeding approaches. As part of agronomic treatments, fertilizers—both organic and inorganic—as well as biofertilizers—are highlighted. Because it has a much lower mineral content and takes longer to release nutrients, organic matter is more difficult to utilize than inorganic fertilizers, which are typically available in sizes larger than 100 nm and are more susceptible to leaching and volatilization. The goal of improving crops' capacity to absorb nutrients through biofortification has been pursued for a long time but with little success. Therefore, the moment has come to use nanotechnology to solve some of these problems.

5. ZnO-NPs absorption and translocation in plants

5.1 Physiological approaches of absorption and translocation of ZnO-NPs

The bioavailability and toxicity of NPs are determined by bio-geotransformations that occur in the soil. NPs translocate to the plant's aerial portions in reaction to root contact, where they gather in cells and organelles (Fig.2). The first stage of bioaccumulation—the absorption of NPs from the soil—is carried out by the plant roots. The distribution of the absorbed component from the roots to the rest of the plant is facilitated by a variety of processes, including crystal phase dissolution, biotransformation, and bioaccumulation, according to several studies that have examined different NPs and suggested that they accumulate in plants. Their toxicity, rate of accumulation, and rate of entry into plant cells are all influenced by size. This is due to the fact that their size affects how they enter cells (plasmodesmata) or plant cell organelles. The reactive nanoparticles (NPs) with respect to their surface area, shape, and aggregation is linked to these factors. When determining the exact location of the point of contact between NPs and plant cells, the size and shape of the NP surface are important factors. The only way the NPs can attach themselves to the plant cells is by their electrical charges since the cell wall is negatively charged. After the charge and particle size, hydrophobicity on the plant surface was the third most significant determinant in absorption and translocation. By closely examining the underlying structures of the nanomaterials, it is possible to gain a better understanding of the absorption, translocation, and aggregation of nanoparticles (NPs) in plants. To determine the precise effect of NPs supported by their physical-chemical characteristics, standardized laboratory tests evaluating NPs in plant tissues at different levels are necessary. Consequently, a detailed analysis and understanding of the composition of the Nps are required in order to completely comprehend and explain the processes of absorption, translocation, and accumulation. Careful tracking and monitoring are necessary to determine the distribution and localization of NPs within plants and their cellular organelles. The translocation through root epidermal cells, osmotic pressure, and capillary forces have all been described as routes by which the small NPs (diameters ranging from 3 to 5 nm) might reach plant roots. Roots epidrmal cells are only partly permeable and unable to let big NPs into the root due to their tiny pore size (Fig 2). The creation of new pores in the epidermal cell wall facilitated the absorption of some NPs. After NPs get through cell walls and extracellular spaces and migrate apoplastically to the central vascular cylinder, the xylem can develop vertically. To access the core vascular cylinder, NPs must, however, symplastically pass through the casparian strip barrier (Fig 2) via attaching to carrier proteins in endodermall cell membranes.

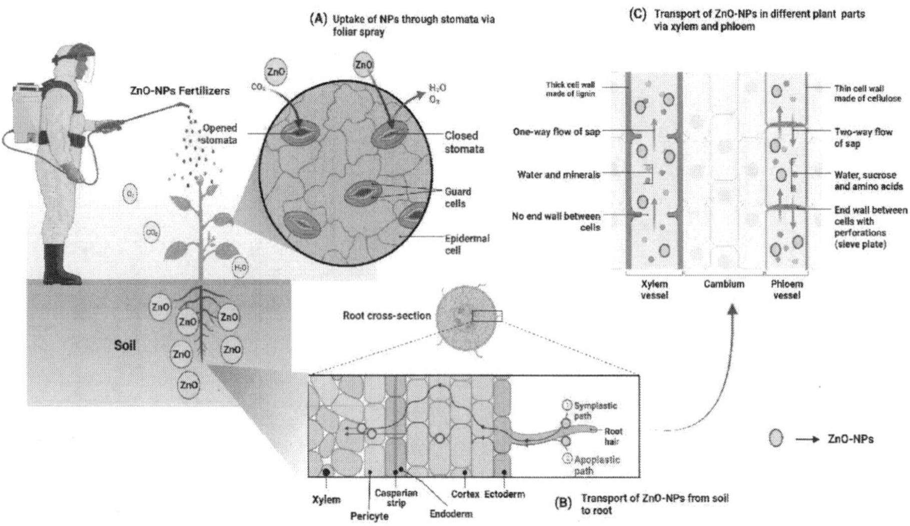

Figure 2. ZnO NPs based Nanofertilizers mode of action, depicted diagrammmatically (A) plant system; (B) from soil to root and (C) transport through xylem and pholem in different parts of plants via symplastic and apoplastic pathways.

This is achieved via inducing endocytosis, pore creation, and transport. Upon internalization into the cytoplasm, NPs have the ability to spread via plasmodesmata to adjacent cells. NPs that are unable to internalize are collected in the Casparian strip; those that do get it the xylem are subsequently carried to the shoots, where the phloem returns them to the roots (Fig 2). Reports have indicated the presence of NPs in the cytoplasm, walls of epidermal cells, and plant nuclei. NPs that are not absorbed in the soil aggregate'ss root surface have the potential to change how nutrients are taken. Through the parenchyma and into the intercellular space, seeds can directly absorb NPs, which are then distributed to the cotyledons. When NPs are applied to leaves, they may be absorbed by the cuticles or stomata. The cuticle serves as a leaf's initial line of protection against foreign particles, keeping those smaller than five nanometers out. Through stomata, they enter the plant's vascular system. From there, they go via apoplastic and symplastic routes to the organs responsible for photosynthesis and transpiration, respectively. The most effective way for NPs (10-50 nm) to go to new sites is through neighbouring cells' cytoplasm (the symplatic pathway). This explains why larger NPs (between 50 to 200 nm) are transported by the apoplastic pathway, which is the body's network of intercellular gaps.

Together with the sugar solution, internalised NPs pass via the phloem sieve tubes. The roots, immature leaves, stems, fruits, and grains of the plant act as an efficient sap sink, allowing NPs to freely travel by pholem transport and accumulate to different degrees. The apoplastic route is a well-known nonselective path for least resistance. The apoplastic pathway is widely acknowledged as the most effective means of translocating several water nutrients and non-essential metal complexes. After foliar application, it was discovered that the following factors largely affected NPs' ability to adsorb: application technique, particle size, concentration, and ambient circumstances. The shape, chemical composition, trichrome content, and the presence of waxes and exudates on the leaf are some of the elements that influence the trapping of nanoparticles on the leaf surface. Synchrotron X-ray absorption spectroscopy was used to investigate the alterations in zinc oxide (ZnO) nanoparticles following their interaction with various plants. Under hydroponic cincumstances, ZnO NPs in various forms, such as Zn-phosphate, aggregate mostly densely in the roots and shoots of Zea mays. It might be the consequence of enhanced translocation, plant absorption, or ionic Zn dissolution in the rhizosphere. Wheat crops cultivated in soil showed comparable zinc buildup and speciation. It was shown that ZnO NPs moved from the soil to Glycine max.

5.2 Molecular approaches of absorption and translocation of ZnO-NPs

Zinc is taken from the soil by micronutrient transporters in plants. ZRT-IRT-like proteins (ZIPS) directly absorb Zn^{2+} upon applications of ZnO NPs (Fig 3). Numerous species, including mammals, plants, microbes, and fungi, have ZIP transposases. Eight transmembrane helices are predicted for ZIP proteins, and in the majority of instances, a loop region containing a histidine-rich motif that may be able to bind metals is found between transmembrane domains 3 and 4, suggesting a function in the transport or control of zinc (Fig. 3).

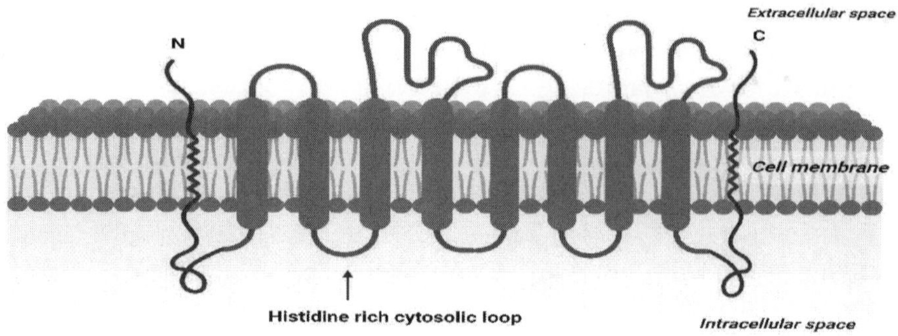

Figure 3: Predicted diagrammatic structure of protein structure of membranes of ZIP family of micronutrient transporter that helps in the transport of zinc at the molecular level in plants.

The proteins produced by ZIP1, ZIP3, and ZIP4 have a strong affinity for zinc and may facilitate the uptake of Cd and Cu by the plant, according to research on gene expression in yeast and root tests. In cases of zinc shortage, ZIP4 is expressed in both the roots and the shoots, whereas ZIP1 and ZIP3 are only expressed in the roots. YSL1, FER1, FER2, and PHT4 are examples of the multigene families that plants normally use in conjuction with ZIPs to manage the transport of ZnO NPs metals (Fig. 4). When there is a zinc shortage, ZIPs transporters are engaged in the movement of external ZnO NPs, which aid in maintaining a steady supply of zinc throughout the plant.

The transfer of zinc into the vacuole affects the overall transport of zinc from the roots to the shoot, and metal chelators such as nicotianamine (NA) are critical to this process (Fig.4). The ZIPs family of proteins participates in the remobilization of sugars from the leaves while the filling of grains as well as the migration of sugars from the root or basal section of the shoot to the problem. In ZnO NPs, the zinc ions are transported in the xylem either as a free cation or as a complex with organic acids like citrate (Fig.4). Because the xylem in certain cereals is discontinuous, phloem serves as the only pathway for the transportation of nutrients. Small proteins or NA complexes with zinc are transported via the phloem. Numerous ZIP and metal tolerance protein (MTP) family members have been suggested to act as transporters from the material tissue into the aleurone, embryo, and endosperm cavity (Fig 4).

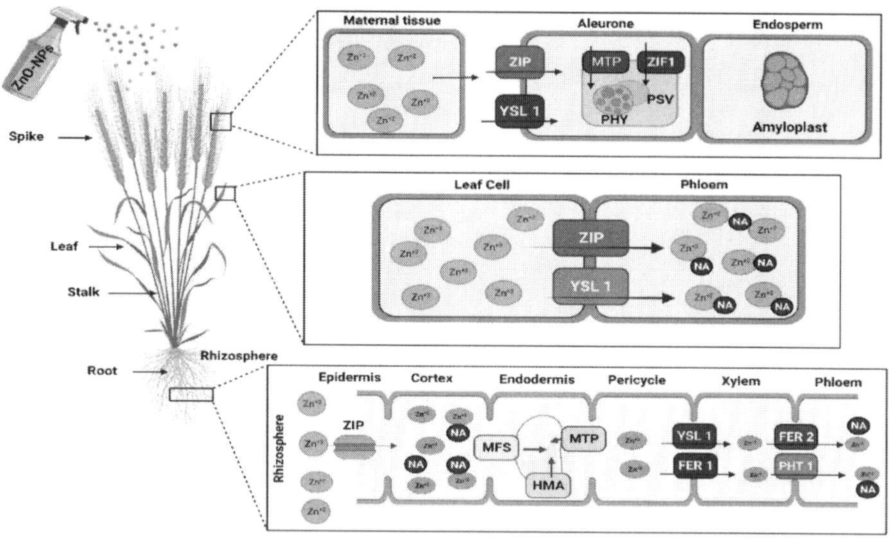

Figure 4: Diagrammatic representation of molecular bases of translocation of ZnO Nps into Zn supplement in Wheat plants. Translocation of Zn from roots to grain via shoots (stalk) assisted by ZIP (ZRT-IRT like protein), YSL (yellow stripe like transporter), MFS (Major facilitator super family transporter), PHT1 (Phosphate transporter 1), NA (nicotianamine), PHY (phytate), PSV (protein storage vacuoles).

6. Nano-Biofortification approaches for Zinc fortification in Crops

6.1 Different Nano-biofortification strategy for Zinc fortification in Crops

6.1.1 Soil Application

The pH of the soil affects how well Zn fertilizer works in the soil; more acidic soils facilitate easier zinc availability. Increases in pH have a hundred-fold detrimental impact on the solubility of soil. There is proof that acidic soil liming can reduce zinc availability. The three mechanisms that induce the zinc deficit in calcareous soil are the development of insoluble calcium zincate, a pH rise brought on by $CaCO_3$, and direct sorption of zinc to the precipitated $CaCO_3$. The alkaline soil that covers around 30% of the world's cropland has a limited plant availability of zinc. Other than pH, Zn fixation is influenced by complexation with organic matter, occlusion in minerals, diffusion into micropores and interparticle space, soild-phase diffusion, and co-precipitation with other metals. Applying 40,80 and 120 ppm of ZnO NPs to sandy loam soil throughout a pot culture study reveals that the optimal concentration for increased zinc content (19.94g) and increases in plant height, seeds per spike, yield, biomass, and 100 seed weight is 80 ppm of ZnO NPs. Applying ZnO NPs at 1,3, and 5 mg Zn/kg to the soil during the drought can increase the zinc content of wheat grains by up to 94%. Additionally, this concentration of ZnO NPs effectively promotes faster plant growth and development, leading to increased output under drought stress and enriched edible grains with critically important nutrients. Different ZnO NPs concentrations (40-400 mg/kg) in soil can raise the Zn content in soybean grains, which in turn boosts seed production. Antioxidant enzyme activity also reduced MDA content and lipid peroxidation. This suggests that ZnO NPs may be used to supplement zinc deficit soils and increase soybean yields. Higher zinc levels would also assist combat hunger on a worldwide scale. Applying ZnO NPs to the soil can increase Zn content while adjusting production parameters for various crops (Table 1).

6.1.2 Seed Nanopriming of ZnO NPs

A pre-planting technique called "seed priming" alters the seed's physical makeup to hasten its germination. NPs-based pre-soaking treatments for seeds, referred to as "seed nanopriming", aid in the creation of nanopores in the shoot, which facilitates water absorption. Additionally, it triggers the seed's antioxidant defenses against reactive oxygen species (ROS), which result in hydroxyl radicals that break down cell walls and speed up the starch's hydrolysis and nutrient absorption. Zinc-rich seeds have been demonstrated to improve crop and seedling health. Nutrient-deficient seeds are produced by plants cultivated in nutrient-poor soil, which lowers the general vigor, development, and yield of succeeding plant generations. Because zinc is absorbed by plant roots by diffusion, low moisture levels reduce the availability of zinc. Therefore, depending on the soil irrigation, the efficiency of Zn added to the soil may change. When soil moisture content in a rainfed state is entirely reliant on precipitation, the results of zinc feeding can be quite varied. When Zn-rich seeds, such as pre-soaked ZnO NPs, are utilized, plant development and yield in Zn-deficient circumstances—especially when rainfed—are enhanced (Table 1). Red bean morphological and phenological characteristics, yield and yield components, maturity, grain filling duration, plant height, and number of seeds per pod can all be enhanced by seed priming with 4% ZnO NPs. By using 10 mg/L ZnO for nanopriming, it is possible to enhance the number of photosynthetic pigments, phenolic compounds, flavonoids, vitamin C, glutathione (GSH), zinc content, and antioxidant activity, including ascorbate peroxidase (APX), catalase (CAT), and glutathione (PAL), as well as the percentage of germination, length, and dry weight of the plumule and radicle. The agronomic profile of rice can be improved by pre-sowing treatment with 25 ppm of ZnO-NPs, which increases yield features with grain Zn nutritive value. ZnO NP priming at a dosage of 10mg/L significantly and favorably affects fresh biomass, water absorption, photosynthetic efficiency, vigor index, and seed germination activity. Pretreating direct-seeded rice with 10 mol ZnO NPs improves plant growth, grain nutrients, and productivity in DSR, as well as seed vigor and early seedlings emergence. But in low-resource and high-stress environments, nanopriming seeds might be a key component in solving the zinc shortage issue.

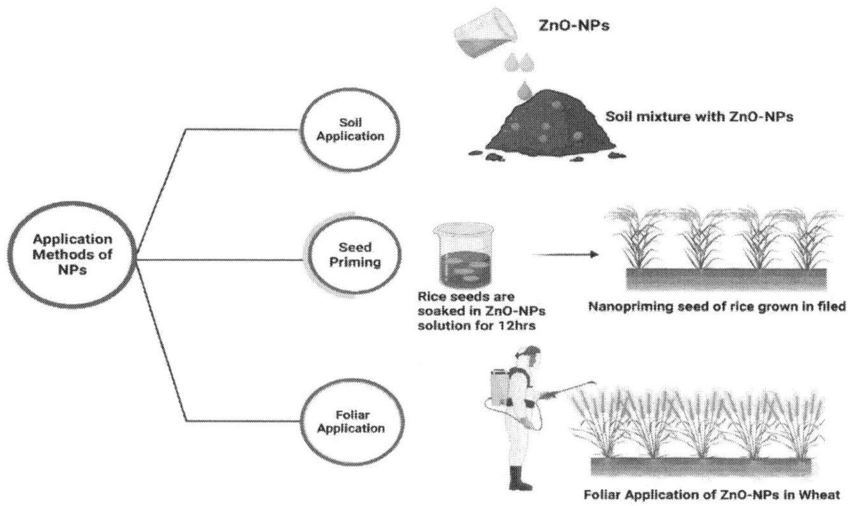

Figure 5: Application methods of ZnO NPs

6.1.3 Foliar Application of ZnO-NPs

The kind of fertilizer used, crop features (especially leaf trails), and crop genetic potential all have a significant impact on how successful foliar Zn delivery is (Table 1). Among the many advantages of applying fertilizer via foliar approach are the reduction of Zn fixing, the avoidance of antagonist nutrient interaction with Zn absorption, and so on (Table 1). Applying zinc to the leaves allows it to pass via the phloem and reach the growing grains. It has been discovered that applying zinc topically increases the concentration of zinc in grains, such as wheat, more effectively than applying zinc topically. Foliar treatments can increase the content of zinc in grains to a greater or lesser extent. When compared to nutrient inputs with greater Zn levels in amaranth, foliar sprays of 10ppm biogenic ZnO NPs significantly improve nutrient recovery, yield, and productivity. Applying 100µMZnO NPs by foliar spraying may enhance cauliflower's zinc biofortification. ZnO NPs applied foliarly to tomatoes at a concentration of 100 ppm can enhance physiological characteristics, yield attributes, yield, and quality qualities. The most favourable outcomes with regard to growth, photosynthetic pigments, protein content, activity, and the quantity of both enzymatic and non-enzymatic antioxidants present are obtained in mung bean seedlings at a dosage of 20ppm of ZnO NP.

Table 1: Extensive information on the various published studies involving the biofortification applications of ZnO-NPs.

Plants	Scientific name	Concentration of ZnO-NPs	Preparation Method of ZnO-NPs	Characterization of ZnO-NPs	Application Methods	Main finding
Common bean	*Phaseolus vulgaris*	10-40 mg/L	Chemical	20nm	Foliar Spraying	ZnO-NPs at 30 ppm produced the highest seed yield.
Wheat	*Triticum aestivum* L.	20-1000 mg/L	Chemical	100 nm	In soil	In comparison to ZnSO4, ZnO-NPs increased Zn in grain while having no effect on Zn in leaf.
Wheat	*Triticum aestivum* L.	25-100 mg/L	Chemical	20-30 nm	Seed primming	ZnO-NPs increased photosynthetic pigments and crop yield
Read beans	*Phaseolus vulgaris* L.	4% ZnO-NPs	Chemical	NA	Seed primming	Increased yield and Zn contents in grain of red beans
Moringa	*Moringa oleifera*L.	10 mg/L ZnO-NPs	Chemical	16.49nm	Seed primming	Improving germination, photosynthesis pigments contents, antioxidant enzyme activity Zn nutrient contents
Rice	*Oryza sativa* L.	25ppm	Chemical	20-30nm	Seed primming	Pre-sowing treatment with 25 ppm ZnO-NPs increases rice yield and grain Zn nutritive value
Wheat	*Triticum aestivum* L.	10mg/L	Chemical	NA	Seed primming	Improved seed germination, vigour index, fresh biomass, water uptake, and photosynthetic efficiency.

Rice	*Oryza sativa* L.	10 µmol	Chemical	10 nm	Seed primming	Increase seed vigour, seedling emergence, and DSR field plant growth and productivity
Amaranth	*Amaranth*L.	10 ppm	Biogenic	15-30 nm	Foliar	Foliar application of biosynthesized ZnO-NPs enhanced fertilization that increases yield, zinc content, and zinc use efficiency
Cauliflower	*Brassica oleracea* L.	100 µM	Chemical	25 nm	Foliar	ZnO-NPs applied foliar to cauliflower improve zinc biofortification.
Tomato	*Solanum lycopersicum*L.	100 ppm	Chemical	NA	Foliar	ZnO-NPs improved tomato growth, physiological traits, yield, and quality.
Mung bean	*Vigna radiate*L.	20 ppm	Chemical	NA	Foliar	ZnO-NP positive effects on growth, photosynthetic pigments, protein content, activity, and enzymatic and non-enzymatic antioxidants in mung bean seedlings
Soybean	*Glycine max*L.	2 mg/kg	Chemical	18 nm	In soil / Pot culture	ZnO-NPs increased grain yield and P uptake by 14% by regulating nutrient availability

Sorghum	*Sorghum bicolor* L.	1, 3, and 5 mg/L	Chemical	18 nm	In soil /Pot culture	ZnO-NPsincreasedgrainnitrogen, potassium and zinc contents under drought stress help in biofortification under abiotic stress
Eggplant	*Solanum melongena* L.	50-100 mg/kg	Chemical	< 200 nm	Foliar Spraying	ZnO-NPs increased crop production under drought
Sesame	*Sesamum indicium* L.	3-10 mg/L	Biogenic	10 nm	In soil /Pot culture	Biogenic ZnO-NPs improved sesame germination and growth.
Loquat	*Eriobutria japonica*L.	12 mg/mL	Biogenic	<50 nm	Petriplate culture	Antimicrobial activity with fortification use
Wheat	*Triticum aestivum* L.	40, 80, 120 ppm	Biogenic	<50 nm	Soil culture	At a dose of 80 ppm, ZnO-NPs caused the greatest increases in height, seed weight, yield, and biomass.

6.2 Nanofertilizers a key approach for Zinc fortification in Crops

Due to their small size, high surface area to volume ratio, and exceptional optical properties, nanomaterials are unlike other materials. Because of their unique qualities, nanofertilizers are beneficial because they enable improvements in plant growth, nutrient security, and a variety of farming techniques (Fig. 6). Recent years have seen significant progress in the study of nanotechnology applications, and numerous industries have begun integrating this cutting-edge technology into their operations to produce cutting-edge goods. Since the human population is growing, it makes sense that agricultural productivity would grow as well. Fertilizers have been helpful in increasing food production, and new improved fertilizers verities, known as nanofertilizers, can be produced with the aid of nanotechnology. They are easily absorbed by the soil, which enhances soil quality and, as a result, promotes plant growth. The bulk of research and development efforts have been directed towards electronics, optical devices, water purification, and healthcare, despite the fact that nanotechnology has many potential applications. There is a lot of potential for developing materials that can carry nanofertilizers, which can be used as an environmentally friendly solution to nutrient-related problems because elements necessary for crop development and nutritional composition.

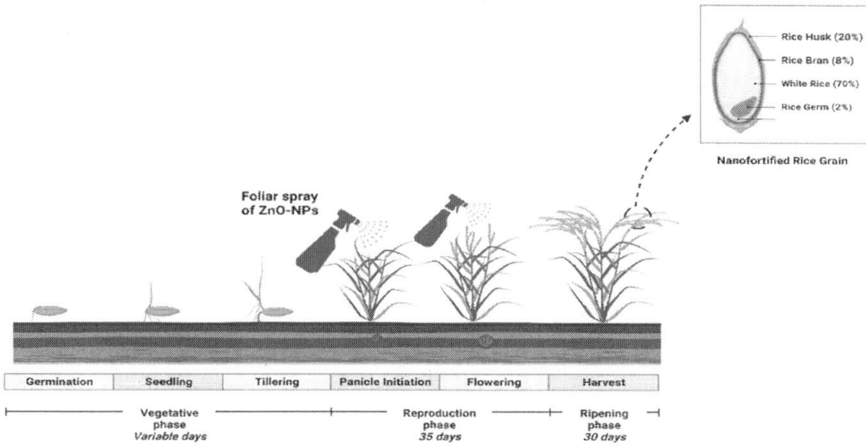

Figure 6: Diagrammatic representation of mechanism of uptake of ZnO-NPs based fertilizer by rice plants through foliar spray or applying in the soil that uptake by root via symplastic or apoplastic pathways and stomata that increased Zn contents into the rice grains.

Utilizing zinc nanoparticles to enhance plant uptake is a relatively new strategy. Particle size also affects how effective zinc fertilizers are. Fertilizers like zinc oxide (ZnO), which dissolves slowly in water, dissolve more quickly when particle size decreases because of an increase in surface area and number of particles per unit weight that occurs. The efficiency of using zinc nanoparticles to raise crop zinc content has been shown by the increased zinc content of shoots in response to ZnO NP supplementation. Granular ZnO was completely ineffective, whereas fine $ZnSO_4$ (0.8 to 1.2 mm) and granule $ZnSO_4$ (1.4 to 2 mm) were equally effective. Zn uptake gradually increased with decreasing granule size. Instead, 1.5 mm granules were utilized because they are lighter than 2.0 mm and 2.5 mm granules. Zn fertilizer has a greater surface area as a result, enabling more uniform distribution and absorption. Better crop uptake and higher yields are made possible by increased fertilizer efficiency. ZnO NPs were used to promote Cicer arietinum root development. It has been observed that ZnO NPs impact the germination and root development of *Cicer arietinum* seeds. When employing ZnO NPs as nano-nutrients, the following factors may be taken into account:

- **Nano-dosing**: when plants are exposed to extremely high doses, it may be hazardous to the physiology and systems of the plants, and consequently to humans who eat the plants. Thus, biofortification cannot start until the ideal nutrient dose is established.
- **Nutrient preparation and application technique**: Foliar application of nanonutrients is a good way to apply nutrients directly to plants in problematic soil types such as saline, alkaline, low or high pH, and others. In particular, biologically produced nutrients are favored because of their low toxicity and environmental friendliness.
- **Techniques for biofortification**: Like seeds priming with engineered nanomaterials, are being studied because they might be useful in the fight against hunger. There are various methods to achieve nutrient enhancements through biofortification, including foliar spraying, seed priming, and soil enrichment with fertilizer containing the candidate nutrient.

According to recent research, nanotechnology might eventually totally change the way we practice agriculture. This makes it possible to use agrochemicals with an advanced delivery system that is easy to use, target specific, and free of risks. When considering even the most advanced polymeric-type conventional fertilizers, the efficiency of nanofertilizers is higher due to their higher surface-to-volume ratios. Moreover, their composition might permit a gradual release, enhancing the plants' ability to assimilate the nutrients. Consequently, this technology offers a basis for creating innovative and long-lasting nutrient delivery systems that capitalize on the nanoporous surfaces of plant components (Fig. 2B and C). Without compromising plant health or the environment, encapsulated nanoparticles, nanoclays, and zeolites increase fertilizer efficiency, restore soil fertility (Fig. 2A), and maintain agroecological balance. Zinc oxide, silica, iron, and titanium dioxide nanoparticles; ZnS/ZnCdSe core-shell quantum dots (QDs); InP/ZnS core -shell QDs; Mn/ZnSe QDs; gold nanorods; AL_2O_3, TiO_2, CeO_2, and FeO; and numerous other materials are possible components of nanofertilizers. However, the effectiveness of applying nanomaterials as fertilizers as fertilizers to a particular plant species depends on a number of aspects due to their nanoscale dimensions, concentrations, compositions, and chemical properties. For long-term food security, then, the development of nanotechnology-based agriculture is crucial. This is only possible with the aid of new technologies, which necessitate a thorough comprehension of biology, engineering, material science, and biotechnology. Traditional fertilizers' nutrient loads are prone to leaching loss and contaminating aquifers that store underground water. People are searching for alternatives, such as nanofertilizers, due to the disastrous effects of chemical fertilizers on the environment (greenhouse gas emissions, hypoxia). Apart from their numerous other advantages, nanofertilizer possesses the special ability to release nutrients gradually, lowering the risk of them being lost through evaporation.

Compared to conventional fertilizers, nanofertilizers offer a number of benefits, such as the capacity to triple crop resistance to disease and drought while also requiring three times less chemical fertilizer. However, the sizes and shapes of the particles have a significant impact on how much of the plants absorb from the soil. Before plants can absorb the nanoparticles and use them as a source of the micronutrients they require to grow, it's possible that they must go through a series of events like oxidation and recombination. The idea of fortifying plants with nutrients has gained interest due to their nanoscale. Such nutrients are stored by the plant, which allows it to grow and fills the void left by their deficiency. Nanofertilizers specifically designed for this purpose could also be used to remedy certain nutrient deficiencies in plants. The atoms on the surface of the nanomaterial can be arranged in a variety of ways to achieve the desired results. Zeolites are mineral materials that protect soil by retaining soil nutrients, which is why they are used for efficient crop production and reducing environmental risks associated with agricultural practices. They recently created natural zeolite, a nanofertilizer made up of over fifty different kinds of aluminosilicates, which are basic alkali and alkaline earth minerals. This is because its affordability and ease of use have made it a popular choice for maize production. In addition to providing a tiny amount of minerals to the plants, it functions as a clever carrier and regulator of major mineral fertilizers. Because zeolite can transfer both potassium and nitrogen, less fertilizer is needed to get the same effects. Crop yields were found to rise when zeolite compound was mixed with humus materials. Zeolite nanocomposites containing nitrogen, phosphorous, and potassium have been reported, and plants have been shown to absorb and grow amino acids, mannose, calcium, iron, and zinc, to mention a few. Zeolite's limited ability to absorb and retain anions is why biopolymers are added to enhance this property. The most common zinc compounds added to fertilizers are oxides and sulphates. There is no way to overstate how crucial zinc is to all living things. It is a constituent of most enzymes, required for the regulation of hormones and chlorophyll, and critical for the metabolism of carbohydrates. ZnO in the form of nanoparticles is relatively easily absorbed, metabolized, and accumulated in plant systems. Although excessive ZnO NP concentrations can stunt plant growth and cause abnormalities in germination, root development, and early-stage seedling biomass, zinc is a micronutrient that is only needed in small amounts. However, if plants don't get enough of it, their growth and development may be hindered. Zinc oxide (ZnO) is mixed with zinc sulphate to make fertilizer because ZnO is not soluble in water. Crop yields are reduced in alkaline soils because plants are more vulnerable to zinc deficiencies. It is difficult to apply these micronutrients to the soil uniformly because a significant amount is required. Zinc oxide and Zinc sulphate, on the other hand, is only marginally effective as micronutrients when taken by themselves. They work well in conjunction with micronutrients, which serve as carriers. While orthophosphate aids in the absorption of only 0.05% of zinc, carriers are crucial to the solubilization and absorption of micronutrients, particularly in aqueous solutions, as demonstrated by the fact that polyphosphate, ammonia, and anhydrous ammonium nitrate can all dissolve 2.0% of zinc. The mineral content, ionic strength, and organic matter of the soil are essential for nutrient uptake and bioavailability. Calcareous soils, characterized by high pH and calcium carbonate ($CaCO_3$) concentrations, have an impact on zinc solubility and absorption. When there is calcium carbonate present and zinc is exchanged for calcium, zinc carbonate precipitates.

Crop nutrient status has been found to be unaffected by the physical application of micronutrients to the soil. Plants attempting to absorb micronutrients from sandy soils may encounter difficulties due to the composition of the soil. Soil pH has an impact on micronutrient availability as well. With the exception of molybdenum, soil pH reduces the availability of micronutrients. Zinc precipitates out as $Zn(OH)_2$, $ZnCO_3$, and Zn_2SiO_4 when the pH of the soil rises, it is then adsorbed on the soil's surface, where it is found near other clay and wound inorganic materials. By doing this, the system becomes less soluble, which hinders plants from effectively desorbing soil minerals. However, nanozinc oxide has a wide range of agricultural uses, such as crop genetic modification, the creation of nanofoods, prescribed diets for certain illnesses, nutrient recovery from the soil, blocking excessive UV radiation during plant growth, and the use of fertilizers in crop production, among many other uses.

7. Challenges of nano-biofortification

7.1 In Covid-19 Era

A worldwide pandemic is the severe acute respiratory syndrome (SARS) caused by COVID-19. Most places in the world have restricted people's freedom of movement as a precaution. The World Health Organization (WHO), the National Institute of Health (NIH), and the Centres for Disease Control and Prevention (CDC) have developed therapeutic guidelines for patients who are infected. As of yet, however, there are no generally available preventive measures beyond hand washing and awareness of frequently occurring oronasal symptoms and fever. It is imperative that a biologically active material be available during a pandemic to be used as a preventive measure. For the vast majority of people, the perfect solution would be affordable, easily assessable, and low-risk. It has been demonstrated that the mineral zinc strengthens the immune system and prevents the spread of viruses. To ascertain whether zinc consumption (within suggested upper safety limits) can be utilized as a preventive measure to guard against the development of Covid-19, clinical investigations are necessary. Even after vaccination, low zinc levels can affect a person's response to vaccinations. Zinc is an essential mineral that helps more than a hundred enzymes function properly. It is also necessary for the synthesis of DNA and RNA, healthy cell integration, and the proper functioning of the immune system. Zinc is a necessary nutrient since the body is unable to produce or store it. An adult Indian must consume 10 to 15mg of zinc daily in order to meet the recommended daily intake (RDI). A higher incidence of Covid-19 has been associated with immune suppression (Fig.7). Eating a nutritious diet is one of these elements since it supports a strong immune system. Zinc, an antioxidant nutrient, has a significant effect on the immune system. The maturation and function of natural killer cells and neutrophils, which mediate innate immunity, depend heavily on zinc. The cells known as macrophages, which are in charge of phagocytosing invasive pathogens, are also impacted by zinc deficiencies. The growth and function of T and B lymphocytes are correlated with zinc deficiency. Similar to SARS-COV-2, an RNA virus causes Covid.

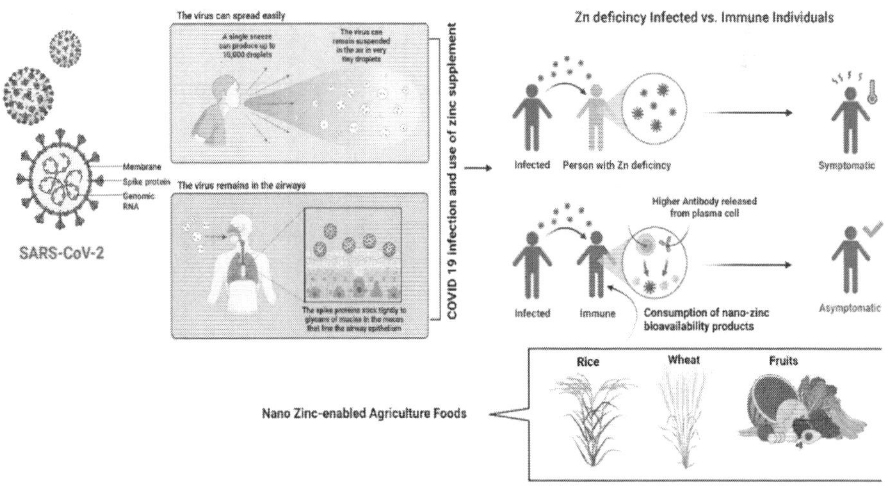

Figure 7. Diagrammatic representation of Covid-19 SARS-COV-2 virus infection mode spreading process and impact of nano zinc enabled agriculture foods to boosting the immune system.

Ribonucleic acid (RNA) is the genetic material that RNA viruses store. RNA-dependent RNA polymerase (RdRp) enzymes are the key players in the RNA-synthesizing machinery of RNA viruses. RdRps are therefore given top priority in the development of antiviral medications.

This week's publication of a new study demonstrates that intracellular zinc ion (Zn^{2+}) concentrations efficiently block RdRp activity, which in turn prevents RNA viruses like Covid and artery viruses from replicating. In cell culture studies, it has been demonstrated that high concentrations of Zn^{2+} and the addition of ionophores—compounds that stimulate cellular import of Zn^{2+}--such as pyrrolidine dithiocarbamate (PDTC), pyrithione (PT), and hinokitiol (HK) inhibit the replication of various RNA viruses. Zn content in horticultural and cereal crops can be increased through nano fortification, meeting humans' daily requirement for zinc. Major illness like COVID-19 can be avoided by having access to zinc content. Crops enhanced with nanotechnology can meet the Zn requirement necessary for human health (Fig.7).

7.2 Toxic impact of Zinc deficiency and accumulation in plants

Because zinc is becoming less accessible in arable soil worldwide, zinc deficiency is becoming a significant farm problem that jeopardizes food safety and crop productivity. The availability of zinc in soils for plant growth is negatively impacted by a number of factors, including low zinc contents, high pH, lower reduction potential, sustained flooding, rhizosphere microbial communities, high organic matter, high iron/manganese oxide contents, high calcium carbonate and bicarbonate contents, high sodium and phosphorous availability, high exchangeable magnesium/calcium ratio, and many others. Protein metabolism is disturbed by zinc deficiency because it causes ribosomes to break down. Because zinc is a necessary component of several enzymes required for protein synthesis, including glutamate ribonuclease, ATPase, and dehydrogenase, low zinc levels inhibit the synthesis of RNA and protein. The amount to which protein and RNA production are decreased by zinc deficiencies varies amongst plant species. When plants are severely stressed by zinc deficiency, the most obvious signs of zinc deficiency are wilted stems, reduced leaf size, and immature leaves turning yellow. However, aging leaves often display wilting, rolling, bronzing, and restricted growth. Plants have evolved defense mechanisms that include root system reorganization, arbuscular mycorrhiza symbiosis, photo-siderophore and organic acid ion production, and increased resistance to oxidative stress brought on by zinc deficiency. These plant modifications increase zinc uptake and transport mediated by zinc transporters and affect the transcription of Zn-required processes. Growing apprehensions about bioavailability, nanoparticle toxicity, and an inadequate regulatory framework are impeding the agricultural sector's ability to fully embrace and adopt nanotechnologies. Remember that NPs are a double-edged sword and that using them incorrectly can have a negative impact on plant growth by preventing the plants from producing proteins and pigments. Plants respond to high NP concentrations in different ways on a molecular, physiobiochemical, and morphological level. A major determinant of NPs' bioavailability and toxicity is their bio/ geotransformations in the soil. Comprehensive evaluations of the factors influencing the phytotoxicity of NPs have already been carried out. Studies have demonstrated that a number of variables affect the phytotoxicity of nanoparticles (NPs), including the type of nanomaterial used, the test plants used, the NP concentration, the NP size distribution, the NP morphology, and the NP surface charge. Moreover, these investigations have demonstrated that the concentrations-dependent mediated effect of nanoparticles (NPs) on plant growth and development can result in a series of issues for plants, animals, and eventually humans when high concentrations of these NPs are used.

8. Conclusion

As a micronutrient, zinc is needed in trace amounts by both humans and plants. But because zinc is an essential cofactor for numerous enzymes, this trace amount plays a crucial role in the healthy operation of both plants and human systems. Food security is impacted because of the low available concentrations of zinc in the grain, which has a detrimental effect on yield potential. To combat this, the Zn concentrations in the seeds effectively rise whenever the crops are sprayed with the appropriate quantity of ZnO NP fertilizer at the appropriate times. The biofortification of grains and other crops, such as pulses, is a great idea for people worldwide, especially for rural populations, to ensure that everyone has access to diets rich in zinc. The strategy of nano-biofortification holds promise as a weapon against malnourishment. This technique shares some of the advantages of nano-fertilizers, including efficiency and lower usage, especially in its biological nano-form. This approach is still in its infancy and needs more testing before it is put into practice globally.

CHAPTER 12

ROLE OF NANOTECHNOLOGY IN PRICISION AGRICULTURE MANAGEMENT

Short Summary

Conventional agriculture frequently relies on an excessive number of pesticides and fertilizers, which have negative effects on the ecosystem and current species. The simplest definition of sustainable agriculture is the application of the fewest agrochemicals possible in order to protect the environment and preserve a variety of species. The main goal of sustainable agriculture is to provide for the current needs of society, including food and textiles, without compromising the ability of future generations to provide for their own needs. A number of agriculture practices, including food security enhancement, poverty reduction, and sustainable increase in agricultural productivity, include geoinformatics, nanoparticles, and nanobionics. The identification of the type of soil and its architecture is aided by geoinformatics techniques. Using nanotechnology in agriculture has advantages, including specific applications such as trace production and nutrient levels to increase productivity without decontaminating soil and water, protection against various insect pests and microbial diseases, and security against various diseases caused by microorganisms. Through the use of nanoparticles and nanobionics, the application of nanotechnology in agriculture can significantly increase in efficacy of agricultural efforts and consequently, suggest a significant way to uphold the sustainable development of agroecosystems. In real time, the use of nanobionics as sensors, fungicides, herbicides, insecticides, and fertilizers can help determine and diagnose a crops needs and health. Therefore, the purpose of this chapter is to critically highlight ideas regarding the application of nanoparticles, nanobionics, and geoinformatics as cutting-edge instruments for sustainable agriculture.

1. Introduction

The production of plant and animal products to meet the needs of the current population and safeguard the resources for future generations is the significance of sustainable agriculture. In essence, sustainable agriculture is a modern method of farming that preserves natural resources for the benefit of public health while maintaining their nutritional value. Using plant breeding and other cutting-edge technologies, numerous new crop varieties were created at the start of the 20th century. However, the agricultural sector is dealing with a number of environmental problems in the twenty-first century, including pollution, global warming, and climate change, which have negatively impacted grain production and even improved crop varieties. Reduced grain production can result in a number of problems, including food security. To address this, it is vital to increase grain production by using three essential steps: diagnosis, determination, and detection. The precise analysis of agricultural production elements or raw materials, such as soil architecture, and the need for fertilizer, herbicides, fungicides, insecticides, and pesticides for each crop are necessary for these three main steps. The precise examination of these components of agricultural production combined with contemporary instruments like nanobionics, nanoparticles, and geoinformatics. These agricultural applications improve food security. Lower poverty, and foster sustainable increases in agricultural productivity. The identification of the type of soil and its architecture is aided by geoinformatics techniques. Conversely, real-time crop requirements and health diagnosis can be facilitated by the application of nanoparticles and nanobionics as pesticides, fungicides, herbicides, fertilizers, and sensors (Figure 1). With the help of these modern instruments, farmers can produce crops with greater efficiency and at a lower cost of production, thereby increasing their income.

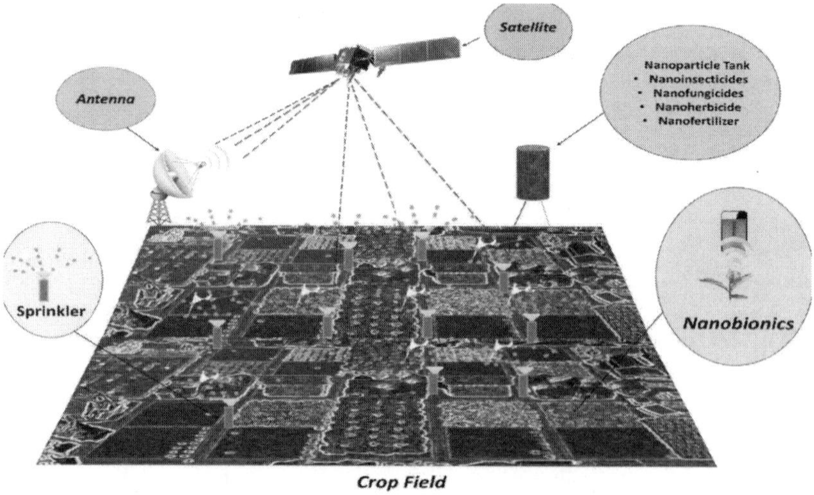

Figure 1. integrated farming system that have geoinformatics, nanoparticles, nanobionics emerging tools for sustainable agriculture.

"3-T" concept for sustainable agriculture

Every year, the effect of climate change on crop patterns and productivity is more pronounced. Drought, floods, salinity, and diseases have impacted crops as a result of climate change (FAO<2019). Crop production is directly impacted by climate change, which means that a nation whose people are entirely dependent on agriculture may also experience food shortages or starvation. Furthermore, this issue might get worse in the future. Agriculture needs technology that can quickly and effectively address climate change-related issues in order to address this issue. In the context of solving this issue, the 3-T concept appears to be useful for agriculture (Figure 2). First T stands for "trace," which refers to tracking or real-time weather, crop, and soil monitoring using geoinformatics tools such as satellite imagery, unmanned aerial vehicles (UVAs), automated weather stations, location-based mobile applications, ground sensors/base station, UGVs (unmanned ground vehicles), etc. Their geoinformatics tools gathered data on disease detection, pest control, fertilizer recommendations, weather prediction, digital soil and crop mapping, and more. These data assist farmers in micromanaging their fields and mitigating the adverse effects of climate change on their crops. The second T stands for test. Validation requires testing after receiving geoinformatics data, and this testing necessitates conformance with ground data. In order to calculate the appropriate dosage of fertilizers, pesticides, and herbicides for each crop plant, we primarily test the soil and each plant individually. Following the test results, prompt action is required for per-diagnosis or treatment. For this action, traditional fertilizers, herbicides, insecticides, and pesticides are combined with nano fertilizers, herbicides, and insecticides. Which are quicker to work and improve the nutrition and health of the crop as well as the soil; all of these elements fall under the third T, or treat.

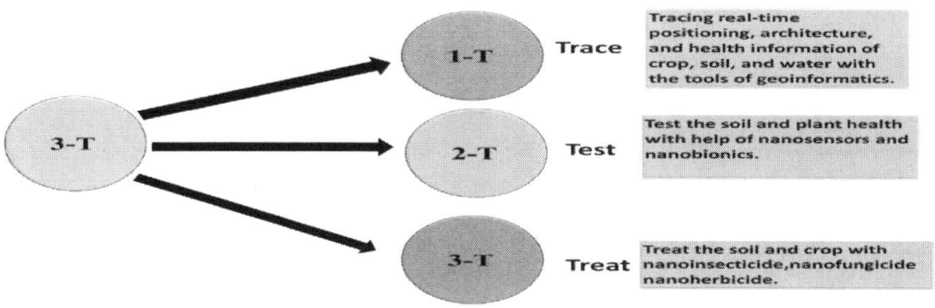

Figure 2: Diagrammatic representation of 3-T concept.

2. Geoinformatics

The term "geoinformatics" combines the terms "earth" and "the study of information processing," which is known as informatics. Consequently, the study of "Earth sciences" and "Informatics" is combined to form geoinformatics. With the aid of information technology, geoinformatics primarily deals with the gathering, analysis, storing, retrieval, representation, and distribution of data about the planet. Agriculture uses a variety of geoinformatics tools and methods, such as the Geographic Information System (GIS), Remote Sensing (RS), and Global Positioning System (GPS). The application and integration of these three instruments could contribute to increased crop yield, decreased water waste, improved soil fertility, and increased use of agricultural resources for crop production.

2.1 Global Positioning System (GPS)

A satellite-based navigation system called GPS (Global Positioning System) can help locate objects anywhere on Earth. Global data, including real-time, 3-D positioning, navigation, and earth timing, is continuously provided around the clock. A GPS system uses signals broadcast by satellites to enable GPS receivers to determine their position in real time. This positioning data can be used to map agricultural, soil, and water measurements. In order to locate farm equipment within a 3-meter radius of a real site in the field, smart framing uses GPS systems. GPS is mostly used in agriculture for mapping yield and varying fertilizer and pesticide application rates.

2.2 Remote Sensing (RS)

An integral part of information science, remote sensing aids in both large and micro-scale crop health, soil, weather, temperature, and humidity management. Through the use of a variety of sensors, including satellites, airplanes, balloons, rockets, and ground-based sensors, remote sensing is a scientific art that analyses the characteristics of the earth's surface. For land surveys, the remote sensing methods use wavelengths between 0.4 and 2.4 nm.

2.3 Geographical Information System (GIS)

The Geographical Information System, or GIS, is a computer-based method that facilitates the mapping, analysis, and observation of events and things that occur on Earth. Data about crop productivity and agronomic factors that are important for mapping to improve understanding of various factors such as yield, fertility, pests, and weeds. This information aids in decision-making and improves farmers' comprehension of sustainable agriculture.

2.4 Role of Geoinformatics in sustainable agriculture

Different pigment systems found in plant leaves are known to strongly absorb visible (red) light wavelengths while strongly reflecting near-infrared light wavelengths, which are invisible to human vision. These reflectance characteristics also shift as a plant canopy processes from early spring growth to late-season maturity and senescence. Among other uses, Vegetation Indices (VIs) derived from canopies based on remote sensing are efficient algorithms for both quantitative and qualitative assessments of vigour, vegetation cover, and growth dynamics. An important method for creating, organizing, and maintaining protected areas of uncontaminated or natural areas is mapping vegetation. Plant leaves contain various pigment systems that are known to strongly reflect near-infrared light wavelengths—which are invisible to human vision—(red) light wavelengths. As a plant canopy develops from early spring growth to late-season maturity and senescence, these reflectance characteristics also change. Vegetation Indices (VIs) derived from remote sensing-based canopies are effective algorithms for both quantitative and qualitative assessments of growth dynamics, vegetation cover, and vigor. Mapping vegetation is a crucial step in the creation management, and preservation of protected areas of pristine or natural areas. The LAI is used to quantify the amount of leaf material present in a given environment to accurately represent radiation, heat, water, and various gas exchanges with the overlying atmosphere or the underlying soil in a wide range of models used in agriculture, ecology, carbon cycling, climate, and other studies. The primary applications of LAI are in biogeochemical cycle models, agro-meteorology from crop assessment, and soil-vegetation-atmosphere-transfer schemes. These applications frequently call for lengthy time series at different temporal and spatial scales.

Figure 3. The LAI, total chlorophyll content of various types of plant in Indore district Madhya Pradesh India.

The current NDVI is compared to the range of values recorded during the same period in prior years using the Vegetation Condition Index (VCI) (Figure 4). The VCI, which is expressed as a percentage, indicates the range of value between which the observed value falls, i.e., the minimum and maximum, in prior years. Bad and good vegetation state conditions are indicated by lower and higher values, respectively. The majority of the time, vegetation condition and growth anomalies—both in terms of extent and intensity—are detected and spatially demarcated through the qualitative application of vegetation color imaging (VCI). This is especially helpful for tracking the current growing season (for early warming purposes, for example). Additionally, the most effective tool for measuring canopy chlorophyll content is the IRECI (Figure 5).

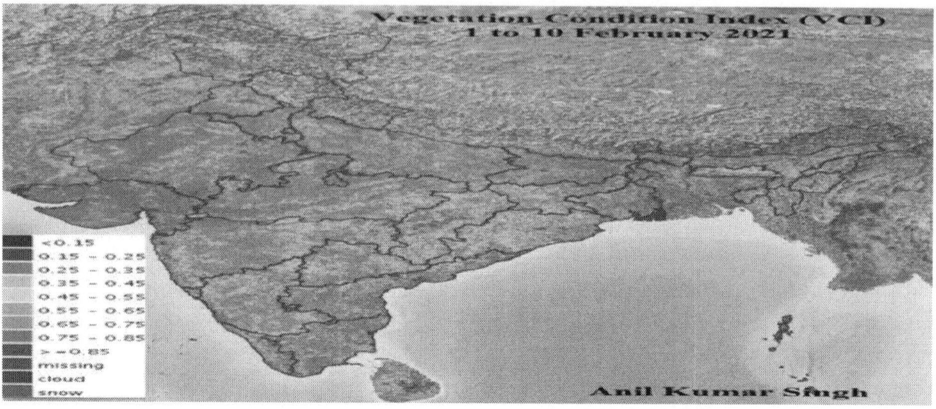

Figure 4. Diagrammatic representation of Vegetation Condition Index (VCI) of India from 1February to 10 February 2021

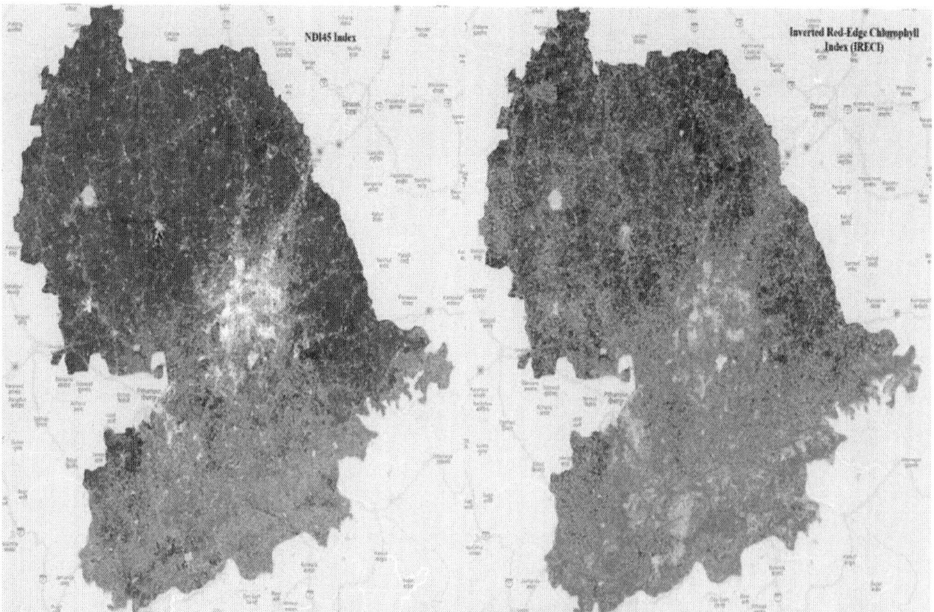

Figure 5. Diagrammatic representation of IRECI of Indore district Madhya Pradesh India.

3. Nanoparticles

The functional components of a larger structure known as a nanotube are the nanoparticles. It is a new form of carbon, similar to a two-dimensional sheet of graphene rolled into a tube. Nanotubes come in two primary varieties: Three types of nanotubes: (i) single-walled (SWNTs) and (ii) multi-walled (MWNTs). It is perfect for creating nanoelectromechanical systems and reinforced composites because of its 200 GPa tensile strength. Moreover, its metallic or semiconducting characteristics present amazing opportunities for creating electronic circuits or even whole nanodevices. The nanotube systems are made up structurally of graphitic layers that are smoothly coiled into cylinders. Recently, plant protein labeling has been made possible with the development of fluorescent nanoparticles (NPs) or quantum dots (QDs). Because of their various properties—mechanical, electrical, thermal, optical, elastic, etc.—the geometrical dimensions of carbon nanotubes, particularly their diameter, determine the applications for which they can be used. Most SWNTs have a diameter of about one nm. This property has a strong correlation with electron orbital re-hybridization, synthesis methods, and maxing of s and a bond. Utilizing these CNT properties will undoubtedly create new opportunities for the development of various nanodevices with unique thermal, optical, and conductive qualities for use in the agricultural sector and the creation of sustainable farming conditions. Delivery systems based on carbon nanotubes (CNTs) can target agrochemicals or other substances to hosts, reducing chemical release into the environment and causing less damage to other plant tissues. According to a report, compared to CNT powder, CNT powder, CNT nano-sponges enhanced the absorption of the hazardous organic solvent dichlorobenzene from water by approximately 3.5 times. Utilizing CNT nano-sponges, pesticides, fertilizers, oil, medications, and other water contaminants are absorbed. Their efficiency is increased by the iron and sulfur they contain. Yet, these CNTs are still in the early stages of development. In rare circumstances, they might enter the cytoplasm and build up there, causing human cells to become vital. Semiconductor QDs feature large effective excitation, strong resistance to photobleaching, remarkable resistance to photochemical degradation, broad absorption spectra with tapered, symmetric fluorescence spectra spanning from ultraviolet to near-infrared, high quantum yield, and molar extinction coefficients. As a result, these have exceptional fluorescence, materials for change carriers that are quantumly confined, and band energies that are size-tunable. Because QDs have unique spectral characteristics compared to traditional organic dyes, they are a new generation of fluorophores with modern applications in bioimaging and biosensing. In order to create solar fuel, QDs also serve as photocatalysts in the chemical reaction that turns water into hydrogen when exposed to light. When used in small quantities, QDs had no discernible cytotoxicity on seed germination or seedling development. Thus, QDs can be used for live imaging in plant root systems to verify established physiological processes based on this transport method. Multifunctional plasmonic materials that can be coupled with MENS, induce particular field responses and connect sensing phenomena and size-tunable energy regulation well. The watermelon plant's physiology was greatly altered by the gold nanorods, which also demonstrated phytotoxicity to plants, especially when present in high concentrations, and the ability to transfer the auxin growth regulator 2,4-D, which had a major impact on the control of tobacco cell culture growth. Small emulsion nanoscale droplets (oil/water system) with sizes less than 100 nm are what create nanoemulsions. The physical characteristics of nanoemulsions can differ significantly from those of microscale emulsions, despite the possibility that there are no fundamentally important differences between the two types of emulsions. The ratio of surface to volume, Laplace pressure, and elastic modulus of nanoemulsions are substantially higher than those of regular emulsions because of the size of the droplets. In addition, most nanoemulsions appear optically transparent, which gives them many advantages over general emulsions, including the ability to be added to beverages. Unfortunately, creating a nanoemulsion requires a lot of energy, so certain equipment—that can produce extremely high shear stress is needed. Researchers described a "low-energy" method for creating nanoemulsions. This method involves bringing two liquid phases into contact with each other: one is a homogenous liquid that contains a lipophilic phase, hydrophilic surfactant, and possibly a polymer or drug as well as a potential solvent. The hydrophilic species present in the oily phase is then quickly dissolved into the aqueous phase, causing the oil to demix into nano-droplets that are immediately stabilize by the amphiphiles. As a result, this approach seems to be the easiest and doesn't call for any energy-intensive special devices.

3.1 Crop protection

Nowadays, sprays are the most popular method for applying nutrients, pesticides, herbicides, fertilizers, and antibiotics to agricultural land. Pesticides are any substance or mixture of substances intended for preventing, destroying, or controlling any pest, including vectors of human or animal disease, unwanted species of plants or animals, causing harm during or otherwise interfering with the production, processing, storage, transport, or marketing of food, agricultural commodities, wood and wood products or animal feedstuffs, or substances that may be administered to animals for the control of insects, arachnids, or other pests in or on their bodies (World Health Organization).

3.1.1 Nanoinsecticides

When applying pesticides, water solubility is a critical consideration. These days, the majority of pesticides are hardly soluble in water. This makes the pesticide more expensive and harmful since it needs organic solvents to dissolve the pesticide entirely in water. Using nanoparticles can make pesticides more soluble, which lowers their toxicity. By effectively incorporating nanoparticles such as modified chitosan and porous silica into insecticides, fewer water-soluble pesticides have been created. Azadirachtin combined with modified chitosan nanoparticles (hydrophobic insecticide) demonstrated favorable reduction of cell growth and prolonged drug release in *S. litura* ovarian cell line. Increased mortality has been seen in insects when anacardic acid is intercalated into LDH (Layered Double Hydroxides) nanoparticles. The mortality rate is increased when mustard leaves or *S. litura* skin is sprayed directly. Preferably, these findings demonstrated the possible advantages of using nanoparticles to increase the solubility. With the use of solid lipid nanoparticles (SLNs), evaporation or volatilization associated with post-insecticide application has been addressed, which was another significant issue. Because pesticides release active chemicals slowly, combining them with nanoparticles can help reduce the toxicity of the insecticides.

3.1.2 Nanofungicides

Chemical or biological agents are known as fungicides, and their purpose is to destroy fungus. Generally speaking, fungicides fall into two main categories:

a. **Preventive fungicides:** These include compounds like sulfur, dichlorocarbamates, organometallics, phthalimide, and benzamides that shield plants against fungal diseases.

b. **Curative fungicides:** These agents penetrate the site of infection and stop the plant pathogen from developing further. Examples of such agents include acetimides, dicarboximids, and sterol inhibitors.

The most studied and practical nanoparticle carriers include polymer blends, silica, and chitosan. A wide variety of fungus species were used to test the nanofungicide's efficacy. However, certain nanofungicides were tested on the plant lately, and no harmful signs were found. The reason behind the unintended impact of generic insecticides upon application is their low water solubility, volatilization, and poor stability. Pyrchlostrobin, a fungicide based on nanoparticles, has been shown to be an effective remedy for these issues. A fungicide with poor water solubility is pyraclostrobin. Chitosan-lactide copolymer is loaded onto nanoparticles in different doses to inhibit *C. gossypii*. An environmentally acceptable substitute for traditional fungicides is nanofungicides. Suppressive characteristics are exhibited by metal nanoparticles including copper (Cu-NPs, CuO-NPs), silver (Ag-NPs), and zinc (ZnO-NPs). These characteristics help protect against a number of pathogens that cause economically significant soil and foliar-bone plant diseases. The characteristics of these metal nanoparticles were evaluated in vitro. ZnO-NPs, which mean inhibition rates ranging from 235 to 848 μg/mL). It was found that CuO-NPs had little effect on the fungus. CuO-NPs essentially had no impact on any fungal species, whereas Ag-NPs only showed a notable inhibitory effect on *B. cinerea*. A comparative analysis of mycelial growth fungitoxicity experiments between the nanoparticles and their bulk-sized counterparts revealed that ZnO- NPs were more toxic to all fungal species tested than $ZnSO_4$, whereas CuNPs were more fungitoxic to $CuSO_4$ in all cases except for *B.cinerea, A. alternate,* and *M.*

fructicola. The toxicity of Cu-NPs and CuO-NPs is positively correlated, although there is no link between NPs and the corresponding bulk metal equivalents. This suggests that bulk and nanosized metals may function differently in terms of mechanism of action. The fungitoxicity of all nanoparticles was shown to dramatically increase upon application to spores rather than fungal hyphae, notwithstanding the notable variance across fungal species. This suggests that these compounds have a significant deal of promise for application as antifungal medicines that provide protection.

3.1.3 Nanoherbicides

A class of insecticides known as herbicides is made specifically to destroy unwanted grasses and weeds. When traditional herbicides are applied, the crop is also impacted. They exhibit significant levels of toxicity and a protracted half-life, and they remain unaffected by standard wastewater treatment plant treatment. Both soil and water pollution levels rise as a result of this. These chemicals damage the crops in the field as the weeds grow alongside them. Herbicides are used to eradicate weeds that pose a threat to crops as well. This is where nanoherbicides come into play since they can help with this issue. Because of their tiny size (between 1 and 100 nm), these particles may dissolve readily in soil particles, eliminating weeds from their roots without harming the primary crops. Herbicides based on nanoparticles could be able to eradicate weeds and increase crop rise to 84% when certain herbicides, such as triazine, ametrine, and atrazine, are combined with them. A common pesticide is atrazine. The nutrients in the soil that were previously dispersed by the weeds may now be used by the crops. Additionally, this makes the crops more resistant to disease. In addition to serving as transporters of molecules, the nanoparticles catalyze processes in which different compounds degrade. Herbicide breakdown efficiency can be further increased by combining nanoparticles such as modified silver with carboxymethyl cellulose.

3.2 Crop improvement

Every year, nearly one-third of agricultural yield is lost due to a variety of stresses, including microbial diseases, weeds, insect stress, natural catastrophes, low soil fertility, and other factors. Every day, new approaches and tactics are created to address this problem; nevertheless, the majority of these technological and associated initiatives have their own set of drawbacks. A wide range of materials, including magnetic materials, ceramics, metal oxides, quantum dots, semiconductors, dendrimers, lipids, emulsions, and polymers, are used to make nanoparticles. Different applications of these nanoparticles are used in agriculture to combat biotic stressors.

3.2.1 Abiotic stresses

Abiotic stressors are those that cause direct disruptions to a plant's physiological processes, such as photosynthesis, respiration rate, metabolism, transport, etc., which hinders the plant's ability to grow and develop. Drought, salinity, submersion (food), and metal toxicity are these causes. They lower agricultural yield and growth. In order to equalize the impact of various pressures, plants adapt. A series of modifications start at several levels, including the molecular, biochemical, and physiological ones, and they lead to the plant's unrestricted development. On several levels, these adjustments serve as a defensive mechanism. The site of expression of several gene-altered pathways that lead to plant adaptation against these stimuli is known as the biochemical or genic level. Numerous investigations have demonstrated that applying nanoparticles to a plant can affect its growth and development while it is involved in the creation of antioxidant compounds. These nanoparticles indirectly catalyze the metabolic processes by increasing the activity of antioxidant enzymes (CAT, SOD, and POD)

(A) Drought stress

It happens as a result of the plant not having access to water. Drought is mostly caused by a number of causes, including higher temperatures, less precipitation, dry winds, intense light, and a rapid pace at which water evaporates from the soil. As a result, the plant cannot get enough water for proper growth and development. Applying varying concentrations of nanoparticles has been shown to increase plant tolerance to drought stress. This stress causes a reduction in leaf size, an increase in stem extension and root multiplication, a crucial disturbance of plant water relations, and a decrease in water absorption efficiency. Per-treatment with SNPs has been shown to have a beneficial effect on the following: relative water content (RWS), photosynthesis, membrane electrolyte leakage channel (ELI), chlorophyll, carotenoid, carbohydrate, and proline (components implicated in a barrier for drought stressors). Uncertainty surrounds the Hawthron plant's adaptation mechanism for preserving essential physiological conditions to prevent drought stress in the presence of silicon nanoparticles. When SiO_2 nanoparticles were added to a plant medium, the plant used a variety of defense mechanism, such as

a. Decrease in the leaf cells' capacity to penetrate their plasma walls, leading to a loss of lipid peroxidation.

b. Plant cell wall defense against heat stress and drought.

Proline levels rose following the administration of silica nanoparticles as compared to conventional silica fertilizer. In stressed plant leaves, silica nanoparticles as compared to conventional silica fertilizer. In stressed plant leaves, silica nanoparticles catalyze the activities of the enzymes peroxidase (POD) and catalase (CAT). Moreover, silica nanoparticles aid in boosting water efficiency under pressure. When plants were treated with TiO_2, the rate of germination, photosynthesis, chlorophyll production, and RuBisco increased.

(B) Salinity stress

Excess salt builds up in the soil, which results in salinity stress. It has an impact on plant growth, which lowers output. Currently, salt stress affects more than 20% of farmed land worldwide, and this number is rising daily. Increased osmotic stress, lowering the leaf water potential, increased production and accumulation of reactive oxygen species (ROS), ion toxicity that modifies biochemical processes, and decreased leaf water potential are all consequences of increased salinity in the soil for plants. In addition to improving growth and development, using silicon fertilizer and silicon nanoparticles under salt stress also improved the quality of proline and chlorophyll in basil (*Ocimum basilicum*). Following the application of SiO_2 nanoparticles, there was an increase in the quantity of proline, the amount of chlorophyll, the dry weight and fresh weight of the leaf, and the upregulation of the gene that synthesized an antioxidant enzyme under salt stress. When Si ions and silica nanoparticles are applied to the rice, the antioxidant defense mechanism is activated and osmolyte synthesis is stimulated.

(C) Metal Stress

Plants that get an excessive amount of metal nutrients are subject to metal stress. These result in oxidative stress and toxicities. Among the important macro- and micronutrients are zinc, copper, and magnesium. They are necessary for healthy plant development and growth. A crucial micronutrient for all living things, including plants, is zinc. It performs a variety of essential functions in the cell. ZnO NPs are crucial because they lessen the absorption of harmful heavy metals by plants, such as cadmium.

(D) Oxidative Stress

Numerous causes, including intense light and other comparable circumstances that result in an excess of free radicals and other reactive oxygen species being produced, can induce oxidative stress in plants. They have a detrimental effect on growth and development because of their extremely reactive character. It has been noted that TiO_2 nanoparticles have a major role in reducing light stress. Superoxide anion radicals and hydroxide are produced when TiO_2 NPs are exposed to light because they catalyze the oxidation-reduction process. In addition, exposure to UV-B rays is another cause of impaired photosynthesis and aberrant leaf structure. Through the inhibition of SOD and APX functions, it causes the formation of ROS in plant cells. UV-B exposure also causes enhanced lipid peroxidation, the production of H_2O_2 and superoxide radicals, and electrolyte loss. It has been discovered that Si NPs boost antioxidant activity to control oxidative damage brought on by UV-B exposure in wheat. Although the exact mode of action is yet unknown, it is hypothesized that Si NPs shield the plant by using its antioxidant defense system to balance off photosynthetic damage brought on by ROS activity. Therefore, NPs are crucial in triggering abiotic stress management responses in plants. They may be viewed as possible instruments in the process of managing abiotic stress in crops as they modify abiotic stress-induced responses at several levels. However, to guarantee little or no toxicity for their use in agriculture, appropriate tuning of the physicochemical, optical, electrical, and biological characteristics of NPs is crucial in NP engineering.

3.3 Soil management

The existence of NPs in the soil environment, such as smectite, imogolite, hallyosite, palygorskite, sepiolite, allophane, hematite, and goethite, is now a well-established fact. "A novel class of clay particles is called nanosols. They are not the same as clay particles and typically range in size from 1 to 100 nm. Nanoparticles have been created to elicit unique features. Their extremely tiny size, which produces extremely high surface charges and specific surface areas (SSAs), is responsible for these characteristics. They can now interact with other particles in the soil matrix rather aggressively as a result. Because artificially produced NPs are very resistant, do not break down rapidly, and accumulate gradually in the soil, they have a substantial influence on the soil. Once in the soil, they do not break down rapidly and hence accumulate. As a result, it shows a detrimental influence on the soil environment, which in turn affects the quality of the soil. To maximize their benefits, these NPs are mixed with clay to create a colloidal solution.

4. Nanobionics

Since the term "nanobionics" combines the terms "nano" and "bionics", it may be characterized as the study of electrical interactions in biological systems at the nanoscale. The overarching goal is to produce a wide variety of wild-type plants that are educated to function as self-powered light sources, self-powered groundwater sensors, infrared communication devices, and photographing things in their surroundings. Because plants can produce energy from sunlight through a process called photosynthesis, they are the only animals suitable for carrying out these functions. Researchers working in the field of nanobiotechnology hope to create bionic plants with the best potential efficiency in photosynthesis and biochemical sensing.

4.1 Plant Nanobionics

A larger degree of advancement has been made in nanotechnology, where materials at the nanoscale are designed to deliver medications to specific cells or tissues. Drug stability, pharmacokinetics, and the development of treatments into nanoparticles with reduced toxicity can all be enhanced by nanoparticles. The creation of nanoparticles for drug delivery systems and the larger-scale manufacturing of these medications are time-consuming processes that represent significant barriers to their effectiveness. An emerging type of biologics utilized in medication delivery systems is plant viruses. The primary benefit of employing these plant sources to make biologics is that the expressed PDs are protected from stomach acids and enzymes by the plant cell wall through a process known as bio-encapsulation. Nevertheless, commensal microorganisms in the gut break down the cell walls and release the proteins when whole plant cells containing PDs enter the stomach. This is because gut microbes have advanced to break down every component of the cell wall. Receptor-binding proteins, or tags, are fused to PDs and then effectively traverse the intestinal epithelium, which is the biggest mucosal region in the body, spanning 1.8-2.7 m2, to reach either the immune system or the circulatory system. Different polysaccharides make up the cell walls of plants. As a result, the cell wall serves as a barrier to prevent foreign particles and nanoparticles from entering the cell and damaging internal components. A unique type of nanoparticles known as layered double hydroxides (LDHs) has been shown to effectively transport biomolecules into plant cells. The primary advantages of using LDH nanoparticles are their little damage upon entry through intact cell membranes and the security of RNA activity. LDHs are often expressed as [M2+1 xM3+x$(OH)_2$] [An—]x/n zH_2O, where M^{2+} and M^{3+} denote interlayer anion and divalent and trivalent metal ions, respectively. The adjustable LDH complex is created by intercalating positively charged ion sheets with charge-balancing anions. The ion sheets may then delaminate from ultrathin nanostructures. LDHs have a number of extremely attractive qualities, including minimal cytotoxicity, easy biodegradability, accurate drug release, and high compatibility. Because of their ultrafine size and strong surface reactivity, ENMs have excellent physiochemical skins that are suitable for a variety of applications in medicine, biotechnology, energy, pharmaceuticals, cosmetics, and electronics. Since engineered nanoparticles (ENPs) have had a significant impact over the past few decades, they often have characteristics like a high absorption capacity, the ability to release molecules more effectively within target cells or to specific organelles, and the ability to pass through cell walls without causing any harm. ENMs contribute by entering plant cells and interacting with intracellular structure and metabolic pathways, which can either negatively impact plant development and growth or positively impact it through a variety of processes.

4.1.1 Application of Plant Nanobionics

(A) Crop production

1. **Improved Photosynthesis Efficiency**: Thylakoid membranes in chloroplasts are the primary location of the photosynthetic machinery in the majority of plant types. Chloroplasts are able to absorb 50% of the incoming solar radiation that falls within the visible region of the light spectrum. Furthermore, only around 10% of the sunlight that is available to plants is actually used by them. Therefore, scientists have worked to increase the range of solar light absorption in an effort to increase photosynthetic efficiency. In photocatalytic complexes based on chloroplasts, nanomaterials with seamless chemical and physical features enhanced unique functional capability. The visible and near-

infrared spectrum of light wavelength can be captured by SWNTs, but the absorption rate of chloroplast antenna pigments is constrained in this scenario. Scientists have successfully created highly charged single-walled carbon nanotubes (SWNTs) that may spontaneously penetrate chloroplasts. These nanotubes are coated with DNA and chitosan, a biomolecule derived from the shells of shrimp and other crustaceans. By encasing CNTs or nanoparticles in highly charged DNA or polymer molecules, the novel lipid exchange envelop penetration (LEEP) method for integrating the nanostructures enables them to pass through the hydrophobic, fatty membrane that encircles chloroplasts. With their characteristic optical characteristics, single-walled carbon nanotubes (SWNTs) inserted within chloroplasts have the potential to enhance photosynthetic light responses. SWNTs can record visible and near-infrared light spectrums, even if the absorption rates of chloroplast antenna pigments are insufficient for this purpose. SWNTs transform this absorbed solar energy into excitations that move electrons to the photosynthetic apparatus. When CNTs were incorporated into plant-extracted chloroplasts, the photosynthetic activity of the chloroplasts increased by 49% above the control. These nanocomposites increased the electron flow associated with photosynthesis by 30% when they were introduced into live leaf chloroplasts. It is also possible to extend SWNT real-time sensing of NO in excised chloroplasts and leaves to recognize a wide variety of exogenous substances, including insecticides, herbicides, and environmental contaminants, as well as plant signaling molecules.

2. **Plant Nanobionic with Higher ROS Scavenging Ability**: Most remarkably, single-molecule free radical dynamics within chloroplasts may be displayed using SWNT-based nanosensors, which can be used to optimize light and CO_2 levels for photosynthetic circumstances. The primary impediment to the utilization of harvested chloroplasts in solar energy applications is their susceptibility to photosynthesis protein deterioration caused by light and oxygen. After being combined with a highly charged polymer (polyacrylic acid), cerium oxide nanoparticles, or nanoceria, are able to pass through the chloroplast's outer membranes and settle in the stroma. They also remarkably prevent damage to photosystems by quenching reactive oxygen species, which are widely dispersed throughout the chloroplast. This allows for the real-time monitoring of free species and environmental pollutants using vivo and ex vivo embedded nanosensors. Furthermore, photosystems I and II, also known as PSI and PSII, are two forms of pigment-protein complexes in which chlorophylls absorb solar energy and convert it into electrochemical energy that produces ATP and NADPH, which are then utilized for CO_2 fixation. PSII carries out the light-induced oxygen evolution process and moves electrons from water to plastoquinone in the membrane. PSI uses the electrons that PSII provides to provide to produces to produce

strong reducing power and lowers NADPH and ferredoxin. Even inside silica nanopores, the high and steady oxygen-evolving capability of *T. vulcanus* PSII was preserved by the conjugation of nanomesoporous silica compound (SBA) with photosystem II (PSII). Under sufficient cycles of light and dark, the movement continued for almost three hours. The negative impacts of electron acceptors and light-induced radicals can be eliminated by combining PSII-SBA conjugates with mediator recycling systems. These conjugates also have the potential to construct photosensors and artificial photosynthetic systems.

(B) The aspect of plant nanobionic in environment

1. **Nanobionic Plant as Nitroaromatics Detector**: Although it seems beautiful, it is actually just a simple plant transpiration mechanism. In addition to absorbing water and other elements from the soil, plants are also able to detect tiny amounts of elements in their tissues. Using the principle in mind, researchers created a nanobionic plant that can detect explosives in groundwater and alert users to their presence nearby. Carbon nanotubes (CNTs)-based IR-fluorescent sensors that respond only to nitro aromatics injected into the leaves of spinach plants. When nitro aromatics develop, the fluorescence intensity of these nanotubes decreases. Next, a reference sensor with a constant signal intensity was created.

   The plant uses its roots to draw nitroaromatics, also known as picric acid (2,4,6-trinitrophenol), which is a common explosives component. Once within the leaves, the suppressed infrared signal is captured by a night-vision camera and sent to the smartphone via a Wi-Fi connection. High contrast pictures are produced by the method, which also involves implanting the reference sensor in the leaf.

2. **Nanobionic Plant as Temperature Detector**: Cyberwood was created by embedding a novel synthetic carbon nanotube—which has precise temperature sensitivity and mechanical and structural qualities similar to wood—into a tobacco plant cell matrix. This kind of carbon-nanotube's electrical conductivity varies with temperature, preserving plant cells' innate ability to sense temperature changes even after they die. Multi-walled carbon nanotubes (MWCNTs) are known for their high electrical conductivity and structural stability, which may be used to connect the samples to an external circuit. Indeed, a critical function of pectin and charged atoms (ions) is to regulate the temperature sensitivity of dry cyberwood and live plant cells. Pectin is a type of sugar molecule found in plant cells that is temperature-dependent and can cross-link to form a gel that contains both calcium and magnetism ions. As the temperature rises, the gel's contacts break down. allowing the ions to move more freely and improving the material's ability to conduct electricity. Several conventional nano-synthesis methodologies were combined to create

cyberwood. In the first phase, undifferentiated tobacco BY-2 cells, which were derived from *Nicotina tabacum* callus seedlings, were cultured in a growth medium containing, MWCNTs. Spontaneous accumulation of cells was observed when tobacco cells were combined with MWCNTs. After 24 hours, a gel-like substance developed; it was gathered and dried for 15 days at 47 C. The material was created with a hierarchical, intricate structure akin to that of wood. The design might find use in distance sensors for consumer goods and security systems, or in thermal sensors, such as those used in thermal cameras. The cyberwood sensor's exceptional temperature sensitivity allows it to detect warm bodies even at a distance of a few dozen centimetres, such as when a hand is approaching the sensor. The hand's distance from the sensor directly affects the conductivity of the sensor. Cyberwood's extreme temperature sensitivity suggests that it might be utilized as a temperature distance sensor. It is possible to deduce the distance of a heated body from the sensor by using temperature data taken under constant ambient circumstances.

5. Conclusion

In conclusion, nanotechnology plays a pivotal role in advancing precision agriculture management, offering innovative solutions for sustainable farming. By integrating nanoparticles, nanobionics, and geoinformatics, agricultural practices can be optimized to enhance productivity while minimizing environmental impact. Nanotechnology allows for the precise application of agrochemicals, reducing the overuse of pesticides and fertilizers, thus protecting soil and water quality. The use of nanobionics as real-time sensors aids in diagnosing crop health and nutrient needs, ensuring targeted interventions that boost crop yields and resilience against pests and diseases. Additionally, geoinformatics facilitates the accurate assessment of soil types and conditions, enabling tailored agricultural practices. These technologies collectively support sustainable agriculture by promoting food security, reducing poverty, and increasing agricultural productivity without compromising future generations' ability to meet their needs. Embracing nanotechnology in precision agriculture represents a significant step toward achieving sustainable development goals and maintaining the health of agroecosystems.

CHAPTER 13

ROLE OF ARTIFICIAL INTELLIGENCE, SENSOR TECHNOLOGY, BIG DATA IN AGRICULTURE: NEXT-GENERATION FARMING

Short Summery

By embracing technology advancements, such as harvesting and tillage machinery, controlled irrigation systems, biotechnological instruments, crop breeding, fertilizers, herbicides, and genetic research for improving traits, modern farming has advanced. These developments made it easier for farmers to produce vast amounts of higher-quality crops. On the other hand, efforts are still ongoing to get the highest yield feasible from different types of soil, and the main losses are associated with food waste, particularly before and after harvest, when the production is not closely examined and handled. The industry requires a clever and accurate solution, which modern technologies make feasible. Utilizing specialized, cutting-edge technology to increase agricultural productivity and product quality is the goal of smart farming. As an illustration, consider precise agricultural data and a crop management strategy tailored to a particular location that tracks and evaluates crop and field variability. The application of artificial intelligence (AI) decision support systems encourages farmers to apply pesticides and fertilizers to crops in accordance with the necessary amounts. Therefore, combining the use of appropriate electronic sensing devices that capture data in the soil, environment, or crops might enable such monitoring. The data might provide insightful information about the needs of the crop to determine how best to use the soil in a specific region, manage crop care, and maximize output after harvest. The development and application of sensors to assist determine the quality of a wide range of horticulture goods, including fruits, require well-informed judgements to be made about allocation. Jumble-sale computer data is used in exemplary, contemporary study to measure fruit quality and identify flaws. Artificial intelligence is therefore recommended as a data driven explanation with several benefits. This method may lessen the amount of produce that is lost laterally in the supply chain from the farm.

1. Introduction

Agriculture has long been thought of as an intuitive environment where knowledge is transmitted from one generation to the next. However, problems like shifting climate and the absence of profitable farming are now significant obstacles. According to UN predictions, there will be 9.8 billion people on the planet by 2050, a 2.2 billion increase over current numbers. This indicates that we must increase agricultural production to fulfil the needs resulting from population growth: sadly, rapid urbanization and climate change have destroyed much of farmland. The overall area of farms in the United States has decreased from 913 million acres in 2014 to 899 million acres in 2018, primarily due to urbanization and climate change. The term "big data" refers to the bulk number collection of soil architecture, weather prediction, fertilizer advice, disease management, climate change, and crop managing data. This notion emerged in the contemporary technology-based period. Numerous resources, including software, online portals, IoT devices, etc., extract these data. However, robotics and some types of AI are used to implement big data. Traditionally, farms have been employed to harvest crops and maintain agricultural today face the difficulty of a labour shortage as a result of society shifting from being an agricultural one with a significant number of people living on farms to one where people live in cities. Artificial intelligence-enabled agricultural robots are one solution to the labour problem. A Forbes report claims that agricultural robots supplement human labourers and are capable of harvesting crops more quickly and in greater quantities than human labourers. Agricultural robotics is actively creating robotic systems to operate in the field and assist producers with labour-intensive jobs, pushing agricultural systems to the new idea of agriculture, but still there are many situations in which robots cannot outpace humans in speed. The introduction of robots into agriculture has significantly cut operational expenses for farms and raised production in a number of nations. Crop yields and agricultural production throughout the world declined in 2015. Agricultural robotics was proposed as a solution to these issues and the growing need for large yields. The worldwide market for agriculture and crop production is growing because of robotic advancements. The Verified Market Intelligence research states that agricultural robots will be able to perform field work more effectively than farmers. Over the past five years, agricultural technology entrepreneurs have raised over 800 million dollars. In 2014, startups using robots and machine learning to untangle agricultural tangles began to gather steam, coinciding with growing interest in artificial intelligence. Over the past five years, venture capital financing for AI has surged by 450%. The Food and Agriculture Organization of the United Nations (FAO) projects that there will be 9.6 billion people on the planet by 2050, so feeding people while growing production sustainability and protecting the environment will be crucial in the years to come. For this reason, this type of new agriculture pretends to do more with less. In agriculture, advanced sensing methods can help reach the task by offering specific utilization of phytosanitary products based on detailed information about soil, crop status, and environmental conditions. This reduces the need for pesticides and herbicides, increases crop yield and quality, and improves water use efficiency.

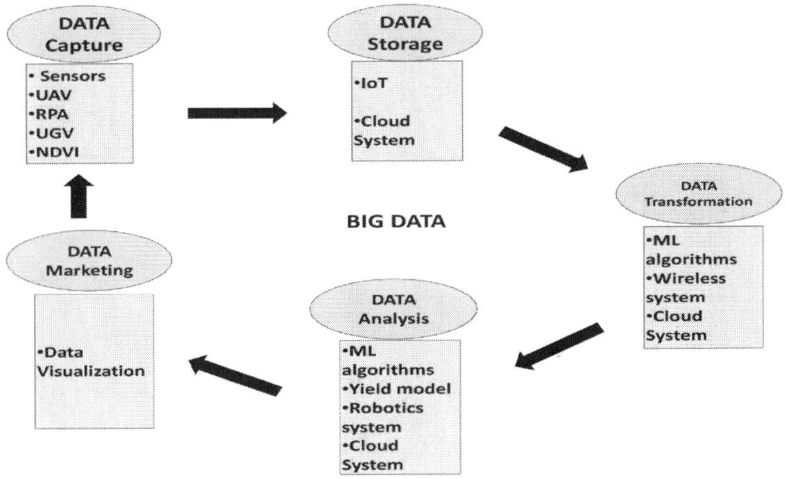

Figure-1: Diagrammatic representation of Big data information cycle

2. Characteristics of Big Data

Metaphorically, big data is like the ocean, and information is like a floating Iceland in it that can be overdone with various software and platform combinations. Four Vs can be used to categorize big data: velocity, volume, veracity, and variety.

2.1 Volume

The volume increases the enormous quantity of data that is collected, stored, managed, and analysed using a wide variety of tools.

2.2 Velocity

The term "velocity" describes data that must be gathered, saved, processed, and evaluated instantly. An example of this would be the real-time traffic, weather, soil, and other data that Google Map provides

2.3 Variety

Variety in data refers to information that comes in a variety of formats, including unstructured text documents, financial transactions, audio, video, email, as well as structural data from traditional databases.

2.4 Veracity

Another word for big data that describes access to massive amounts of data created from many sources with little latency and data flows that fluctuate significantly with periodic peaks and baselines is veracity.

3. Big Data and Smart Agriculture

Both big data and smart agriculture are relatively novels ideas in the field of agriculture. The precision Agriculture idea is an extension of Smart Agriculture, based on big data, the farmer takes the choice and achieves the situation as per information offered by big data. The ability to provide real-time assistance in the form of crop management systems such as weather alert systems, pest spraying UAVs, crop sensors, and other systems to carry out agile actions in response to unexpected changes in operational conditions or other circumstances, such as weather or disease alerts, is a crucial feature of big data. Intelligent support for big data also aids in the application, upkeep, and use of agricultural technologies. The primary use and utilization of big data in intelligent farming is to guarantee low costs, increased profits, and sustainability. The present era's digital revolution in agriculture has been driven by the use of artificial intelligence, sensors, and smart machinery. In agriculture, data refers to a collection of information about different kinds of soil mapping that includes details about their physical and chemical properties, previous management practices, weather, etc. Because of this data, farmers are notified in advance of any bad weather or diseases, which lessens their suffering. A comprehensive level collection of good agriculture practices data is included in big data for smart agriculture. These include the following:

3.1 Digital Soil and Crop Mapping

The creation of digital maps for soil types and their physio-chemical characteristics is the focus of digital soil and crop mapping. Farmers manage so many acres of land in industrialized and some emerging nations that is nearly impossible to get real-time reports and alerts from their planted areas without using up all of the available technology. Many nations, like Ireland, employ satellite-based soil and crop monitoring in order to manage and check these vast tracts of land. These approaches are quicker and less expensive than traditional ones.

3.2 Weather Prediction

Weather affects crop growth, developmental yield, and overall agricultural production. The agricultural systems of many other nations, including China, Japan, Korea, Bangladesh, and Pakistan, are likewise impacted by the weather. Put another way, the agricultural systems of India and other nations rely on the weather. Weather anomalies have the potential to cause physical harm to crops and soil erosion. Sudden variations in the weather do serious harm to crops. The weather affects all economic activity, including produce quality from farm to market and storage and transportation. Weather that is debauched affects all of the aforementioned factors, leading to significant financial losses. The following elements are conjectured by agricultural weather:

1. Cyclones, tornadoes, depressions, and low-pressure zones
2. Snowfall and rain
3. The kind and extent of cloud cover over the sky
4. Wind direction and speed
5. Conditions such as wind squalls, fog, frost, hail, and thunderstorms
6. The kind and extent of cloud cover over the sky
7. The dew point, maximum, and minimum temperatures

Agriculture weather forecasting aspects are monitored by Earth observation satellites, Unmanned Aerial Vehicles (UAVs) and Automatic Weather Stations.

3.3 Fertilizers Recommendation

Precisely calculating the right amount of fertilizer for a crop is a science that necessitates nanoscale examination of several characteristics. Crop nutrient absorption rates, land type, irrigation systems, interaction between fertilizers and crops, characteristics of fertilizers, methods for evaluating soil, physical, chemical, and biological features of soil, and research data are some of these aspects. More fertilizers are applied to agricultural fields, and these excess fertilizers build up as hazardous substances in the soil that result in a variety of pollutants. With the use of big data techniques, farmers may now receive advice on the proper amount of fertilizer.

3.4 Disease Detection and Pest Management

In contemporary agriculture, sophisticated algorithms are employed to detect patterns and behaviours in a range of microorganisms and pests, with the aid of big data technologies. This process aids in the prediction of pest incursions and the dissemination of microscopic illness. Agricultural pests may quickly reduce a farmer's income, but improper application and excessive pesticide use can have negative impacts like cancer on humans, plants, and other living beings. UAVs and agricultural sensors help in mid-season crop health monitoring and pest management by lowering the need for pesticides.

3.5 Adaptation to Climate Change

Global warming-related climate change is a serious issue that has an impact on the agriculture industry. In one big data initiative, Taiwanese rice farmers receive IoT (internet of things) devices so they may gather the essential data on their crops. Farmers will be able to improve their production cycles even in climate-based unfavourable situations, thanks to the combined data from IoT sensors. Traditional farmers suffer enormous financial losses as a result of their inability to assess these climate changes. Big data has the power to transform farming in the future by providing answers to all of these questions.

3.6 Automated Irrigation System

At present, every nation on the planet is forced to utilize water in an extremely creative way. Recent research indicates that water is becoming increasingly scarce globally, with over one-third of the world's population expected to experience a complete scarcity by 2025. Water scarcity is a major issue for farmers in agriculture as well. To address this, two automated irrigation systems have been developed: one uses drip irrigation, which is used as an automated irrigation system for small-scale farms, and the other uses weather prediction.

4. Sources of Big Data

Big data may come from a wide range of sources and on a large scale. Large data sources include government agencies, non-governmental organizations (NGOs), and other private organizations that gather data (government statistical yearbook, for example). Ground sensors, which monitor farmers and provide data, are also important sources of big data. Big data is sourced from online storage and web services that use data from aerial sensors such as UAVs, satellites, and small aircrafts. Big data is sourced via the cloud system, which combines wireless sensor networking and Internet of Things systems. This gives farmers access to real-time online data (data about weather, plants, yield, etc.) via commercial enterprises on their mobile devices: media (such as audios and videos from social platforms, research papers, and other articles) which is also a source of big data. Big data from altered sources is needed to address the different issues facing agriculture, and these applications leverage big data. For instance, research on crops, soil, and animals uses field-installed ground sensors: weather station data is used for climate change; satellite data is used to map land. Various government agencies use this data to create policies that help farmers increase productivity and predict agricultural yields for export and import. For the purpose of providing services to consumers, a variety of public and commercial service providers have gathered agricultural big data information.

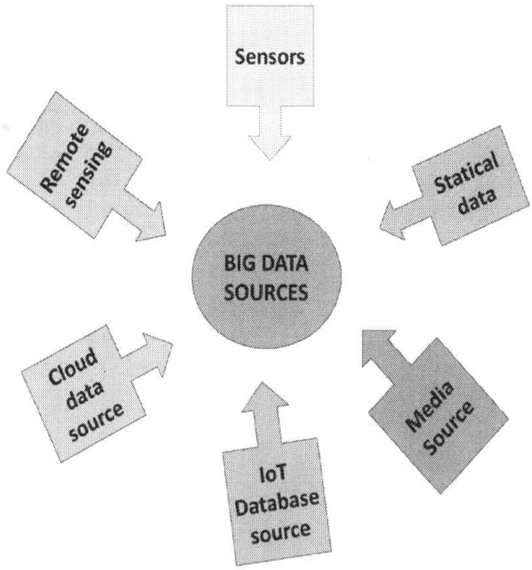

Figure 2. Diagrammatic representation of Big Data Sources.

Table-1: Overview of Big Data sources, service providers

Agriculture area	Big data source	Big data innovative service providers
Weather and climate	Earth observation satellites UAV, Automatic weather stations	Fertilizer Calculator Crop Water Needs Estimation Yield models
Soil	Mobile Apps Location-based datasets Ground Sensors/ Base Stations, UAV	Soil Indicators for Scottish Soils,SoilInfo, SoilWeb mKrishi (India), CiAgriculture (China), Fujitsu (Japan) Batian (China), RedBird, etc.
Crop and yield	Fertilizer Calculator Crop Water Needs Estimation Yield models	BaiKhao

Source: Swiss Re Centre for Global Dialogue Switzerland

4.1 Sensors

The innovative tools that keep an eye on crops and gather unbiased data from them are called sensors. Sensors can be categorized based on the platform they are used on, such as satellites, ground-based, and aerial or airborne like airplanes, unmanned aerial vehicles, and balloons. Cameras on airplanes, satellites, and unmanned aerial vehicles (UAVs) are mostly used to gather pictures, while ground-based optical sensors can gather reflectance data and store it in a text file.

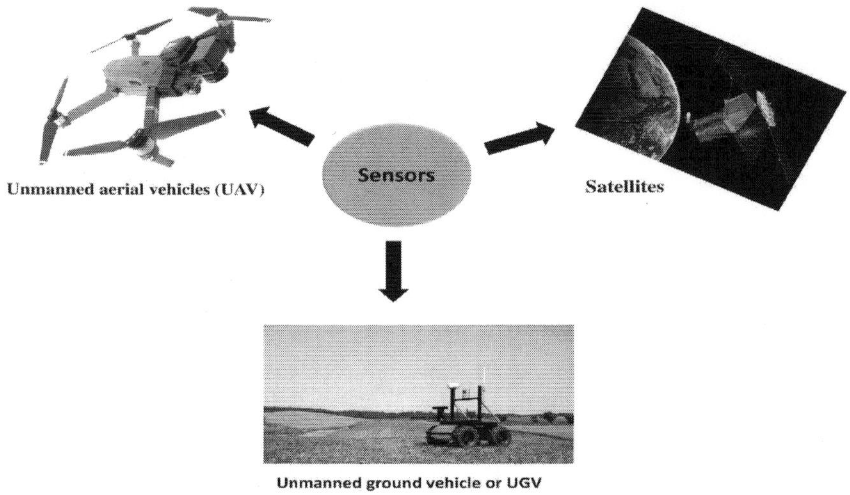

Figure 3:Diagrammatic representation of sensors

4.1.1 Remote Sensing Platforms: Satellites

With photos taken from satellites, remote sensing has played a vital role in gathering information about the geographical features of a certain region and mapping these geographical areas without requiring direct contact with the places that need to be measured. Every nation in the world is launching satellites to gather real-time agricultural data. Eight Landsat satellites in America gather gross spectrum data from the Earth every sixteen to eighteen days, while two Landsat satellites in Europe are part of European Sentinel, which provides multispectral data at a resolution of 10 m pixels for NDVI (Normalized DiTerence Vegetation Index) imaging, soil, and water cover every ten days. Furthermore, the German geospatial information provider Rapid Eye used a five-satellite constellation to provide multispectral RGB imagery, as well as red-edge and NIR bands at a resolution of five meters. Another satellite, GooEye-1, was launched on September 6, 2008, and collected multispectral RGB data as well as NIR data at a resolution of 1.84 meters. The World View-3 satellite was launched on August 13, 2014, and it collected multispectral data at a nadir of 1.24 meters using the RGB bands, two NIR bands, and eight SWIR bands. Numerous researches have examined the nutritional condition of field crops and concentrate on the possible uses of thermal technology through remote sensing.

4.1.2 Airborne platform Systems: Unmanned Aerial Vehicles (UAV) and remotely-piloted aircraft (RPA)

A crop and a satellite can be separated by 700 kilometres, and when sensors are placed closer to their objectives, more in-depth information can be obtained. 100 meters separates the aerial system from ground, allowing it to readily survey the area and gather data regarding crops, fertilizers, soil, crop disease, etc. Fixed wing aircraft and multi-rotar aircraft are two primary categories into which unmanned aerial vehicles (UAV) and remotely piloted aircraft (RPA) fall. Although fixed-wing UAVs are more expensive and prone to breaking after repeated landings, they are able to cover greater agricultural territory in a single flight and carry larger payloads than rotating-wings. Because they can take off and land vertically, UAVs are more stable when in the air; on the other hand, thy are slower and have a smaller field of vision when the battery runs out. UAVs outperform remote sensing in a number of ways, including greater spatial resolution and frequency flexibility, which facilitate the collection of large amounts of data for agricultural purposes. UAVs are able to collect data from hard-to-reach places where traditional equipment is unable to, in contrast to ground vehicles. However, they need meticulous professional planning of the flight route and some machine vision applications that would need fly in the afternoon to avoid ground-level foliage shadows that could skew the picture.

4.1.3 Ground platform Systems: Unmanned ground vehicle or UGV

Unmanned ground vehicles, or UGVs, are sensors used in ground platform systems. Advanced positioning, orientation, navigation, planning, and sensing technologies are fitted into this UGV. These technologies are used in agriculture for a variety of tasks, including tilling, seeding, weed removal, pest control, soil analysis, transplanting, crop scouting, and harvesting. The Clearpath Robotics firm created the medium-sized robotic unmanned ground vehicle (UGV) called Husky. Its distinctive characteristics, which include stereo cameras, LIDAR, IMUs, GPS, and manipulators, can be used for monitoring field crops and gathering data.

4.2 Statical data

Governmental agencies, non-governmental organizations, and other private organizations compile vast amounts of agricultural data that are then published. Examples of this data include statistical yearbooks, soil-mapping, water-related data, crop-mapping, soil property types, and government weather survey data. These organizations are a major source of big data.

4.3 Remote sensing

With the use of satellite imagery gathered, remote sensing has been essential in gathering data pertaining to the geographical features of a certain region. This data is charted without requiring direct personal contact with the places that need to be measured. Satellites used in remote sensing gather information on drought and flood forecasting, water availability, agricultural and soil mapping.

4.4 Cloud data source

A new paradigm for the supply of computer infrastructure and big data processing techniques for diverse resources is being linked to cloud computing, which is currently emerging as a marketable infrastructure. It does away with the requirement to maintain costly computer gear, software, personnel, infrastructure, resources, and information technology. Information on weather, crops, farmers, soil, pesticides, fertilizers, and agricultural marketing is stored in a single location on the cloud via a variety of digital instruments used in agriculture, such as sensors, UAVs, remote sensing, and others. With the aid of different techniques in the form of online portals, mobile applications, and other applications, end users such as researchers, farmers, consultants, and specialists may readily access this big data via the cloud.

4.5 IoT Database source

The Internet of Things (IoT), is a new and revolutionary technological age that allows objects, like plants, and gadgets, like sensors, to be connected to massive databases through the use of the internet. Interaction between agricultural items is facilitated by IoT. Since different agricultural equipment produced output data in different forms, it is crucial to gather all of this information into a single protocol so that networked objects and devices may communicate with one another. Following the one-to-one attention given to the crops throughout production, agricultural goods must also be monitored following harvest. Wireless sensor networks (WSN) and radiofrequency identification (RFID) are fundamental components of the Internet of Things (IoT) used in agriculture. WSN is crucial to the intensive management of crop storage and logistics facilities. It is possible to use these WSN sensors to monitor temperature, water, moisture contentment, pH, evaporation, fertilizers, and pests of soil. Using RFID to capture plant pictures, these various types of sensors gather data to detect changes in plant growth patterns, such as leaf area index, chlorophyll measurements, evaporation rate, plant height measurements, etc. Farmers may better comprehend plant health, chlorophyll and plant height measurements, the administration of irrigation pesticides, fertilizers, and herbicides, and the weed pressure charting for micro-level management with the use of big data information gathered by IoT base sources.

4.6 Media Source

Big data may also be found in media, which is equivalent to social media platforms like agricultural channels, Facebook, Youtube, Instagram, and Twitter. Big data can also be found in research papers, audios, videos, and articles.

5. Techniques and Tool use in Big Data analysis

Large data sets are highly dimensional and diverse, require complicated tools and approaches to extract information due to their complex content. Big data analysis is done with a variety of methods, either alone or in conjunction with two other approaches, such as Image processing, remote sensing, cloud computing, machine learning, GIS (Geographic Information System), Vegetation Index (VI), and Normalized Difference Vegetation Index (NDVI).

5.1 Machine learning (ML)

These days, machine learning is a popular technology that may be used in agriculture to increase sustainability. Because it manages micro-labels, machine learning in agriculture makes the industry more profitable. Artificial machine learning, or ML, is rapidly advancing in applied agricultural science. In the agriculture industry, artificial approaches are being employed to increase accuracy and provide answers to problems.

5.1.1 Livestock Management

Livestock production and animal livestock are the two categories under which livestock is divided. Animal welfare addresses the wellbeing, health, and illness of animals. Monitoring the early identification of animal illnesses and behaviour is the primary use of machine learning. Along with losses and the creation of commodities connected to animals, it also scans the economic profit.

a. **Livestock Production:** Since the dawn of human civilization, domesticated cattle have been an essential component of human culture, society, and most importantly- the world economy. By supplying sources of meat, fertilizer, milk, and pulling ploughs, domestic livestock has increased the quantity of food and nourishment accessible to people, contributing to the development of human cultures and civilizations. Livestock has also been used historically for transportation, wool, leather, and other raw products. The diets, fatty acid content in milk, and milk output of animals were recommended by the precise data prediction of rumen fermentation, which was made possible by machine learning (ML) analysis. The poultry industry benefits from the early detection and warning of issues in the commercial production of eggs through the application of support vector machines (SVM) in ML models. These machines aid in the analysis of data pertaining to hen and egg production. Convolutional neural networks (CNNs), a machine-learning algorithm-based approach, are successfully used in digital images for animal face recognition, such as pigs. This method solves the issue of distressing activity for radio frequency identification (RFID) tagging, which has many drawbacks like low range and time consuming.

b. **Animal Welfare:** The ties that people have with animals and their need to treat the animals in their care with compassion and responsibility are referred to as animal welfare. Prior to events like the oestrus and the detection of nutritional changes in animals, data from collar sensors equipped with magnetometers and three-axis accelerometers is gathered and subjected to machine learning modelling. ML uses behavioural data from calves' chewing behaviours in conjunction with nutritional supplements like ryegrass and hay that were gathered by optical FBG sensors and subjected to ML analysis.

5.1.2 Water Management

Water is a basic component of agriculture, demonstrating the significance of water in farming and the need for micro-management. Water also plays a major part in hydrological, climatological, and agronomical balance. The estimation of daily, weekly, or monthly evapo-transpiration is done using machine learning. Every plant has a highly complicated evaporation process, however by comprehending this process, the irrigation system can be designed to supply water based on each plant's needs in the agricultural field. This is crucial in the 2020 scenerio as well since, based on current research, over one-third of the world's population will experience complete water scarcity by 2025 if groundwater usage continues. Thus, this large data analysis model will aid in agricultural water management. Temperature sensors supplied data, which was analyzed by the ELM model of ML to predict the exact weekly evapo-transpiration in the dry region of India.

5.1.3 Soil Management

The best medium for a plant's growth and development is its soil. Plants need healthy soil to flourish, and ML application is needed to identify and forecast soil variables including moisture content, temperature, drying time, and condition. The soil in every location on earth is a diverse natural resource with intricate processes and mechanisms that are challenging to comprehend because of the planet's abundant biodiversity, varied temperature, and uneven geographic distribution of lands. A technique for providing remote agricultural management decisions was created by combining machine learning (ML). using data on evapo-transpiration and precipitation, this clever technique precisely measures the amount of soil drying in an area near Urbana, Illinois, in the United States. Big data also refers to an extensive compilation of information on soil characteristics, such as composition, nutrient availability, mineral availability, mineral content, and soil type. For this analysis, a visible-near infrared (VIS-NIR) spectrophotometer was used to collect soil spectra from 140 unprocessed and wet samples were collected from an arable field in Premslin, Germany in August 2013, following the harvest of wheat crop. Various modelling aspects of ML with combination soil data predicted soil organic carbon (OC), moisture content (MC), total nitrogen (TN). Additionally, the ML ANN (Artificial Neural Networks) model estimates soil moisture using data gathered from force sensors.

5.2 Cloud platforms

Through the internet, consumers can access shared pools of reconfigurable system resources via the cloud computing paradigm in information technology. Big data sources such as crops, growth progress and patterns, soil, weather, and climate are collected via cloud platforms. These sources require to be combined into a single, readily available platform that can be pre-processed, visualized, and analyzed. Agricultural cloud computing is utilized to monitor and evaluate moisture content in soil, temperature, and water delivery operations. Using gadgets connected to the cloud, farmers may examine information on their farms or fields. In addition, the microcontroller in the system enables farmers to remotely operate farm machinery, including turning on and off motors and lights. In the context of big data, the cloud system plays the following roles:

1. A unified cloud system houses all the agriculture-related data from big data sources, making it accessible to all users- farmers, agricultural businesses, etc, anytime and anywhere.
2. Centralized decision support systems for cloud system administration of all large data pertaining to land, its characteristics, soil, area, and location.

5.3 GIS (Geographic information systems)

Computer hardware and software known as geographic information system (GIS) employ feature qualities and location data to create maps. The storage of layers of data, including yields, maps from soil surveys, data from remote sensing, crop scouting reports, and soil nutrient levels, is one of the key features of GIS in agriculture. A significant amount of data is gathered by the GIS tool in conjunction with big data sources from earth observation satellites, unmanned aerial vehicles, and ground sensors like temperature and moisture sensors. This data is then analyzed by GIS analysis software and presented in a variety of formats for weather forecasting.

5.4 Vegetation indices

Data pertaining to determine the wavelengths of light absorbed and reflected by green plants were gathered by combining UAVs equipped with thermal cameras and satellite sensors. Every plant leaf in the biological system has a pigment system that substantially reflect near-infrared light- which is undetectable to human vision -also powerfully absorbs visible (red) light wavelengths. These reflectance characteristics also shift as a plant canopy progresses from early spring development to late- season maturity and senescence. Among other users, vegetation indices (VIs) derived from canopies based on remote sensing are useful algorithms for both quantitative and qualitative assessments of vigour, plant cover, and growth dynamics. There are several types of vegetation indices (VIs) including the Soil Adjusted Vegetation Index (SAVI), Green Normalized Difference Vegetation Index (NDVI), and Normalized Difference Vegetation Index (NDVI). However, the NDVI is a commonly used VI for gathering and analysing large amounts of data related to agriculture. A biological system must recognize the various hues (wavelengths) of visible and near-infrared sunlight reflected by the plants in order to regulate density of green plants on lands. The spectrum of colours known as VIBGYOR (Voilet-Indigo-Blue-Green-Yellow-Orange-Red) is visible when sunlight travels through a prism. Certain wavelengths of this spectrum are absorbed by objects while other wavelengths are reflected. For usage in photosynthesis, the pigment chlorophyll in plant leaves significantly absorbs visible light (between 0.4 and 0.7 μm). conversely, the near-infrared light (0.7-1.1μm) is significantly reflected by the leaf cell structure. The influence of various light wavelengths increases with the number of leaves on a plant. The range of NDVI values in areas of bare rock, snow, or sand are typically extremely low (0.1 or less, for example). Moderate NDVI levels (between 0.2 and 0.5) can be caused by sparse vegetation, such as senescing crops, bushes, and grasslands. High NDVI values (between 0.6 and 0.9) are indicative of thick vegetation, as that seen in tropical and temperate forests, or of crops that are at the height of their development. A crucial instrument for gathering and analyzing large amount of crop health data is the NDVI. To calculate the density of plant growth on Earth, all sensors, including satellites and unmanned aerial vehicles (UAVs), gathered vast amounts of data that were then analyzed using a different mathematical formula. This process produced the Normalized Difference Vegetation Index (NDVI), which is calculated as follows:

$$\text{NDVI} = \frac{NIR - VIS}{NIR + VIS}$$

Where, NIR is near infrared radiation

VIS is visible radiation

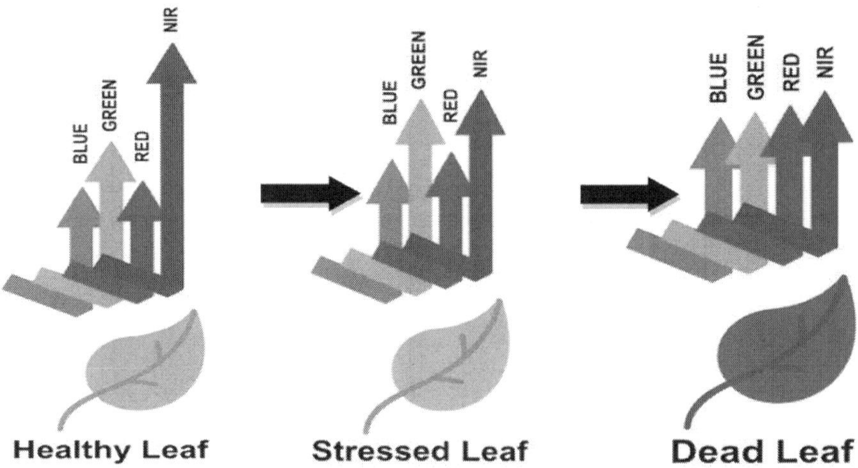

(A)

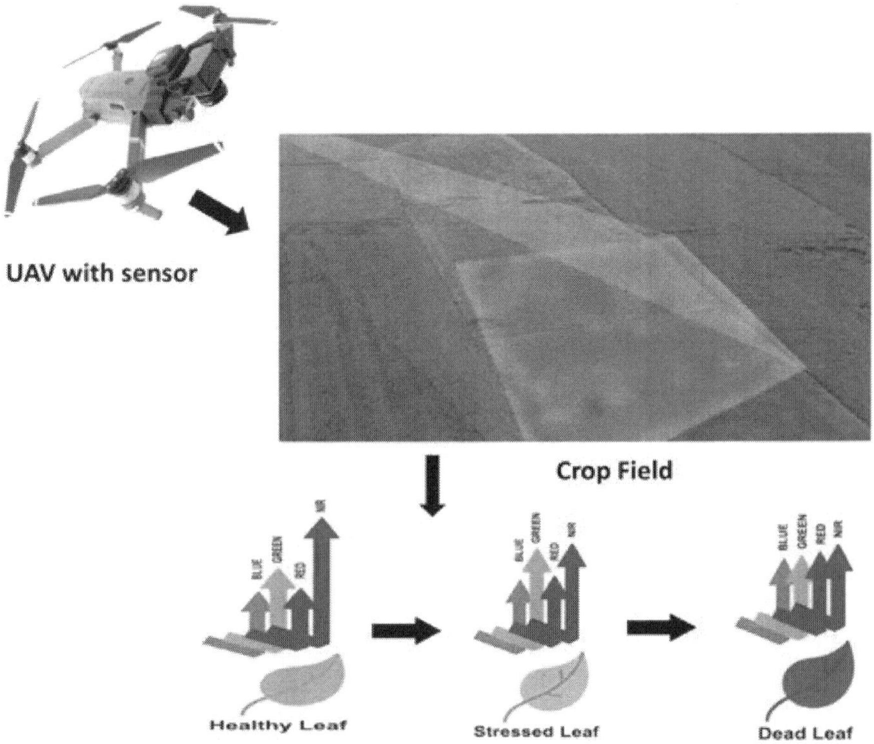

(B)

Figure 4. Figure (A) and (B) is a diagrammatic representation of NDVI.

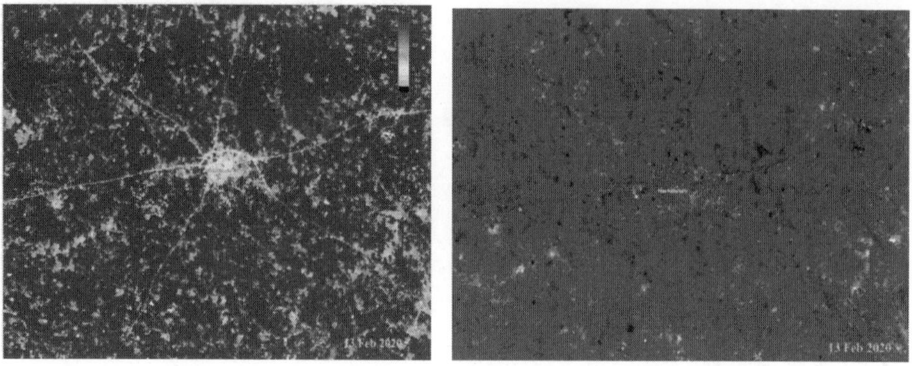

(A) (B)

Figure-5. Satellite image of Machhlishahr, Jaunpur, (A) Normalized Difference Vegetation Index (NDVI) (B) False-color composite (FCC)

Table 2. Small overview of big data source, analyzer tools, software, and service provider company

S. No	Agriculture field	Bigdata source	Big data analyzer tools	Software	Service providers
1.	Weather and Climate	Weather stations, surveys, static historical information (weather and climate data, earth observation data), remote sensing (satellites), geospatial data.	Machine learning modeling algorithms, Image processing, Remote sensing, Cloud platforms, GIS	Weather and Climate Toolkit (WCT), ArcGIS, Google Earth, MatLAB, QGIS, ModelVis	Weather Decision Technologies, Inc (WDT), IBM The Weather Company, CropProphet Enterprise, World Climate Service, AccuWeather, Aeris-Weather, Atmograph, Hurricane Mapping
2.	Soil Management	Mobile Apps Location-based datasets Ground Sensors/ Base Stations, UAV	Machine learning modeling algorithms, GIS, Remote sensing, NDVI vegetation indices	SOILMAP,Soil Data Viewer 5.1, SOTER ArcGIS, Autodesk	Soil Indicators for Scottish Soils,SoilInfo, SoilWeb mKrishi (India), CiAgriculture (China), Fujitsu (Japan) Batian (China), RedBird

3.	Crop and yield	Fertilizer Calculator Crop Water Needs Estimation Yield models	Machine learning modeling algorithms ANN, VIS-NIR, GIS, Remote sensing, NDVI vegetation indices	AgroMetShell, SCOPUS, BaiKhaoNK, SOCiT, Pock-etLAI, RaGPS, MapIT, m-Sa-hayak, mKRI-SHI's	Soil Indicators for Scottish Soils, *SIFSS, ICAR, IASRI,*BaiKhao
4.	Animal research and man-agement	Livestock production	Machine learning modeling algorithms SVM CNNs, RFID, Face detection sensors	Ranch manager, Chetu, Cattle Max, Livestock	SCR Dairy, **Lely Qwes**

6. Role of Big data in Agriculture ecosystem: For Smart Farming

After carefully considering all of the information in the Big Data Source and Big Data Analyzed Sauce tool sections, it was discovered that big data will enable common farming to be transformed into smart farming, allowing farmers to make excellent profits at cheap cost of labour. After taking into account everything in the Big Data Source and Big Data Analyzed Sauce tool sections, it was discovered that big data will enable common farming to be transformed into smart farming, generating high profits at cheap cost of labour for farmers. A choice for optimal resource use must be made promptly and with more precision in order to produce the crop in a sustainable manner. Big data is crucial for many aspects of smart farming, including GPS mapping, navigation, and Internet of Things (IoT) connections to distant sensors and monitoring systems, autonomous driverless tractors, robots and smart autonomous machines for seedbed preparation and reseeding, and helicopters. For bigger lands, digital planting and sowing management by automation and robotics machine monitoring by IoT and Cloud system is more cost-effective when done through smart planting from air to field, which is handled by GPS and IoT system. UAV, RPA, and UGV technology are also prevalent in many smart farms for the weather forecasting and spraying of pesticides, herbicides, and fertilizers. Smart ways to regulate crop health through Artificial intelligence (AI) and machine learning (ML), both are integrated and implemented to render easier for farmers to recognize pest, weed management. Through image processing like NDVI, along with all the information of crop yield analysis, NDVI prepares data on crop production that is stored in a cloud platform and estimates the yield of a particular place using crop mapping via satellite UAV. Field robotics-based harvesting is a mart way of harvesting that aims to efficiently analyze crop maturity and harvest without compromising the grains. The robotic system has a sensor that analyzes fruit and crop ripening and maturing prior to harvesting, ensuring that the fruit and crops are perfectly ripened and mature before they are harvested.

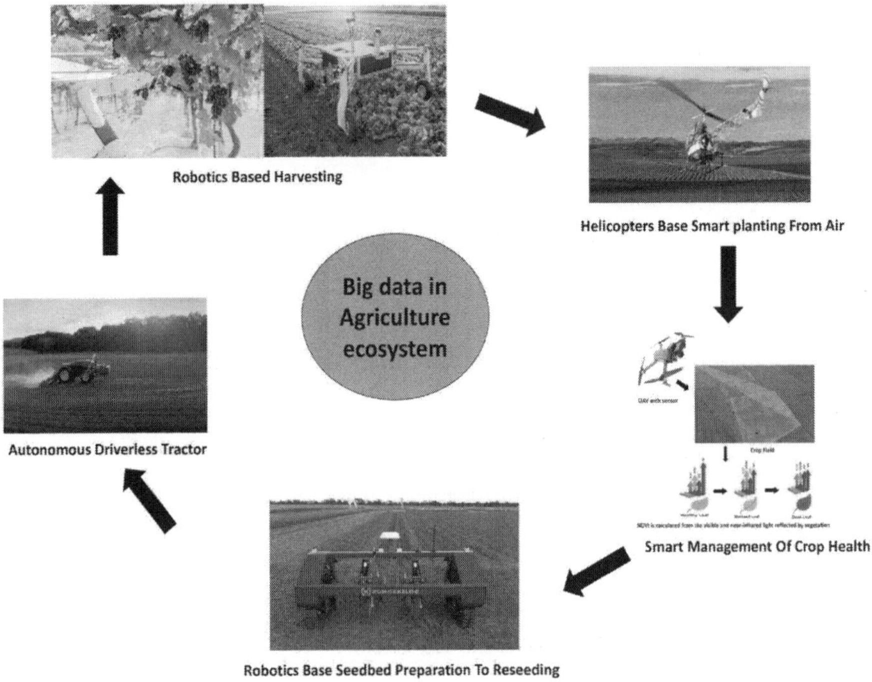

Figure 6. Diagrammatic representation the role of Big data in agricultural ecosystem.

7. Conclusion

In conclusion, the integration of artificial intelligence (AI), sensor technology, and big data is revolutionizing next-generation farming, addressing critical challenges and enhancing agricultural productivity. AI decision support systems enable farmers to apply precise amounts of pesticides and fertilizers, optimizing crop management and reducing waste. Sensor technology provides real-time data on soil, environment, and crop conditions, facilitating informed decisions to improve yield and quality. Big data analytics helps in understanding field variability, tailoring crop management strategies to specific locations, and predicting crop needs accurately. These technologies collectively enhance the efficiency and sustainability of farming practices, reducing food waste and improving post-harvest management. The use of AI and sensors also aids in quality assessment of horticultural products, ensuring better allocation and reducing losses in the supply chain. Embracing these advanced technologies is essential for achieving sustainable agricultural development, meeting global food demands, and minimizing environmental impact.

GLOSSARY

Agrochemcials: Agrochemicals are chemical substances used in agriculture to enhance crop production, control pests, and improve soil fertility.

Amperometric measurement: Amperometric refers to a type of electrochemical measurement where the current is measure as a function of the concentration of an analyte in a solution.

Antioxidants: Antioxidants are substances that inhibit the oxidation of other molecules, protecting cells from damage caused by free radicals.

Bioaccumulation: Bioaccumulation is the accumulation of substances, such as toxins or chemicals, in an organism over time, typically through ingestion or absorption.

Biomarkers: Biomarkers are measurable indicators of biological processes, conditions, or diseases, often used for diagnosis or monitoring health.

Biomolecules: Biomolecules are organic molecules, such as proteins, nucleic acids, lipids, and carbohydrates, that are essential for the structure and function of living organisms.

Biosensors: Biosensors are analytical devices that detect biological substances by converting a biological response into an electrical signal for analysis.

Capillary electrophoresis: Capillary electrophoresis (CE) is a high-resolution analytical technique that separates ionic species based on their charge-to-size ratio under an applied electric field within a narrow capillary tube.

Chelation: Chelation is the process by which a single ligand forms multiple bonds with a central metal ion, creating a stable ring-like coordination complex.

Chromatography: Chromatography is a technique used to separate, identify, and analyze components of a mixture based on their differential distribution between a stationary phase and a mobile phase.

Coercivity: Coercivity is a measure of the intensity of the applied magnetic field required to reduce the magnetization of a ferromagnetic material to zero after it has been magnetized to saturation.

Crop mapping: Crop mapping is the process of identifying and visualizing the spatial distribution of crops in a specific area using tools like satellite imaginary, GIS, and remote sensing.

Electroplating: Electroplating is the process of depositing a thin layer of metal onto the surface of a conductive object using an electrical current passed through an electrolyte solution containing the metal ions.

Eutrophication: Eutrophication is the process by which a water body becomes enriched with nutrients, particularly nitrogen and phosphorous, leading to excessive growth of algae and depletion of oxygen, which disrupts aquatic ecosystems.

Fluorescence: Fluorescence is the emission of light by a substance that has absorbed electromagnetic radiation, typically ultraviolet or visible light, occurring as the substance transitions from an excited electronic state to its ground state.

Foliar: Foliar refers to anything related to or applied to the leaves of a plant, such as foliar sprays or nutrients absorbed directly through leaf surfaces.

Gene sequencing: Gene sequencing is the process of determining the precise order of nucleotides (adenine, thymine, cytosine, and guanine) in a DNA molecule, which provides detailed information about an organism's genetic makeup.

Genome: A genome is the complete set of genetic material (DNA or RNA) present in an organism, encompassing all of its genes and non-coding sequences.

Homeostasis: Homeostasis is the process by which living organisms maintain a stable internal environment, despite external changes, to ensure optimal functioning or cells and physiological systems.

Homo-aggregation: Homo-aggregation is the process in which similar or identical particles or molecules aggregate or clump together, often due to intermolecular forces like van der Waals forces or hydrophobic interactions.

Horticulture: Horticulture is the branch of agriculture focused on the cultivation, management, and study of fruits, vegetables, nuts, seeds, herbs, sprouts, mushrooms, algae, and non-food crops like grass and flowers.

Hydathodes: Hydathodes are specialized structures in plants that excrete water through pores at the leaf margins, typically during guttation.

Hydrodynamic radius: The hydrodynamic radius is a measure of the effective size of a particle in solution, representing the radius of a hypothetical sphere that moves through a solvent with the same velocity as the particle in question under a given flow condition.

Hypoxia: Hypoxia is a condition in which there is a deficiency of oxygen in the tissues, leading to impaired cellular function and potentially causing damage to organs and systems.

Immobilization: Immobilization is a process of restricting the movement of molecules, cells, or enzymes, typically by physically trapping them in a solid matrix or on a surface to enhance their stability, reusability, or interaction in biochemical applications.

Leaching: Leaching is the process by which soluble substances, such as nutrients or contaminants, are dissolved and carried away by water from a solid material, typically soil or ore.

Lignification: Lignification is the process by which lignin, a complex polymer, is deposited in the cell walls of plants, providing structural support, rigidity, and resistance to decay.

Lithography: lithography is a printing process based on the principle that water and oil do not mix, where an image is etched onto a flat surface, typically a stone or metal plate, and then inked to transfer the image onto a medium such as paper.

Metabolomics: Metabolomics is the study of the complete set of metabolites, including small molecules and biochemical intermediates, present in a biological sample, and their changes in response to various conditions or treatments.

Nanobionics: Nanobionics is the integration of nanotechnology with biological systems to enhance or develop new functions in living organisms or devices.

Nanofertilizers: Nanofertilizers are fertilizers engineered at the nanoscale to improve nutrient delivery, enhance plant absorption, and reduce environmental impact by controlling the release of nutrients in a more efficient and targeted manner.

Nanosheets: Nanosheets are thin, two dimensional structures with a thickness on the nanoscale, often made of materials like graphene, transition metal dichalcogenides, or other nanomaterials, and exhibit unique properties due to their high surface area and atomic thickness.

Nanotubes: Nanotubes are cylindrical nanostructures made of materials like carbon, with unique mechanical, electrical, and thermal properties, used in various scientific and industrial applications.

Pathogen: A pathogen is any microorganism, such as bacterium, virus, fungus, or parasite, that causes disease in a host organism.

Pest: A pest is an organism that harms crops, livestock, or human health by causing damage or spreading diseases.

Pesticide: A pesticide is a chemical substance used to kill or control pests, including insects, weeds fungi or bacteria.

Photostability: Photostability is the ability of a substance to remain stable and retain its chemical properties when exposed to light, particularly ultraviolet radiation.

Phytohormones: Phytohormones are chemical substances produced by plants that regulate growth, development, and responses to environmental stimuli.

Polymer: A polymer is a large molecule composed of repeating structural units called monomers, which are chemically bonded to form long chains or networks.

Probe: A probe is a specialized tool or molecule, often labelled with a detectable marker, used to identify, bind or measure specific targets in biological or chemical systems, such as DNA, RNA, or proteins.

Proteomics: Proteomics is the large-scale study of proteins, particularly their functions, structures, and interactions within a biological system.

Quantum dots: Quantum dots are nanoscale semiconductor particles that exhibit unique optical and electronic properties, such as size-dependent fluorescence, due to quantum confinement effects.

Salinization: Salinization is the accumulation of soluble salts in soil or water to levels that can negatively affect plant growth and soil health.

Spectrometry: Spectrometry is an analytical technique used to measure the mass-to-charge ratio of ions, allowing the identification and quantification of chemical substances based on their mass and molecular structure.

Tomography: Tomography is an imaging technique that creates detailed cross-sectional views of an object by reconstructing data from multiple projections, often used in medical and scientific fields.

Transcription: Transcription is the process by which an RNA molecule is synthesized from DNA template, copying the genetic information to produce messenger RNA (mRNA) for protein synthesis.

Transcriptome: The transcriptome is the complete set of RNA molecules, including mRNA, non-coding microRNA, transcribed from the DNA of an organism at a specific time or under specific conditions.

Translation: Translation is the process by which a messenger (mRNA) molecule is decoded by ribosomes to synthesize a corresponding polypeptide or protein.

Transposon: A transposon, or jumping gene, is a DNA sequence that can change its position within a genome, potentially altering gene function and contributing to genetic variation.

Xenobiotic: A xenobiotic is a foreign substance or compound, typically a drug, chemical, or pollutant, that is not naturally produced or expected to be present in an organism.

SCIENTIFIC ARTICLES FOR FURTHERS READING

Al Tawaha, A. R. M., Singh, A., Rajput, V. D., Varshney, A., Agrawal, S., Ghazaryan, K., ... & Shawaqfeh, S. (2024). Green Nanofertilizers–The Need for Modern Agriculture, Intelligent, and Environmentally-Friendly Approaches. *Ecological Engineering & Environmental Technology, 25.*

Azarin, K., Usatov, A., Minkina, T., Alliluev, I., Duplii, N., Mandzhieva, S., ... & Ghazaryan, K. (2024). Impact nano-and micro-form of CdO on barley growth and oxidative stress response. *Journal of King Saud University-Science, 36*(11), 103493.

El-Ramady, H., Prokisch, J., Sári, D., Singh, A., Ghazaryan, K., Rajput, V. D., & Brevik, E. C. (2024). Nanotechnology in the soil system: An ecological approach towards sustainable management. *Applied Soil Ecology, 204,* 105669.

Fathi, A., Shiade, S. R. G., Parmoon, G., Yaghoubian, Y., Pirdashti, H., Rajput, V. D., ... & Minkina, T. (2024). Bioremediation of heavy metals contaminated soils using nanotechnology. In *Bio-organic Amendments for Heavy Metal Remediation* (pp. 611-628). Elsevier.

Ghazaryan, K., Agrawal, S., Margaryan, G., Harutyunyan, A., Rajput, P., Movsesyan, H., ... & Singh, A. (2024). Soil pollution: an agricultural and environmental problem with nanotechnological remediation opportunities and challenges. *Discover Sustainability, 5*(1), 1-33.

Ghazaryan, K., Pandey, D., Singh, Ş., Varagyan, V., Alexiou, A., Petropoulos, D., ... & Singh, A. (2024). Enhancing Crop Production: Unveiling the Role of Nanofertilizers in Sustainable Agriculture and Precision Nutrient Management. *Egyptian Journal of Soil Science, 64*(3), 981-1007.

Ghazaryan, K., Verma, A., Rawat, S., Rajput, P., Minkina, T., Mandzhieva, S., ... & Singh, A. (2024). Implementing Sustainable Nano-Bioremediation for Emerging Pollutants: An Environmentally Friendly Remediation Strategy. In *Nanotechnology Applications and Innovations for Improved Soil Health* (pp. 316-332). IGI Global.

Pandey, D., Singh, A., Darbinyan, N., Chakhmakhchyan, A. D., Parmar, S. S., & Ghazaryan, K. (2024). Revolutionizing Sustainable Agriculture with Nano-Priming Technology: A Leap Towards Resilient and High-Yield Crops.

In *Nanotechnology Applications and Innovations for Improved Soil Health* (pp. 305-315). IGI Global.

Prokisch, J., Nguyen, D. H., Muthu, A., Ferroudj, A., Singh, A., Agrawal, S., ... & Rai, M. (2024). Carbon Nanodot–Microbe–Plant Nexus in Agroecosystem and Antimicrobial Applications. *Nanomaterials, 14*(15), 1249.

Rajput, P., Singh, A., Agrawal, S., Ghazaryan, K., Rajput, V.D., Movsesyan, H., Mandzhieva, S., Minkina, T. and Alexiou, A. (2024) "Effects of environmental metal and metalloid pollutants on plants and human health: exploring nano-remediation approach," Stress Biology, 4(1).

Rajput, V. D., Minkina, T., Feizi, M., Kumari, A., Khan, M., Mandzhieva, S., ... & Choudhary, R. (2021). Effects of silicon and silicon-based nanoparticles on rhizosphere microbiome, plant stress and growth. *Biology, 10*(8), 791.

Rajput, V. D., Minkina, T., Kumari, A., Shende, S. S., Ranjan, A., Faizan, M., ... & Kızılkaya, R. (2022). A review on nanobioremediation approaches for restoration of contaminated soil. *Eurasian Journal of Soil Science, 11*(1), 43-60.

Rajput, V. D., Singh, A., Ghazaryan, K., Alexiou, A., Al-Tawaha, S., & Mohammad, A. R. (Eds.). (2024). *Harnessing NanoOmics and Nanozymes for Sustainable Agriculture*. IGI Global.

Rajput, V. D., Singh, A., Ghazaryan, K., Minkina, T. M., & Al-Tawaha, A. R. M. (Eds.). (2024). *Sustainable Agriculture: Nanotechnology, Biotechnology, Management and Food Security*. Walter de Gruyter GmbH & Co KG.

Rajput, V. D., Singh, A., Ghazaryan, K., Movsesyan, H. S., Minkina, T. M., Al-Tawaha, A. R. M., ... & Rajput, P. (2024). Nanotechnology and biotechnology: transforming agriculture for food security and sustainability. *Sustainable Agriculture: Nanotechnology, Biotechnology, Management and Food Security*, 213.

Rajput, V. D., Singh, A., Lalotra, S., Ghazaryan, K., Movsesyan, H. S., Minkina, T. M., ... & Rajput, P. Nanobiotechnology combined approaches for sustainable agriculture.

Rajput, V. D., Singh, A., Minkina, T. M., Shende, S. S., Kumar, P., Verma, K. K., ... & Sindireva, A. (2021). Potential applications of nanobiotechnology in plant nutrition and protection for sustainable agriculture. *Nanotechnology in plant growth promotion and protection: recent advances and impacts*, 79-92.

Rajput, V. D., Singh, A., Minkina, T., Rawat, S., Mandzhieva, S., Sushkova, S., ... & Upadhyay, S. K. (2021). Nano-enabled products: challenges and opportunities for sustainable agriculture. *Plants, 10*(12), 2727.

Rajput, V. D., Singh, A., Rawat, S., Sharma, R., Minkina, T. M., Singh, A. K., ... & Nazarenko, O. (2023). Nanotechnology as the cutting edge for sustainable agriculture. In *Nanotechnology for Sustainable Agriculture* (pp. 3-15). Apple Academic Press.

Rajput, V. D., Singh, A., Tomar, B., Minkina, T., Movsesyan, H. S., Elshikh, M. S., ... & Ghazaryan, K. (2024). Nanoparticle-Mediated Approaches in Agriculture Addressing Abiotic Stress From Soil to Plant Cells. In *Nanotechnology Applications and Innovations for Improved Soil Health* (pp. 72-89). IGI Global.

Rajput, V.D., Singh, A., Minkina, T.M., Verma, K.K. and Singh, A.K. (2023) Nanotechnology for Sustainable Agriculture, Apple Academic Press eBooks.

Rawat, S., Singh, A., Rajput, V. D., Sharma, R., Mandzhieva, S., & Singh, A. K. (2023). Nanobiotechnological approaches for improved plant breeding. In *Nanotechnology for Sustainable Agriculture* (pp. 307-323). Apple Academic Press.

Singh, A., Agrawal, S., Rajput, V. D., Ghazaryan, K., Movsesyan, H. S., Minkina, T., ... & Gupta, S. K. (2023). Seed Nanopriming: An Innovative Approach for Upregulating Seed Germination and Plant Growth Under Salinity Stress. In *Nanopriming Approach to Sustainable Agriculture* (pp. 290-313). IGI Global.

Singh, A., Agrawal, S., Rajput, V. D., Ghazaryan, K., Yesayan, A., Minkina, T., ... & Alexiou, A. (2024). Nanoparticles in revolutionizing crop production and agriculture to address salinity stress challenges for a sustainable future. *Discover Applied Sciences*, 6(6), 317.

Singh, A., Agrawal, S., Rajput, V.D., Ghazaryan, K., Yesayan, A., Minkina, T., Zhao, Y., Petropoulos, D., Kriemadis, A., Papadakis, M. and Alexiou, A. (2024) "Nanoparticles in revolutionizing crop production and agriculture to address salinity stress challenges for a sustainable future," Discover Applied Sciences, 6(6). Available at: https://doi.org/10.1007/s42452-024-06009-7.

Singh, A., Ghazaryan, K., Margaryan, G., Khachatryan, H., Rajput, V. D., & Minkina, T. M. (2024). Enhancing crop food security in the Ararat Plain region of Armenia: harnessing arid soil resources and biogenic nanoparticles for salinity-resilient cultivation. *Sustainable Agriculture: Nanotechnology, Biotechnology, Management and Food Security*, 49.

Singh, A., Margaryan, G., Harutyunyan, A., Movsesyan, H.S., Khachatryan, H., Rajput, V., Minkina, T., Alexiou, A., Petropoulos, D., Kriemadis, A., ElRamady, H. and Ghazaryan, K. (2024) "Advancing Agricultural Resilience in Ararat Plain, Armenia: Utilizing Biogenic Nanoparticles and Biochar under Saline Environments to Optimize Food Security and Foster European Trade," Egyptian Journal of Soil Science, 64(2), p. 0. Available at: https://doi.org/10.21608/ejss.2024.257000.1703.

Singh, A., Rajput, P., Rawat, S., Sushkova, S., & Rajput, V. D. (2023). Nanobionics aid in agriculture. In *Nanotechnology for Sustainable Agriculture* (pp. 325-332). Apple Academic Press.

Singh, A., Rajput, V. D., Agrawal, S., Ghazaryan, K., Minkina, T., Al Tawaha, A. R. M., ... & Alexiou, A. (2024). Nanoparticles mediated salt stress resilience:

a holistic exploration of physiological, biochemical, and Nano-omics approaches. *Reviews of Environmental Contamination and Toxicology, 262*(1), 19.

Singh, A., Rajput, V. D., Chakrawarti, N., Ghazaryan, K., Minkina, T., & Singh, S. (2023). Nano-Enabled Approaches for Biofortification Strategies to Enhance Crop Production with Micronutrient Enrichment. In *Nano-Biofortification for Human and Environmental Health* (pp. 1-14). Cham: Springer International Publishing.

Singh, A., Rajput, V. D., Chakrawarti, N., Ghazaryan, K., Minkina, T., & Singh, S. (2023). ZnO-NP-based biofortification to enhance crop production with micronutrient enrichment to combat malnutrition. In *Nano-Biofortification for Human and Environmental Health* (pp. 99-108). Cham: Springer International Publishing.

Singh, A., Rajput, V. D., Ghazaryan, K., Gupta, S. K., & Minkina, T. (Eds.). (2023). *Nanopriming approach to sustainable agriculture.* IGI Global.

Singh, A., Rajput, V. D., Ghazaryan, K., Lalotra, S., Minkina, T. M., Al-Tawaha, A. R. M., ... & Rajput, P. (2024). Revolutionizing agriculture: harnessing phytonanotechnology for sustainable food security. *Sustainable Agriculture: Nanotechnology, Biotechnology, Management and Food Security*, 1.

Singh, A., Rajput, V. D., Ghazaryan, K., Sharma, A., Minkina, T. M., Al-Tawaha, A. R. M., ... & Rajput, P. Nanotechnology-based genome editing: empowering sustainable agriculture for enhanced food security.

Singh, A., Rajput, V. D., Lalotra, S., Agrawal, S., Ghazaryan, K., Singh, J., ... & Alexiou, A. (2024). Zinc oxide nanoparticles influence on plant tolerance to salinity stress: insights into physiological, biochemical, and molecular responses. *Environmental Geochemistry and Health, 46*(5), 148.

Singh, A., Rajput, V. D., Pandey, D., Sharma, R., Ghazaryan, K., & Minkina, T. (2023). Nano zinc-enabled strategies in crops for combatting zinc malnutrition in human health. *Frontiers in Bioscience-Landmark, 28*(8), 158.

Singh, A., Rajput, V. D., Rawat, S., Sharma, R., Singh, A. K., Kumar, P., ... & Singh, S. (2022). Geoinformatics and nanotechnological approaches for coping up abiotic and biotic stress in crop plants. *Sustainable Agriculture Systems and Technologies*, 337-359.

Singh, A., Rajput, V. D., Sharma, R., Ghazaryan, K., & Minkina, T. (2023). Salinity stress and nanoparticles: Insights into antioxidative enzymatic resistance, signaling, and defense mechanisms. *Environmental Research*, 116585.

Singh, A., Rawat, S., Rajput, V. D., Ghazaryan, K., Minkina, T., Al Tawaha, A. R. M., & Varshney, A. (2023). Impact of Nanofertilizers for the Mitigation of Multiple Environmental Stresses. In *Nanofertilizers for Sustainable Agroecosystems: Recent Advances and Future Trends* (pp. 431-454). Cham: Springer Nature Switzerland.

Singh, A., Rawat, S., Rajput, V. D., Ghazaryan, K., Minkina, T., Al Tawaha, A. R. M., ... & Sandhu, K. S. (2023). Nanoparticle-Based Priming-Enhanced Agricultural Practices: Promoting Sustainable Plant Growth and Development. In *Nanopriming Approach to Sustainable Agriculture* (pp. 142-157). IGI Global.

Singh, A., Sengar, R. S., Rajput, V. D., Agrawal, S., Ghazaryan, K., Minkina, T., ... & Habeeb, T. (2023). Impact of zinc oxide nanoparticles on seed germination characteristics in rice (Oryza sativa L.) under salinity stress. *Journal of Ecological Engineering*, *24*(10).

Singh, A., Sengar, R. S., Rajput, V. D., Minkina, T., & Singh, R. K. (2022). Zinc oxide nanoparticles improve salt tolerance in rice seedlings by improving physiological and biochemical indices. *Agriculture, 12*(7), 1014.

Singh, A., Sengar, R. S., Sharma, R., Rajput, P., & Singh, A. K. (2021). Nano-Priming Technology for Sustainable Agriculture. *Biogeosystem Technique,* (8), 79-92.

Singh, A., Sengar, R. S., Singh, R., Shahi, U. P., & Yadav, M. K. (2022). Effects of zinc oxide nanoparticles for promoting seed germination of rice (Oryza sativa L.) under salinity stress. *Environmental Nanotechnology, Monitoring & Management*, *28*, 254-259.

Singh, A., Sharma, A., Singh, O., Rajput, V. D., Movsesyan, H. S., Minkina, T., ... & Ghazaryan, K. (2024). In-depth exploration of nanoparticles for enhanced nutrient use efficiency and abiotic stresses management: present insights and future horizons. *Plant Stress*, 100576.

Singh, A., Sharma, R., Rajput, V. D., Ghazaryan, K., Minkina, T., Al Tawaha, A. R. M., & Varshney, A. (2023). Green synthesis of nanofertilizers and their application for crop production. In *Nanofertilizers for Sustainable Agroecosystems: Recent Advances and Future Trends* (pp. 205-231). Cham: Springer Nature Switzerland.

Singh, A., Sharma, R., Rawat, S., Singh, A. K., Rajput, V. D., Fedorov, Y., ... & Chaplygin, V. (2022). Nanomaterial-plant interaction: Views on the pros and cons. In *Toxicity of nanoparticles in plants* (pp. 47-68). Academic Press.

Singh, A., Singh, A. K., Rajput, V., Mehrotra, R., Rawat, N. P. S., & Sadera, G. (2020). Nanotechnology in sustainable agriculture: A big opportunity from small particles. *book Advance Research in Agricultural and Veterinary Sciences. Vidya kutir ISBN*, 978-81.

Singh, A., Singh, P., Rajput, P., Chopra, R., & Singh, S. (2021). Application of nanomaterials in agriculture and environment management.

Singh, A., Tomar, B., Harutyunyan, A., Rajput, P., Minkina, T., Mandzhieva, S., ... & Ghazaryan, K. (2024). Utilizing Nanotechnology in Agriculture: A Balancing Approach Between Environmental Health and Risks.

In *Nanotechnology Applications and Innovations for Improved Soil Health* (pp. 380-399). IGI Global.

Singh, S., Pradhan, A., Singh, R. K., Singh, A., Minkina, T., & Rajput, V. D. (2022). Nanoparticles: physiology, chemistry, and biochemistry. In *The Role of Nanoparticles in Plant Nutrition under Soil Pollution: Nanoscience in Nutrient Use Efficiency* (pp. 21-42). Cham: Springer International Publishing.

Singh, S., Singh, O., Singh, S., Singh, A., Rajput, V. D., Ghazaryan, K., ... & El-Ramady, H. (2024). Impact of Nano-Enzyme and Nanomics for Sustainable Agriculture: Current Status and Future Prospective. In *Harnessing NanoOmics and Nanozymes for Sustainable Agriculture* (pp. 1-18). IGI Global.

Tawaha, A. R. M., Singh, A., Rajput, V., Al Zoubi, O. M., Ghazaryan, K., & Minkina, T. Impact of Salinity Stress and Zinc Oxide Nanoparticles on Macro and Micronutrient Assimilation: Unraveling the Link between Environmental Factors and Nutrient Uptake. *Journal of Ecological Engineering*, *25*(2).

Upadhyay, P., Navvuru, S., Yadav, P. K., Lalotra, S., Singh, A., Rajput, V. D., ... & Ghazaryan, K. (2024). Nano-Omics-Based Abiotic and Biotic Stresses Management. In *Harnessing NanoOmics and Nanozymes for Sustainable Agriculture* (pp. 348-371). IGI Global.

Verma, A., Singh, A., Rawat, S., Rajput, P., Minkina, T., Mandzhieva, S., ... & Ghazaryan, K. (2024). Unveiling Nanomaterial-Induced Toxicity: Navigating Challenges to Modern Cultivation and Soil Management Practices. In *Nanotechnology Applications and Innovations for Improved Soil Health* (pp. 236-253). IGI Global.

INDEX

A

Abiotic 20-21, 70, 73-75, 77, 81, 86, 109, 137, 141, 180, 232, 235

Absorption 24, 50, 53-54, 60-63, 67, 70, 87-89, 92, 120, 122, 125, 163, 182, 201, 205-207, 210, 218, 229, 236-237

Adaptation 233

Adenine 144

Aerial 248-249

Aerogel 45

Aestivum 84, 86, 90-91, 143

Aircraft 249

Airplanes 248

Alkaloids 189

Allele 143

Anhydrase 202

Antenna 237

Antiporter 104, 155

Apical 16

Apoplast 75

Apoptosis 171

Aquatic 163

Aquilaria 107

Arginine 175

Arietinum 216

Aromatics 238

Arsenic 25, 117, 121

Atrazine 232

B

Bacillus 186

Barcode 24

Basmati 167

Bigger 87

185-186, 189, 192-194, 267

Biological 13, 28, 32, 39, 46-47, 50, 130, 180, 185, 188-189, 235, 261, 263-264

Biomarkers 261

Biomass 62-64, 67, 73, 78, 123, 143, 162, 173, 182

Biosensors 23, 261

Blooming 175

Brassica 86

Breeding 130, 141, 143, 149, 151

C

Cadmium 116, 118, 122, 125

Calcium 5-6, 39, 60, 201, 218, 220

Canopy 227

Carbonate 218

Cations 188

Cellular 60, 82-83, 92, 105, 110, 155, 166, 178, 182, 184-185, 203

Cerium 86, 93

Cesium 93

Channels 54, 87-88, 109, 158

Chelation 261

Chlorosis 121

Chromatin 175

Chromosome 171

Cleavage 133, 136

Components 12, 184, 187, 189

Copper 66, 90, 159, 173
Cucumber 65
Cyamopsis 173
Cyberwood 238-239
Cytidine 142
Cytokines 184
Cytoplasm 56, 206
Cytosine 144

D

Density 254
Diffusion 88, 209
Digital 244
Dismutase 78, 91, 97, 126
Dispersion 193
Domains 36, 132-133, 203

E

Electrode 24
Electron 47-49, 52, 75, 82, 91, 95, 97, 102, 168, 193
Electrons 237
Elements 60, 117, 238
Emulsions 230
Endodermis 165
Energy 2, 51, 66, 74-75, 91, 187, 230, 237
Enzymes 62-64, 76-78, 83, 86, 89-97, 103, 105, 107-109, 111, 126, 158-159, 168, 170, 179, 186, 191, 220, 233
Epidermis 54, 87-88
Ethylene 82, 141
Expression 86, 103-105, 107, 111-112, 125-126, 141, 154-156, 158, 172, 175

F

Family 107, 175, 208
Farmers 241, 244, 246, 250, 253, 257
Fertility 226
Fibrosis 184
Fixation 63
Flavonoids 93, 102, 109, 159, 170, 189
Flooding 20
Foliar 6, 55-56, 63, 69, 71-72, 91, 180, 199-200, 211, 216, 262
Fullerenes 40-41, 163
Fungicides 12, 224, 231
Fusarium 14, 191

G

Genomes 22
Genotoxic 28
Glycolate 102
Glycolysis 20, 107, 159
Growth 1, 5-6, 13-14, 18-20, 29, 32, 37-38, 42-43, 60-62, 64-67, 70, 72-73, 86, 90-93, 113, 120-123, 145, 150, 161, 164-165, 167, 173, 178, 182, 203, 210, 215, 218, 220, 227, 230-231, 233-234, 254, 266

H

Herbicides 13-14, 145, 224-225, 232
Heritable 172
Histones 174-175
Homologous 132
Hormones 27
Hydathodes 262
Hydrogen 94
Hypoxia 262

I

Immunity 184
Infrared 49, 51
Intensity 51, 238

L

Leakage 28, 75, 168
Lesions 185
Lignin 143
Lipids 163, 168, 170
Livestock 251
Lysine 174

M

Magnesium 60, 63, 120
Maturity 227
Membrane 54, 56-57, 82, 87, 95, 104, 125, 155, 168, 178, 237
Mercury 116
Meristems 16
Mesophyll 85
Mesoporous 16
Messenger 264
Metabolic 77, 82, 90, 107, 110, 158-159
Metabolism 20, 63, 66, 109, 168, 198, 202
Metabolome 109, 159
Metaphase 181
Micelles 4
Microbes 14, 186, 190
Microbial 14, 186
Microscopy 47-49, 52, 193
Minerals 217
Mitotic 171, 181
Moisture 43, 210, 253
Morphology 52
Mortality 231
Movement 57
Muscles 203
Mutations 136, 144, 146

N

Nanoscale 5, 39, 46, 59, 217, 235, 263
Nanosensor 26
Nanotubes 4, 16, 40, 45, 68, 86, 163, 229, 237-238, 263
Nanozymes 270
Nitrogen 2, 4-6, 60-61, 63, 66, 109, 217
Nucleotide 135
Nutrients 5-6, 30, 40-43, 58-60, 69-70, 116, 119, 199, 205, 210, 216-217, 262-263
Nutrition 92, 198, 204

O

Organelles 76, 166, 205
Oxidases 75
Oxygen 20-21, 67, 75, 82, 84-85, 91, 94, 97, 102, 106, 112, 120, 156, 158-159, 168, 178-179, 234, 237, 262
Oxysporum 191

P

Pathogens 37, 41
Pectin 238
Peptide 27
Peroxidase 76-77, 84, 91, 94, 112, 126, 158
Peroxisome 102
Pesticides 12, 28, 41, 122, 224-225, 230-231, 250
Phytate 197
Pivotal 60
Plasma 4, 54, 75, 87, 104, 187
Pollutants 38, 116, 127
Pollution 24, 115-119
Polymer 237, 264
Polyphenol 90

Potassium 6, 17, 29, 60, 62, 67, 92, 155, 217
Precursors 45
Preventive 219
Proline 105-106, 126, 156, 179, 233-234
Proteins 16, 20, 72, 104-105, 107, 126, 133-134, 145, 154-155, 157, 163, 170, 175, 191, 204, 207-208, 236, 264
Proteome 204
Proteomics 107, 157, 264
Proton 104, 155
Pulmonary 184
Pyrroline- 105

R

Radiation 51, 254
Retention 40
Robotics 242

S

Salicylic 27
Saline 84, 109
Sargassum 193
Satellites 226, 248
Sativus 164, 169
Senescence 227
Sensors 39, 226, 239, 244, 246, 248-250, 252-253
Signals 27, 111
Silica 16, 43, 67, 233, 238
Silicon 234
Skeletal 203
Soil-borne 14, 37
Solanum 126
Speciation 89
Spectrum 254
Spoilage 41
Starch 142, 202
Stomach 54, 236
Strains 191
Strawberry 91
Stressors 41, 73, 232
Symplastic 54-57

T

Thaliana 133
Tissues 52, 54, 57, 87, 123, 163
Titanium 5, 37-38
Transgenic 106, 151, 156
Transposon 264
Triticum 84, 86

U

Uracil 144

V

Vanilla 167
Variations 25, 142, 172
Vegetation 227-228, 251, 254
Velocity 244
Versatile 38
Viruses 220
Vitamin 102, 204
Vulgaris 169

W

Wastewater 119
Weather 225, 244-246, 253

X

Xenobiotic 124, 264
X-rays 51

Z

Zeolite 217